Methods in Molecular Biology

Series Editor
John M. Walker
School of Life and Medical Sciences,
University of Hertfordshire, Hatfield,
Hertfordshire, UK

For further volumes:
http://www.springer.com/series/7651

For over 35 years, biological scientists have come to rely on the research protocols and methodologies in the critically acclaimed *Methods in Molecular Biology* series. The series was the first to introduce the step-by-step protocols approach that has become the standard in all biomedical protocol publishing. Each protocol is provided in readily-reproducible step-by step fashion, opening with an introductory overview, a list of the materials and reagents needed to complete the experiment, and followed by a detailed procedure that is supported with a helpful notes section offering tips and tricks of the trade as well as troubleshooting advice. These hallmark features were introduced by series editor Dr. John Walker and constitute the key ingredient in each and every volume of the *Methods in Molecular Biology* series. Tested and trusted, comprehensive and reliable, all protocols from the series are indexed in PubMed.

Modelling the Microbiome

Edited by

Karthik Raman

Centre for Integrative Biology and Systems Medicine, Wadhwani School of Data Science and AI, Indian Institute of Technology Madras, Chennai, India; Department of Data Science and AI, Wadhwani School of Data Science and AI, Indian Institute of Technology Madras, Chennai, India

Gayathri Sambamoorthy

Department of Chemical Engineering, Indian Institute of Science, Bengaluru, India; Department of Molecular Cell Biology, Institute for Cancer Research, Oslo University Hospital, Oslo, Norway; Centre for Cancer Cell Reprogramming, Institute of Clinical Medicine, Faculty of Medicine, University of Oslo, Oslo, Norway

Humana Press

Editors
Karthik Raman
Centre for Integrative Biology
and Systems Medicine
Wadhwani School of Data Science
and AI
Indian Institute of Technology Madras
Chennai, India

Department of Data Science and AI
Wadhwani School of Data Science and AI
Indian Institute of Technology Madras
Chennai, India

Gayathri Sambamoorthy
Department of Chemical Engineering
Indian Institute of Science
Bengaluru, India

Department of Molecular Cell Biology
Institute for Cancer Research
Oslo University Hospital
Oslo, Norway

Centre for Cancer Cell Reprogramming
Institute of Clinical Medicine
Faculty of Medicine
University of Oslo
Oslo, Norway

ISSN 1064-3745 ISSN 1940-6029 (electronic)
Methods in Molecular Biology
ISBN 978-1-0716-5079-0 ISBN 978-1-0716-5080-6 (eBook)
https://doi.org/10.1007/978-1-0716-5080-6

This Humana imprint is published by the registered company Springer Science+Business Media, LLC, part of Springer Nature.
The registered company address is: 1 New York Plaza, New York, NY 10004, U.S.A.

Preface

This book is a comprehensive resource for navigating the rapidly evolving field of microbiome research, with a focus on key methodologies essential for understanding microbes and their interactions in various communities. High-throughput sequencing has revolutionized the study of microbes across environments; analysis of metagenomic data emerging from various sequencing studies is challenging, yet has had a significant impact on enabling a deeper understanding of the role of microbes in a variety of ecosystems. Parallel to the evolution of experimental approaches to studying microbial ecosystems, there have been numerous advances in the development of sophisticated computational in silico approaches to interrogate these communities and disentangle the complex web of interactions that underlie many ecosystems.

This book aims to equip students, researchers, and practitioners across disciplines such as microbiology, bioinformatics, and systems biology with key tools and practical insights to embark on microbiome studies. A key aspect of this book is the nature of the writing, which is protocol-like, of course, in sync with the rest of this book series and, importantly, presents a practitioner's viewpoint and key insights and challenges that one may likely encounter along the way in such studies. Given the pace of evolution of the field, no book can do enough justice to the breadth of the topics or keep up with the frenetic pace of scientific advances; yet, this book presents a systematic end-to-end overview of approaches, from basic foundational concepts in metagenomic data analysis workflows to important wet-lab protocols highlighting the need for controls and data quality issues, to core bioinformatics and data processing approaches, ultimately leading to community interactions and insights derived from predictive modelling tools leveraging metabolic networks and a variety of modelling paradigms.

Chapter 1 delves into detailed workflows from raw sequencing data to the generation of taxonomic and functional profiles, highlighting various classic tools, guiding the readers through their application for specific analytical requirements. Chapter 2 details key practical wet-lab aspects of sample handling, extraction, quality control, and library preparation. Chapter 3 details comprehensive sampling strategies, DNA/RNA extraction, and importantly, the use of both 16S rRNA amplicon sequencing and whole metagenome shotgun sequencing. It also emphasizes standardization and quality control measures vital in marine samples, which are generalizable to other complex environments. Chapter 4 reinforces best practices introduced in Chap. 3 regarding site selection, sample collection, DNA extraction, and incorporation of positive and negative controls for reproducibility and data quality. It also introduces real-world applications like detecting antimicrobial resistance genes (ARGs) and pathogens. Chapter 5 describes a specific experimental technique ("dilution-to-stimulation") for enriching desired microbial consortia from environmental samples. Following this, Chap. 6 provides a comprehensive, step-by-step guide to the computational pipeline for analyzing both microbial and viral metagenomes and viromes. It covers various important elements of the analysis pipelines, for specifically identifying and analyzing viral sequences.

The next part of the book looks at elucidating community interactions to derive ecological insights. Chapter 7 focuses on building and comparing microbial association networks, identifying "key driver microbes" that influence community composition and interactions. Chapter 8 delves deeper into community structure prediction, going beyond mere pairwise interactions, focusing on predicting higher-order interactions from experimental data. Chapter 9 presents a network-biology approach to evaluate pairwise interactions in microbial communities, marking the transition from inferring associations to understanding the metabolic basis of the interactions.

The remaining chapters focus on advanced aspects of metabolic modelling and engineering. Chapter 10 provides a nice overview of genome-scale metabolic model (GEM) reconstruction, curation, and analysis within a community context, including Flux Balance Analysis (FBA), Flux Variability Analysis (FVA), and Flux Sampling (FS), setting the stage for more specialized modelling approaches. Chapter 11 builds on the constraint-based modelling paradigm to predict two-species microbial communities optimized for producing specific metabolites, and identifies cross-fed metabolites and interaction behaviors, illustrating a concrete industrial application of the modelling principles. Following this, Chap. 12 illustrates mgPipe, a tool enabling the integration of omics data into predictive metabolic models to perform mechanistic simulations of the interactions between the diet, microbiome, and the host. Chapter 13 introduces Flux Coupling Analysis combined with SteadyCom to identify causal relationships and metabolic interdependencies among reactions across different species in a community, enabling a deeper understanding of the metabolic interactions happening in a community. Chapter 14 presents FLYCOP, describing the construction of condition-specific genome-scale models for community members, the generation of community-based metabolic models, and the analysis of community-wide metabolic capabilities and interactions. Specific applications, including an industrial use case, are detailed. Chapter 15 has an increased focus on microbial growth dynamics, considering substrate competition and metabolic cross-feeding. This again has applications for industrial bioprocesses and metabolic engineering. Chapter 16 presents a niche advanced modelling approach that integrates high-resolution mass spectrometry data with thermodynamics to assess energy flows and microbial growth kinetics.

While we recognize that there is so much more to cover in this field, this book can provide a broad overview and useful highlights, which can hopefully spark more research and perhaps even dedicated books on some of the specific topics covered herein, such as wet-lab data generation, metagenomics pipelines, metabolic modelling, or even the various applications.

We are also grateful to Dr. John M. Walker, the Series Editor, for first mooting this idea, and to Anna Rakovsky (Springer Nature) and Thanushri Ajitchandran (Straive) for their help and assistance during the editorial process. We are also grateful to researchers at the Computational Systems Biology Lab at IIT Madras, especially Dr. Pratyay Sengupta, who have given useful comments and suggestions to enhance the content presented here.

Finally, we hope the readers appreciate the breadth of content in this book and also the practical guidance from experienced practitioners, who have already published their work in many reputed journals. I am very grateful to all the authors for their contributions and for taking special efforts to make it accessible to less experienced researchers, perhaps even from diverse backgrounds, providing step-by-step guidance on complex protocols. We hope that this repertoire of methods will help the readers navigate through many complex microbial ecosystems, and sow the seed for many novel discoveries.

Chennai, India — ***Karthik Raman***
Oslo, Norway — ***Gayathri Sambamoorthy***

Contents

Preface . . . . *v*
Contributors . . . . *ix*

1 Unlocking the Metagenome: Pipeline for Microbiome Data Analysis 1
Aarthi Ravikrishnan

2 Targeted Metagenomics Using Next-Generation Sequencing Methods 25
B. S. Yugandhar Reddy, S. Sripradha, and Achyut Kumar

3 Exploring the Ocean's Microbial World: Techniques and Protocols for Microbiome Research 33
Vijaya Raghavan Rangamaran, T. J. Sushmitha, Karthikeyan Kathan Tamilmani, Hinduja Murugesan, and Dharani Gopal

4 Shotgun Metagenomic Analysis of Microbial Community Dynamics in Wastewater Treatment Through Constructed Wetlands 47
Georgios Miliotis and Anna Tumeo

5 Dilution-to-Stimulation: A Method for Selecting Polymer-Transforming Microbial Consortia 75
Diego Javier Jiménez, Laura Díaz-García, Lila Aldakheel, and Alexandre Soares Rosado

6 Computational Microbial and Viral Ecology Analysis 83
James C. Kosmopoulos and Karthik Anantharaman

7 Identifying Differential Network Properties and Driver Microbes in Microbial Association Networks Using CompNet and NetShift 143
Kuntal Kumar Bhusan, Tungadri Bose, and Anirban Dutta

8 Estimating Effective Pairwise Interactions to Predict the Structures of Microbial Communities (EPICS) 167
Gayathri Sambamoorthy, Aamir Faisal Ansari, and Narendra M. Dixit

9 Evaluating Metabolic Support in Pairwise Microbial Communities Using MetQuest 175
Pratyay Sengupta, Sandhya Vasudevan, and Karthik Raman

10 Constraint-Based Metabolic Modeling Approach for Microbial Communities 191
Satyajit Beura, Sayan Saha Roy, Amit Kumar Das, and Amit Ghosh

11 Constraint-Based Modeling of Microbial Communities for Metabolite Production 221
Maziya Ibrahim and Karthik Raman

12 Personalized Constraint-Based Modeling of Microbial Communities from Metagenomic Data 233
Jordi Roma Pi and Almut Heinken

13 Predicting Interspecies Metabolic Dependencies in Microbial Communities by Integrating Flux Coupling Analysis with SteadyCom 261
Steve Zhang, Hugh C. McCullough, and Hyun-Seob Song

14 Describing and Designing Microbial Community Metabolic Models In Silico: A Comprehensive Protocol Utilizing FLYCOP 277
Ana del Ramo, David San León Granado, and Juan Nogales

15 Dynamic Simulation of Growth and Cross-Feeding in Microbiomes with μbialSim.. 317
Ali Nawaz, Jessye L. Schaefer, and Florian Centler

16 LambdaPy and LambdaR: Thermodynamics-Based Biogeochemical Reaction Modeling Packages for Integrating High-Resolution Mass Spectrometry Data.. 331
Manokaran Veeramani, Sanjog Kharel, Hugh C. McCullough, Xingyuan Chen, Jianqiu Zheng, James C. Stegen, Timothy D. Scheibe, and Hyun-Seob Song

Index *353*

Contributors

LILA ALDAKHEEL • *Biological and Environmental Sciences and Engineering Division (BESE), King Abdullah University of Science and Technology (KAUST), Thuwal, Kingdom of Saudi Arabia*

KARTHIK ANANTHARAMAN • *Department of Bacteriology, University of Wisconsin-Madison, Madison, WI, USA; Department of Integrative Biology, University of Wisconsin-Madison, Madison, WI, USA*

AAMIR FAISAL ANSARI • *Department of Chemical Engineering, Indian Institute of Science, Bengaluru, India*

SATYAJIT BEURA • *Department of Bioscience and Biotechnology, Indian Institute of Technology Kharagpur, Kharagpur, West Bengal, India*

KUNTAL KUMAR BHUSAN • *TCS Research, Tata Consultancy Services Ltd., Pune, India*

TUNGADRI BOSE • *TCS Research, Tata Consultancy Services Ltd., Pune, India*

FLORIAN CENTLER • *Department of Chemistry and Biology, School of Science and Technology, Siegen University, Siegen, Germany*

XINGYUAN CHEN • *Pacific Northwest National Laboratory, Richland, WA, USA*

AMIT KUMAR DAS • *Department of Bioscience and Biotechnology, Indian Institute of Technology Kharagpur, Kharagpur, West Bengal, India*

ANA DEL RAMO • *Department of Systems Biology, Centro Nacional de Biotecnología CSIC, Madrid, Spain*

LAURA DÍAZ-GARCÍA • *Department of Chemical and Biological Engineering, Advanced Biomanufacturing Centre, University of Sheffield, Sheffield, UK*

NARENDRA M. DIXIT • *Department of Chemical Engineering, Indian Institute of Science, Bengaluru, India; Department of Bioengineering, Indian Institute of Science, Bengaluru, India*

ANIRBAN DUTTA • *TCS Research, Tata Consultancy Services Ltd., Pune, India*

AMIT GHOSH • *Department of Bioscience and Biotechnology, Indian Institute of Technology Kharagpur, Kharagpur, West Bengal, India; School of Energy Science and Engineering, Indian Institute of Technology Kharagpur, Kharagpur, West Bengal, India; P.K. Sinha Centre for Bioenergy and Renewables, Indian Institute of Technology Kharagpur, Kharagpur, West Bengal, India*

DHARANI GOPAL • *Marine Biotechnology Group, National Institute of Ocean Technology (NIOT), Ministry of Earth Sciences (MoES), Government of India, Chennai, India*

DAVID SAN LEÓN GRANADO • *Department of Systems Biology, Centro Nacional de Biotecnología CSIC, Madrid, Spain; CNB DNA Biofoundry (CNBio), CSIC, Madrid, Spain*

ALMUT HEINKEN • *Inserm UMRS 1256 NGERE, University of Lorraine, Nancy, France*

MAZIYA IBRAHIM • *Department of Biotechnology, Bhupat and Jyoti Mehta School of Biosciences, Indian Institute of Technology Madras, Chennai, India; Centre for Integrative Biology and Systems Medicine, Wadhwani School of Data Science and AI, Indian Institute of Technology Madras, Chennai, India*

DIEGO JAVIER JIMÉNEZ • *Biological and Environmental Sciences and Engineering Division (BESE), King Abdullah University of Science and Technology (KAUST), Thuwal, Kingdom of Saudi Arabia*

SANJOG KHAREL • *University of Nebraska-Lincoln, Lincoln, NE, USA*
JAMES C. KOSMOPOULOS • *Department of Bacteriology, University of Wisconsin-Madison, Madison, WI, USA; Microbiology Doctoral Training Program, University of Wisconsin-Madison, Madison, WI, USA*
ACHYUT KUMAR • *Unilever R&D Center, Unilever India Pvt Ltd, Bengaluru, India*
HUGH C. MCCULLOUGH • *University of Nebraska-Lincoln, Lincoln, NE, USA*
GEORGIOS MILIOTIS • *Discipline of Bacteriology, School of Medicine, University of Galway, Galway, Ireland; Centre for One Health, Ryan Institute, University of Galway, Galway, Ireland*
HINDUJA MURUGESAN • *Marine Biotechnology Group, National Institute of Ocean Technology (NIOT), Ministry of Earth Sciences (MoES), Government of India, Chennai, India*
ALI NAWAZ • *Department of Chemistry and Biology, School of Science and Technology, Siegen University, Siegen, Germany*
JUAN NOGALES • *Department of Systems Biology, Centro Nacional de Biotecnología CSIC, Madrid, Spain; CNB DNA Biofoundry (CNBio), CSIC, Madrid, Spain*
KARTHIK RAMAN • *Centre for Integrative Biology and Systems Medicine, Wadhwani School of Data Science and AI, Indian Institute of Technology Madras, Chennai, India; Department of Data Science and AI, Wadhwani School of Data Science and AI, Indian Institute of Technology Madras, Chennai, India*
VIJAYA RAGHAVAN RANGAMARAN • *Marine Biotechnology Group, National Institute of Ocean Technology (NIOT), Ministry of Earth Sciences (MoES), Government of India, Chennai, India*
AARTHI RAVIKRISHNAN • *Genome Institute of Singapore (GIS), Agency for Science, Technology and Research (A*STAR), Singapore, Republic of Singapore; A*STAR Skin Research Labs (A*SRL), Agency for Science, Technology and Research (A*STAR), Singapore, Singapore*
JORDI ROMA PI • *Inserm UMRS 1256 NGERE, University of Lorraine, Nancy, France*
ALEXANDRE SOARES ROSADO • *Biological and Environmental Sciences and Engineering Division (BESE), King Abdullah University of Science and Technology (KAUST), Thuwal, Kingdom of Saudi Arabia*
SAYAN SAHA ROY • *School of Nano Science and Technology, Indian Institute of Technology Kharagpur, Kharagpur, West Bengal, India*
GAYATHRI SAMBAMOORTHY • *Department of Chemical Engineering, Indian Institute of Science, Bengaluru, India; Department of Molecular Cell Biology, Institute for Cancer Research, Oslo University Hospital, Oslo, Norway; Centre for Cancer Cell Reprogramming, Institute of Clinical Medicine, Faculty of Medicine, University of Oslo, Oslo, Norway*
JESSYE L. SCHAEFER • *Department of Chemistry and Biology, School of Science and Technology, Siegen University, Siegen, Germany*
TIMOTHY D. SCHEIBE • *Pacific Northwest National Laboratory, Richland, WA, USA*
PRATYAY SENGUPTA • *Department of Biotechnology, Bhupat and Jyoti Mehta School of Biosciences, Indian Institute of Technology Madras, Chennai, India; Centre for Integrative Biology and Systems mEdicine (IBSE), Wadhwani School of Data Science and AI (WSAI), Indian Institute of Technology Madras, Chennai, India; Department of Data Science and AI (DSAI), Wadhwani School of Data Science and AI (WSAI), Indian Institute of Technology Madras, Chennai, India*
HYUN-SEOB SONG • *University of Nebraska-Lincoln, Lincoln, NE, USA*
S. SRIPRADHA • *Unilever R&D Center, Unilever India Pvt Ltd, Bengaluru, India*
JAMES C. STEGEN • *Pacific Northwest National Laboratory, Richland, WA, USA*

T. J. SUSHMITHA • *Marine Biotechnology Group, National Institute of Ocean Technology (NIOT), Ministry of Earth Sciences (MoES), Government of India, Chennai, India*

KARTHIKEYAN KATHAN TAMILMANI • *Marine Biotechnology Group, National Institute of Ocean Technology (NIOT), Ministry of Earth Sciences (MoES), Government of India, Chennai, India*

ANNA TUMEO • *Discipline of Bacteriology, School of Medicine, University of Galway, Galway, Ireland*

SANDHYA VASUDEVAN • *Department of Biotechnology, Bhupat and Jyoti Mehta School of Biosciences, Indian Institute of Technology Madras, Chennai, India; Centre for Integrative Biology and Systems mEdicine (IBSE), Wadhwani School of Data Science and AI (WSAI), Indian Institute of Technology Madras, Chennai, India*

MANOKARAN VEERAMANI • *University of Nebraska-Lincoln, Lincoln, NE, USA*

B. S. YUGANDHAR REDDY • *Unilever R&D Center, Unilever India Pvt Ltd, Bengaluru, India*

STEVE ZHANG • *University of Nebraska-Lincoln, Lincoln, NE, USA*

JIANQIU ZHENG • *Pacific Northwest National Laboratory, Richland, WA, USA*

Chapter 1

Unlocking the Metagenome: Pipeline for Microbiome Data Analysis

Aarthi Ravikrishnan

Abstract

Metagenomic technologies have revolutionized our understanding of microbes in different spheres of life, revealing the massive diversity and complex functionalities of microbial communities across various environments. Shotgun metagenomics, which involves sequencing the DNA of all the organisms in a sample, is emerging as a powerful tool in assessing the microbial content. Unlike the traditional culturing approach, the shotgun metagenomic technology provides a comprehensive view of the entire microbial community, including potential functions that the organisms could be performing. In this chapter, we describe a typical bioinformatics workflow to generate the taxonomic profiles from metagenomic sequencing data and demonstrate a few basic statistical analyses that can be performed from this data to generate insights. In addition, we discuss the experimental and analytical considerations that must be taken into account while generating and making inferences from metagenomic data. Lastly, we provide insights on automating the workflow for consistent and reproducible large-scale analyses.

Key words Shotgun metagenomics, Taxonomic profiles, Statistical analyses, Microbial communities

1 Introduction

Shotgun metagenomics is a high-throughput technology that unveils the vast microbial diversity and their functionalities across various ecosystems. Shotgun metagenomic sequencing is used for studying the microbial genetic material recovered from various samples, and this usually includes genomes that are not readily cultivable in laboratories. The potential of shotgun metagenomics to identify the extensive "dark matter" in several communities gives it an edge over the traditional culturing and targeted sequencing approaches that either focus on cultivating specific species or focusing on particular regions of genome (Fig. 1). Further, such a comprehensive sequencing-based workflow allows us to capture the complete spectrum of microbial diversity including archaea, bacteria, viruses, and unicellular eukaryotes, recover whole

Karthik Raman and Gayathri Sambamoorthy (eds.), *Modelling the Microbiome*, Methods in Molecular Biology, vol. 3006, https://doi.org/10.1007/978-1-0716-5080-6_1,

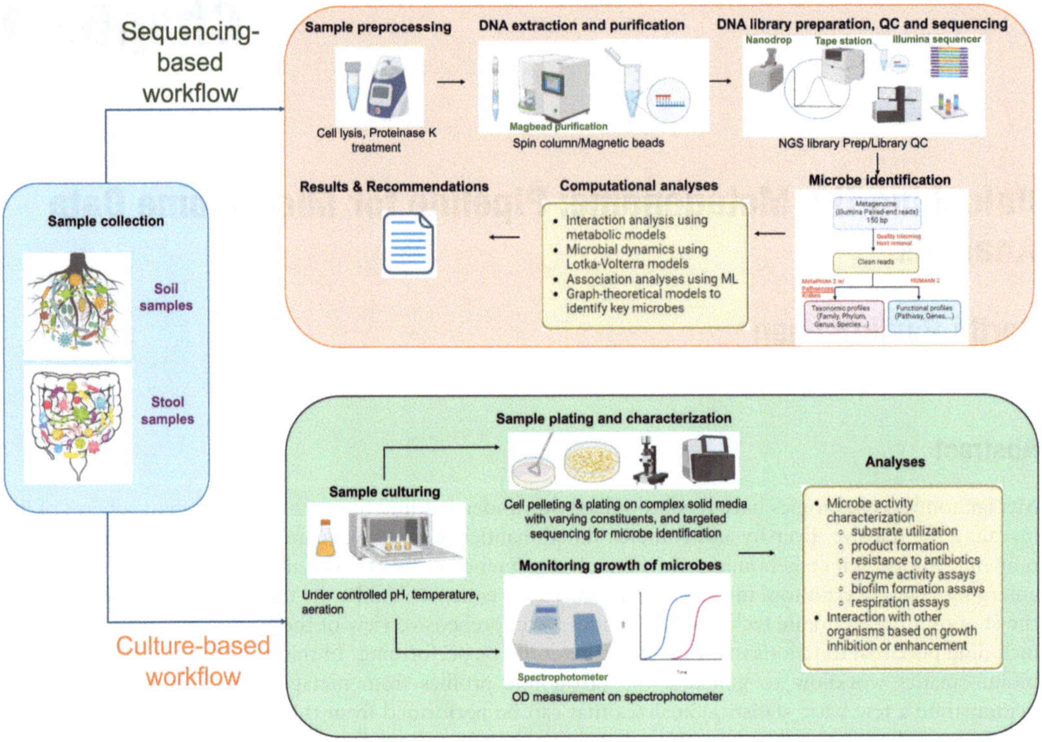

Fig. 1 Overview of workflows to study microbial communities. The top panel illustrates the steps in the sequencing-based workflow, while the bottom panel shows the steps in the culture-based workflow. The inset icons in the top and bottom panel were obtained from BioRender.com

genomes, and characterize the functional potential through identification of genes and pathways [1].

Typically, a shotgun metagenomics study has a combination of experimental and computational workflows, usually consisting of the following steps: (i) Framing hypotheses and study design; (ii) Sample collection, processing, and sequencing; (iii) Preprocessing of sequence reads; (iv) Generation of taxonomic and functional profiles from sequencing; (v) Statistical analyses and postprocessing to derive biological insights; and (vi) Experimental validation of these insights either through *in vivo* or *in vitro* experiments. In experiments where human subjects are involved, often biological insights are first obtained from "discovery cohort" and then tested on "validation cohort".

The process of sample collection, storage, and processing is one of the critical steps that determines the quality of the metagenomic data generated. The variations in protocols used at various stages such as DNA extraction and DNA sequencing can have profound effects on the microbiome data that is generated [2–6]. Indeed, the experimental variations could confound the effects of variables in question; therefore, additional precautions must be taken, and optimization must be done while testing the experimental protocols. Further, care has to be taken to reduce contamination, which can be potentially introduced by laboratory reagents or kits. In general, it is advisable to include positive and negative controls to ensure that the DNA extraction method works well and that the kits are not contaminated [7].

Based on the sequencing platform, libraries are appropriately constructed from the extracted DNA. While Illumina sequencing technology is widely used for generating short-reads, various other sequencers are available, each differing in total output, maximum read length, and sequencing chemistry. Given the well-known throughput of these sequencers, it is often advantageous to preset the sequencing depth according to the specific questions the study aims to address, since the sequencing depth is known to affect the inferences of taxonomic compositions [8]. The sequences thus generated are used to derive insights about the underlying microbial communities.

This chapter aims to offer a comprehensive overview of the essential steps and bioinformatics tools required to generate taxonomic profiles from Illumina short reads. Additionally, it outlines commonly performed microbiome and statistical analyses used in exploratory analyses and derive preliminary insights from the data. We will also provide guidance on visualization of the results wherever appropriate. By providing a robust computational framework, this chapter seeks to empower readers to decipher microbial ecology and formulate testable hypotheses for further validation.

We will divide this chapter into two major sections. In the first section, we provide a detailed outline of all the steps required to generate taxonomic profiles from raw reads. We also elaborate on the various statistical and machine-learning analyses that can be performed with metagenomic sequencing data. Additionally, we provide insights into other types of analyses, including metagenomic assemblies and functional analyses. In the second section, we illustrate the basic steps to obtain taxonomic profiles from raw sequencing reads and further elaborate on microbiome and statistical analyses. To elucidate these steps, we use publicly available MetaPhlAn 3.0 [9] taxonomic profiles obtained from curatedMetagenomicData3 [10], which hosts standardized and curated microbiome data from a wide range of shotgun metagenomics studies. Both sections will utilize data from a previously published study examining the Indian gut microbiome [11].

2 Materials and Methods

2.1 Necessary Software and Packages

1. RStudio (v2023.03.1+446 “Cherry Blossom” Release).
2. R (v4.3.0).
3. Ubuntu operating system (Ubuntu 22.04.3 LTS) or equivalent.

4. R packages and Bash utilities:
   (i) **Microbiome analyses**—`vegan (v2.6.4)`, `curatedMetagenomicData (v3.8.0)`, `ExperimentHub (v2.8.1)`, `Maaslin2 (v1.14.1)`.
   (ii) **Data science, manipulation, and visualization**—`tidyverse (v2.0.0)`, `tidyr (v1.3.0)`, `janitor (v2.2.0)`, `reshape2 (v1.4.4)`, `ggpubr (v0.6.0)`, `scales (v1.2.1)`.
   (iii) **Housekeeping operations**—`pacman (v0.5.1)`.
   (iv) **Read quality control and host DNA decontamination**—`fastp (v>=0.20.0)` [12], `BWA (>=0.7.17)` [13], `samtools (>=1.7)` [14].

2.2 Databases and Reference Sequences Needed

1. **Databases**—MetaPhlAn [15] reference database—this is automatically made available during installation.
2. **Reference sequence**—Host DNA (e.g., Human genome assembly GRCh37/hg19).

3 Methods

This section aims to provide an overview of the general steps used to process and analyze metagenomic paired-end short sequencing reads generated from Illumina sequencer.

3.1 Upstream Procedure for Generating Microbiome Profiles from Metagenomic Data

1. **Quality screening of raw reads**
   The initial step involves screening the raw reads for quality. This process uses the respective FASTQ files to trim low-quality bases and remove adapters.
2. **Host DNA removal**
   Next, based on the sample source, the host DNA is filtered out by mapping the reads to the corresponding host genome (e.g., *Homo sapiens* for human samples) to retain the microbial reads (*see* **Note 1**).
3. **Taxonomic profiling**
   The microbial reads are then analyzed to determine their taxonomy using reference-based mapping tools. Popular tools for this purpose include MetaPhlAn (v4: [15], v3: [9]), Kraken (v2: [16]), and Bracken [17]. Each tool employs different methodologies:
   - MetaPhlAn relies on marker genes.
   - Kraken uses a k-mer-based approach to classify sequences.

   These methodological differences result in varying sensitivities [18]; Kraken tends to report more taxa, whereas

MetaPhlAn is more conservative due to its use of a curated marker gene database.

In this chapter, we will generate and analyze the taxonomic profiles using MetaPhlAn. For analyses using the Kraken tool suite, readers can refer to protocols published elsewhere [19].

4. **Functional and resistome analyses**
   Additionally, microbial reads can be analyzed to gather insights into:
   - **Functional pathways**—using the HMP Unified Metabolic Analysis Network (HUMAnN 3) [9]
   - **Resistomes**—using SRST2 [20]

These upstream steps are usually computationally intensive and are carried out in High-Performance Computing (HPC) clusters (*see* **Note 2**).

3.2 Downstream Procedure for Microbiome and Statistical Analyses

1. **Quality assessment and control**
   First, the sequence quality is assessed by examining the total number of reads in each sample. This step helps identify whether all samples have a uniform number of reads and detect any outliers. Discrepancies in read counts can arise from various factors, such as inconsistent library concentrations and issues with DNA library pooling. Special care should be taken when making predictions using such atypical samples. It is usually recommended to either re-sequence these samples or treat them as outliers during the analysis.
2. **Kitome identification and removal**
   Kitomes, the contaminant bacterial DNA commonly found in laboratory reagents, have a profound influence on downstream statistical analyses [21]. Therefore, it is crucial to identify and remove these contaminant bacteria from our analysis. Identification can be achieved by analyzing negative controls and correlating the abundances of bacteria present in these controls with library concentrations. Detecting kitomes is especially important in samples with low bacterial biomass and high host reads, such as those from skin and environmental sources. Once identified, kitomes should be removed from the taxonomic profiles before proceeding with further analysis.
3. **Filtering**
   After the removal of kitomes, the taxonomic profiles are further filtered to remove very low-abundant taxa to ensure that false positives do not affect the downstream analysis.

4. **Normalization**
   The most commonly used normalization is known as Total Sum Scaling (TSS), where the taxonomic profiles are normalized so that the sum of proportions is 100. Depending on the nature and type of data, other types of normalization can be done, including Cumulative Sum Scaling (CSS) [22], Trimmed Mean of M-values (TMM), Upper Quartile Normalization (UQ), and Median Normalization (MED).
5. **Transformation**
   Since taxonomic profiles are inherently compositional, it is useful to transform the data from the positive simplex to Euclidean vector space. This transformation can be effectively accomplished using log-ratio-based approaches including Additive Log Ratio (ALR), Centered Log Ratio (CLR), and Isometric Log Ratio (ILR) transformations.
6. **Batch correction**
   When conducting meta-analysis with taxonomic profiles, it is crucial to ensure uniform processing of data across batches. Consistency in experimental procedures and sequencing technology helps prevent variations in microbiome profiles. If samples originate from different protocols, correcting for batch effects becomes essential. Statistical methods such as MMUPHin [23] and ComBat [24] are available for this purpose; however, no batch correction tool is flawless. Therefore, caution is advised when interpreting results derived from such analyses.
7. **Microbiome analyses**
   The microbiome profiles are initially filtered for kitomes, followed by appropriate normalization and transformation procedures. Subsequently, these taxonomic profiles are analyzed to assess alpha diversity, beta diversity, microbial community structure, and identification of core organisms. Additionally, functional analyses can be done using the pathway and gene abundance tables. Further, depending on the sequencing depth, read length, and quality, metagenome assemblies can be constructed from these reads. These assemblies help in several ways, including recovering microbial genomes, performing functional analyses, and identifying novel genes and pathways.
8. **Statistical analyses**
   Statistical methods serve various purposes, such as identifying differentially abundant taxa, exploring correlations between organisms, and assessing associations of microbes with metadata containing clinical parameters, lifestyle conditions, diet, and other relevant factors. There are several tools available to identify differentially abundant taxa, with some of the most commonly used being MaAsLin2 [25], LEfSe [26], and

ANCOM-BC [27]. Each of these tools employs distinct methodologies to pinpoint taxa that differ significantly between groups. MaAsLin2 operates within a generalized linear model framework, LEfSe utilizes linear discriminant analysis, and ANCOM-BC adjusts for sampling fractions before employing absolute abundances in a linear regression framework. Additionally, alternative statistical models such as Linear Mixed Models (LMM), which consider the interactions between the variables, can be employed. Given the sparse nature of taxonomic profiles, we may also opt for models tailored to zero-inflated data, such as Zero-Inflated Beta Regression (ZIBR), Zero-Inflated Gaussian Mixed Models (ZIGMM), and Negative Binomial Mixed Models (NBMM), depending on the specific case. The selection of a model primarily hinges on the research questions and the characteristics of the data obtained from the study.

9. **Machine-learning analyses**
   Taken together with the metadata, taxonomic profiles can further be used in developing classifiers using techniques like random forests and Least Absolute Shrinkage and Selection Operator (LASSO). These models are pivotal in addressing fundamental questions, such as whether classifiers can predict underlying conditions using microbiome data. For detailed information on machine learning-based tools, readers are encouraged to explore resources and examples discussed elsewhere [28].

Overall, for better reproducibility of results and portability, it is advisable to run all the applications in a container on Docker platform (*see* **Note 3**).

4 Example Workflow

In this section, we will illustrate how to conduct exploratory microbiome and statistical analyses following the steps outlined above, using an example from a publicly available dataset.

4.1 Generation of Taxonomic Profiles from Raw Reads

Here, we will look at commands to generate taxonomic profiles from the paired-reads generated. We will provide the commands for one example to illustrate all the steps from decontamination to generation of taxonomic profiles.

4.1.1 Decontamination and Removal of Host Reads

In the first step, the paired end reads are trimmed for low-quality bases using fastp. Following this, the trimmed reads are mapped against a reference genome (e.g., hg19) using BWA MEM to identify host reads. Subsequently, samtools is utilized to filter out the host reads, resulting in microbial reads from the sample.

Do note that the reference genome has to be indexed before using it here. This can be done using:

```
bwa index hg19.fa
fastp -i sampleA_1.fastq.gz -I sampleA_2.fastq.gz --
stdout -j sampleA.json -h sampleA.html | bwa mem -p -t
20 hg19.fa - | samtools fastq -f12 -F256 -1 sam-
pleA_dc_1.fastq.gz -2 sampleA_dc_2.fastq.gz -
```

In the command above, we run `fastp` on paired-end reads of `sampleA`. The flags in the command follow the nomenclature from the `fastp` tool. The resulting JSON file from `fastp` provides information about bases in the paired-end `fastq` reads. The filtered reads are then mapped to the `hg19` reference sequence to identify the host reads. `samtools` uses the input from bwa mem and removes all the host reads. The decontaminated and filtered sequences obtained from this output are used for further analyses.

4.1.2 Generation of Taxonomic Profiles

The microbial reads are then classified to respective taxonomies using reference-based approaches. Here, we show the command for obtaining taxonomic profiles using MetaPhlAn 4.

```
metaphlan metagenome_1.fastq,metagenome_2.fastq --bow-
tie2out
metagenome.bowtie2.bz2 --nproc 5 --input_type fastq -o
profiled_metagenome.txt
```

The profiled_metagenome.txt is obtained for all the samples, and the results are merged as a CSV file, which is later used in downstream analyses. These steps are usually written in pipeline such as `snakemake` [29] or `nextflow` [30], since this will enable high-throughput processing of all the sequences (*see* **Note 4**).

4.2 Data Download and Quality Control

In this section, we will elaborate on the basic steps for microbiome data analysis to enable exploratory data analysis. We will be doing all these steps in R.

4.2.1 Obtaining Example Data from curatedMetagenomicData

In the first step, we create an R script using the following commands to access `curatedMetagenomicData` and retrieve the data for the study we are interested in.

We begin by clearing all the variables in the current environment, loading the necessary packages and setting up the workspace:

```
rm(list = ls())
pacman::p_load(tidyverse, tidyr, janitor, here, ggpubr,
scales, reshape2)
```

```
pacman::p_load(curatedMetagenomicData,    ExperimentHub,
ccrepe)
pacman::p_load(Maaslin2)
color_palette  <-  c("#1F77B4",  "#393B79",  "#B5CF6B",
"#D62728",
 "#AEC7E8", "#FF7F0E", "#843C39", "#D6616B",
 "#FFBB78", "#2CA02C", "#98DF8A", "#FF9896",
 "#9467BD", "#C5B0D5", "#8C564B", "#66C2A5",
 "#E377C2", "yellow", "#7F7F7F", "#7B4173")
sampletypecol  <-  c('Bhopal'  ='#D81B60',  'Kerala'  =
'#FFC107')
```

In this work, we will be extracting metagenomic data from a previously published study [11], where the gut microbiome of 110 individuals was analyzed. As a general procedure, we begin by examining the metadata to ensure that all necessary information is available. In addition to adhering to the study design criteria, exploring the metadata, particularly while analyzing publicly available datasets, helps determine whether further stratification of subjects is possible.

```
sample_metadata <- curatedMetagenomicData::sampleMeta-
data
metadata <- sample_metadata %>% filter(study_name == 'Dha-
kanDB_2019')
age_dist <-
 ggplot(metadata, aes(x = age, fill = location)) +
 geom_histogram(bins = 50) +
 facet_wrap(. ~ location, scales = 'free') +
 scale_fill_manual(values = sampletypecol) +
 xlab('Age') + ylab('Count') + theme(legend.position =
"none")
print(age_dist)
```

From the age distribution (Fig. 2), we can clearly notice two groups of subjects from both the locations and hence classify them into two groups (i) Young (<30) and (ii) Old (>50) and get the number of samples in each group:

```
metadata <- metadata %>%
  mutate(age_g = case_when(age < 30 ~ 'Young', TRUE ~
'Old')) %>%
   select(c("sample_id",  "body_site",  "number_reads",
"BMI",
```

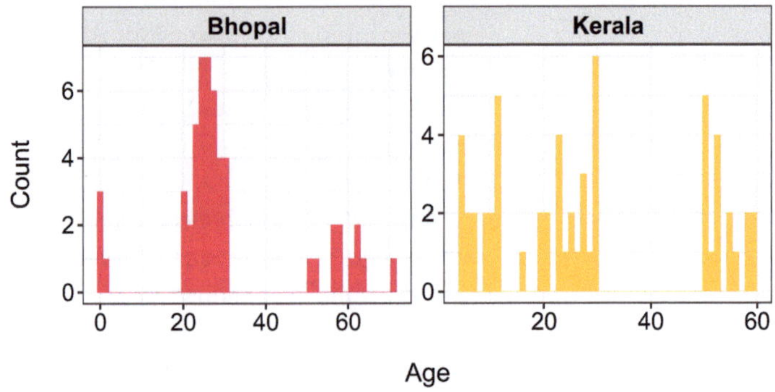

Fig. 2 Age distribution: Histogram showing the number of samples in both locations

```
"location", "age", "gender", "age_g"))
metadata %>% group_by(age_g) %>% summarise(n = n())
```

Next, we obtain the corresponding identifier from ExperimentHub to download the taxonomic profiles for this dataset. We first retrieve the taxonomic profiles and then extract genus and species data into separate data frames for analysis. We also verify the column sums of both genus and species data frames to ensure that all entries are complete and that the total is 100. It is important to note that the taxonomic data obtained from MetaPhlAn is often compositional, meaning the sum of proportions should be 100 (or 1).

```
eh <- ExperimentHub()
ddb_taxa <- eh[["EH3141"]]
species <-
 ddb_taxa %>%
 as.data.frame() %>% t() %>% as.data.frame() %>%
 rownames_to_column('Orgs') %>%
 filter(grepl('s__', Orgs)) %>%
 filter(!grepl('t__', Orgs)) %>%
  separate(col = "Orgs", into=c("Kingdom", "Phylum",
"Class", "Order",
 "Family", "Genus", "Species"),
 sep = '\cr.') %>%
 select(-c("Kingdom", "Phylum", "Class", "Order",
 "Family", "Genus")) %>%
 mutate(Species = str_replace_all(Species, 's__', '')) %>
%
 column_to_rownames("Species") %>%
 mutate_all(as.numeric) %>%
```

```
  select(metadata$sample_id)
colSums(species)
```

4.2.2 Basic Quality Control (QC) Steps

We begin the analysis by performing quality control, starting with an examination of read counts across the samples.

```
num_reads <-
  ggplot(metadata, aes(x = age_g, y = number_reads / 10 ^
6,
  fill = location)) +
  geom_boxplot() +
  facet_wrap( ~ location, scales = 'free') +
  scale_fill_manual(values = sampletypecol) +
  theme(legend.position = "none") +
  xlab('Age group') +
  ylab('# Reads (X 10^6)') +
  theme(legend.position = "none")
print(num_reads)
```

From Fig. 3, we can infer several aspects about the sequencing depth: (a) consistently across both age groups and locations, and young subjects have a higher number of reads compared to older subjects; (b) samples from one location (Bhopal) exhibit higher sequencing depths; and (c) there are a few outliers in both groups across the two locations considered.

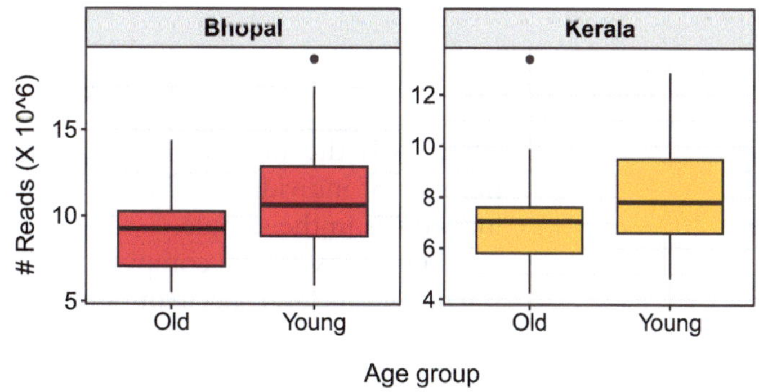

Fig. 3 Number of reads in each group: Boxplot showing the number of reads in each group. Median is represented by the line at the center, upper and lower quartiles are represented by the limits of the box, and the minimum and maximum values are represented by whiskers. Outliers are represented as dots outside the whiskers.

4.2.3 Kitome Identification and Removal

We skip this step in our current example since the taxonomic data is derived from stool samples, which have a higher biomass compared to human DNA reads [31]. Additionally, we lack the necessary data to accurately identify kitomes.

4.3 Microbiome Analysis

In this section, we will discuss some preliminary analyses that can be performed with taxonomic profiles. We will begin by computing alpha- and beta-diversity indices, followed by identifying core taxa, analyzing microbial communities, and constructing microbial correlation networks.

4.3.1 Alpha-Diversity Indices

Alpha-diversity indices are used to quantify the diversity within a particular sample. Commonly used metrics include species richness, Shannon diversity, and Simpson diversity.

Richness

Richness refers to the number of unique species found in an ecosystem or sample. It is typically computed by counting the number of distinct species present in the samples. The `specnumber` function from the `vegan` package in R facilitates the computation of species richness.

```
richness <-
 specnumber(x = t(species)) %>%
 data.frame() %>%
 setNames('Richness') %>%
 rownames_to_column('sample_id')
```

Simpson Diversity

Simpson diversity measures the probability that two individuals randomly selected from a sample will belong to the same species.

Mathematically, the Simpson diversity index (D) is calculated as follows:

$$D = \frac{\sum_{i=1}^{S} n_i(n_i - 1)}{N(N-1)}$$

where S is the number of different microbial species, n_i is the number of individuals of species i, and N is the total number of individuals in the sample.

This index can be computed using the diversity function from vegan package and specifying index as Simpson.

```
simpson <-
 diversity(x = t(species), index = 'simpson') %>%
 data.frame() %>%
 setNames('Simpson') %>%
 rownames_to_column('sample_id')
```

Shannon Diversity

Shannon diversity considers both species richness and evenness. It is often regarded as entropy and quantifies the uncertainty of predicting the identity of species randomly selected from a microbial community. The more the number of species, the higher the uncertainty.

$$H' = -\sum_{i=1}^{S} p_i \ln(p_i)$$

where S is the number of different microbial species, and p_i is the proportion of the ith species relative to the total abundance of all species.

```
shannon <-
  diversity(x = t(species), index = 'shannon') %>%
  data.frame() %>%
  setNames('Shannon') %>%
  rownames_to_column('sample_id')
```

Plotting all alpha-diversities

```
all_alpha <-
  merge(richness, simpson, by = 'sample_id') %>%
  left_join(shannon, by = 'sample_id') %>%
  left_join(metadata %>%
   select('sample_id', 'age_g', 'location'), by = 'sam-
ple_id') %>%
:melt()
alpha_div_plot <-
  ggplot(all_alpha, aes(x = age_g, y = value, fill = loca-
tion)) +
  facet_wrap( ~ variable, scales = 'free', ncol = 3) +
  geom_boxplot() +
  scale_y_continuous(breaks = pretty_breaks(n = 8)) +
  stat_compare_means(label = "p.format") +
  xlab('Age group') + ylab('Value') +
  labs(fill = 'Location') +
  scale_fill_manual(values = sampletypecol)
print(alpha_div_plot)
```

Based on the alpha diversity plot (Fig. 4), we can notice that, except for richness in the young age group, all the other comparisons between samples from the two locations are significantly different from one another. We also note that consistently, samples from Kerala exhibit higher values of alpha-diversity metrics.

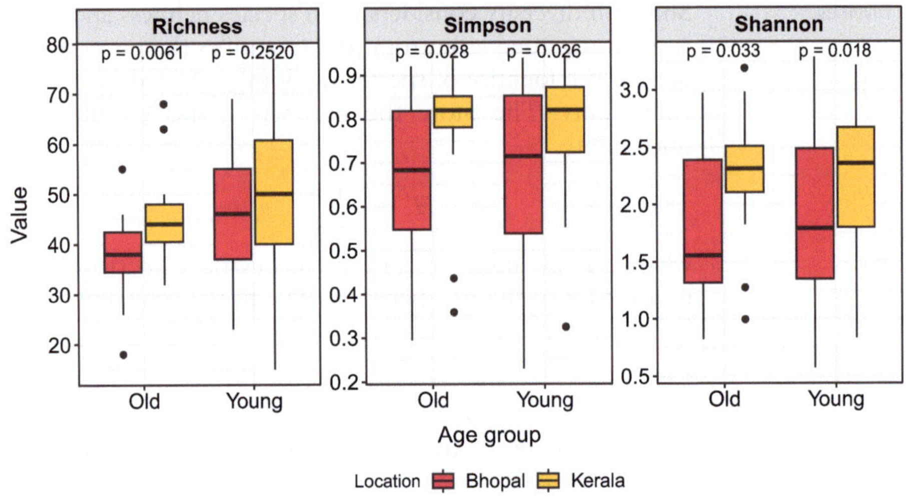

Fig. 4 Alpha diversity differences between two groups: Boxplots showing the various alpha diversity metrics (Richness, Simpson and Shannon diversities (from left to right)). The p-values shown are derived from Kruskal-Wallis test between the samples derived from two locations within every age group. Median is represented by the line at the center, upper and lower quartiles are represented by the limits of the box, and the minimum and maximum values are represented by whiskers. Outliers are represented as dots outside the whiskers

4.3.2 Beta-Diversity Indices

Beta-diversity indices quantify the dissimilarity or distances between different samples based on the relative abundances of microorganisms in each sample. The most commonly used metrics include (a) Bray–Curtis dissimilarity, which utilizes the relative abundance table, and (b) Jaccard distance, which is based on binary presence–absence values.

Bray–Curtis Dissimilarity

Bray–Curtis dissimilarity computes the compositional dissimilarity between two sites based on the abundances at each site.

The Bray–Curtis dissimilarity between two samples A and B is calculated as follows:

$$\text{Bray} - \text{Curtis}(A, B) = \frac{\sum | a_i - b_i |}{\sum a_i + b_i}$$

where a_i and b_i are the abundances of species i in samples A and B, respectively.

Bray–Curtis dissimilarity ranges from 0 to 1, where 0 indicates identical composition between the two samples, and 1 indicates completely different compositions.

Jaccard Distance

Jaccard distance is used to quantify the dissimilarity between samples in terms of the proportion of shared species to the total species present.

The Jaccard index between sets A and B is computed as follows:

$$J(A, B) = 1 - \frac{|A \cap B|}{|A \cup B|}$$

To compute (a) Bray–Curtis dissimilarity index, we use `bray` function from `vegan` package, and to compute (b) Jaccard dissimilarity, we use `jaccard` function from `vegan` package. We visualize the distances on the Principal Coordinate Analysis (PCoA) plot.

```
bray <- vegan::vegdist(t(as.matrix(species)), method =
"bray")
cmds_bray <- cmdscale(bray, k = 3, eig = TRUE)
eigen_bray <- cmds_bray$eig / sum(cmds_bray$eig) * 100
cmdvals_bray <- as.data.frame(cmds_bray$points)
bray_curtis <- cmdvals_bray %>%
 rownames_to_column('sample_id') %>%
 left_join(metadata)
bray_curtis$index <- 'Bray-Curtis dissimilarity'
bray_pcoa_plot <-
 ggplot(bray_curtis, aes(x = V1, y = V2, colour = loca-
tion, shape = age_g)) +
  labs(x = paste0('PCoA1 (', round(eigen_bray[1], 1),
'%)'),
 y = paste0('PCoA2 (', round(eigen_bray[2], 1), '%)')) +
 facet_wrap(~index, scales = 'free') +
 geom_point(data = subset(bray_curtis,location == "Bho-
pal"),
 aes(color = location), size = 4) +
 geom_point(data = subset(bray_curtis, location == "Ker-
ala"),
 aes(color = location), size = 4) +
 scale_color_manual(values = sampletypecol) +
 labs(shape = 'Age group', colour = 'Location')
jaccard <- vegan::vegdist(t(species), method = 'jaccard',
binary = TRUE)
cmds_jaccard <- cmdscale(jaccard, k = 3, eig = TRUE)
eigen_jaccard <- cmds_jaccard$eig / sum(cmds_jaccard$eig)
* 100
cmdvals_jaccard <- as.data.frame(cmds_jaccard$points)
jaccard_distance <- cmdvals_jaccard %>%
 rownames_to_column('sample id') %>%
 left_join(metadata)
jaccard_distance$index <- 'Jaccard distance'
jaccard_pcoa_plot <-
 ggplot(jaccard_distance, aes(x = V1, y = V2, colour =
location, shape = age_g)) +
 labs(x = paste0('PCoA1 (', round(eigen_jaccard[1], 1),
```

```
'%)'),
 y = paste0('PCoA2 (', round(eigen_jaccard[2], 1), '%)'))
+
 facet_wrap(~index, scales = 'free') +
 geom_point(data = subset(jaccard_distance, location ==
"Bhopal"),
 aes(color = location), size = 4) +
 geom_point(data = subset(jaccard_distance, location ==
"Kerala"),
 aes(color = location), size = 4) +
 scale_color_manual(values = sampletypecol) +
 labs(shape = 'Age group', colour = 'Location')
ggpubr::ggarrange(bray_pcoa_plot, jaccard_pcoa_plot,
 common.legend = TRUE, legend = "bottom")
```

From the beta-diversity plots (Fig. 5), we can observe that despite some overlaps, the samples from two locations cluster separately; however, no such clear separation is seen based on the age groups.

PERMANOVA Analysis

To assess significant variations in microbiome composition between the two groups and locations, we conduct PERMANOVA (Permutational Multivariate Analysis of Variance), a nonparametric statistical test capable of detecting differences across multiple conditions. This analysis utilizes a distance matrix to quantify dissimilarities based on a chosen distance measure. Here, we employed the Bray–Curtis dissimilarity metric to evaluate the differences between age groups and locations.

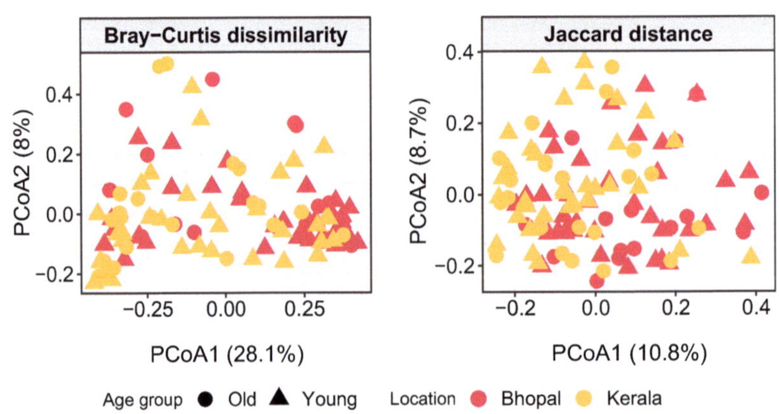

Fig. 5 Beta diversity between two groups: Principal Coordinates Analysis (PCoA) plot showing Bray-Curtis distances (on left) and Jaccard distance (on right). The colour of the points represent the location, while the shape of the points represent the age groups

```
set.seed(1)
fit_adonis <- vegan::adonis2(bray ~ age_g + location, data
= metadata)
print(fit_adonis)
## Permutation test for adonis under reduced model
## Terms added sequentially (first to last)
## Permutation: free
## Number of permutations: 999
##
## vegan::adonis2(formula = bray ~ age_g + location, data
= metadata)
## Df SumOfSqs R2 F Pr(>F)
## age_g 1 0.6272 0.02053 2.3600 0.024 *
## location 1 1.4909 0.04880 5.6104 0.001 ***
## Residual 107 28.4339 0.93067
## Total 109 30.5519 1.00000
## ---
## Signif. codes: 0 '***' 0.001 '**' 0.01 '*' 0.05 '.' 0.1
' ' 1
```

The PERMANOVA results reveal significant differences between age groups and locations, highlighting that location exhibits stronger statistical significance compared to age group.

4.3.3 Microbial Community Analysis

Microbial community analysis is often useful for obtaining a basic overview of the microbes in the community. We start by examining the top 10 species (by mean relative abundances) present in both age groups across the two different locations. This approach helps us identify any dominant organisms within the community and discern any differences between the groups.

```
orgstoplot <-
 species %>%
 rownames_to_column('taxa') %>%
 melt(variable.name = 'sample_id') %>%
 left_join(metadata, by = 'sample_id') %>%
 group_by(age_g, location, taxa) %>%
 summarise(meanval = mean(value)) %>%
 slice_max(order_by = meanval, n = 10) %>%
 mutate(taxaplot = forcats::fct_reorder(taxa, meanval,.
fun = mean,.desc = TRUE)) %>%
 mutate(taxaplot = str_replace_all(taxaplot, '_', ' '))
stack_plot <-
 ggplot(orgstoplot, aes(x = age_g, y = meanval, fill =
taxaplot)) +
```

```
  geom_bar(stat = "identity",
  colour = 'black',
  width = 0.5) +
  scale_fill_manual(values = color_palette) +
  facet_grid(. ~ location, scales = 'free') +
  labs(x = NULL, y = "Relative Abundances (Mean)") +
  theme(
  panel.grid.major = element_blank(),
  panel.grid.minor = element_blank(),
  legend.spacing = unit(0.35, "cm"),
  legend.key = element_rect(colour = "black"),
  legend.text = element_text(size = 9, face = "italic"),
  legend.title = element_blank(),
  legend.position = "bottom"
 ) +
  coord_flip() +
  guides(fill = guide_legend(ncol = 3)) +
  scale_y_continuous(breaks = pretty_breaks(n = 5))
print(stack_plot)
```

From the plot (Fig. 6), we note that across both the groups, the most common species present at a higher mean abundance is *Prevotella copri*.

4.3.4 Core Taxa Identification

We define core taxa as those present in at least 70% of the samples across different groups and locations. To identify these core taxa, we calculate the prevalence of each taxon across both groups and select those that meet the specified criteria.

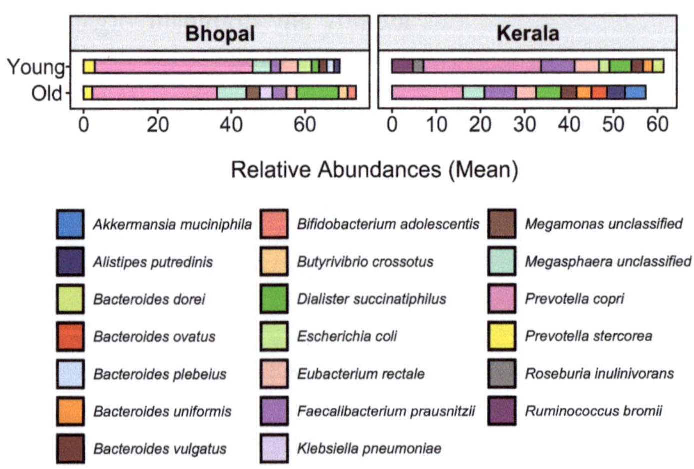

Fig. 6 Relative abundances of top 10 species: Stacked bar chart showing the top 10 most abundant species in each group. Mean relative abundances of each organism in every location and group are plotted. The colours in the bar chart represent different species as represented in the legend

```
num_samples_pergroup <- metadata %>%
  group_by(location, age_g) %>%
  summarise(n=n())
core <-
  species %>%
  t() %>%
  data.frame %>%
  rownames_to_column('sample_id') %>%
  left_join(metadata %>% select('sample_id', 'location',
'age_g')) %>%
  melt(variable.name = 'taxa') %>%
  group_by(location, age_g, taxa) %>%
  summarise(val = sum(value != 0), med = median(value)) %>
%
  left_join(num_samples_pergroup, by = c('location',
'age_g')) %>%
  mutate(prop = val / n) %>% filter(prop >= 0.7) %>%
  mutate(taxa = str_replace_all(taxa, '_', ' '))
core <- core %>% arrange(location, age_g, prop)
dotplot <-
  ggplot(core, aes(y = reorder(taxa, prop), x = age_g)) +
  geom_point(aes(size = prop, colour = location)) +
  facet_grid( ~ location, scales = 'free') +
  labs(color = '', size = 'Proportion') +
  theme(axis.text.y = element_text(face = 'italic')) +
  ylab('Taxa') + xlab('Age group') +
  scale_color_manual(values = sampletypecol) +
  guides(color = FALSE)
print(dotplot)
```

From the dot plot (Fig. 7), it is evident that the core set of organisms varies significantly. Only a small number of species ($n = 6$) are consistently conserved across both locations and age groups.

4.4 Statistical Analysis

In this section, we will demonstrate steps to perform differential abundance analysis and microbial correlation analysis.

4.4.1 Differential Abundance Analysis

In this example, our goal is to identify differentially abundant highly prevalent taxa between the two locations. Considering that age group has a significant impact on taxonomic profiles, we will treat it as a confounding factor. To achieve this, we will use Linear Model (LM) within the MaAsLin2 framework, a tool designed for such analyses.

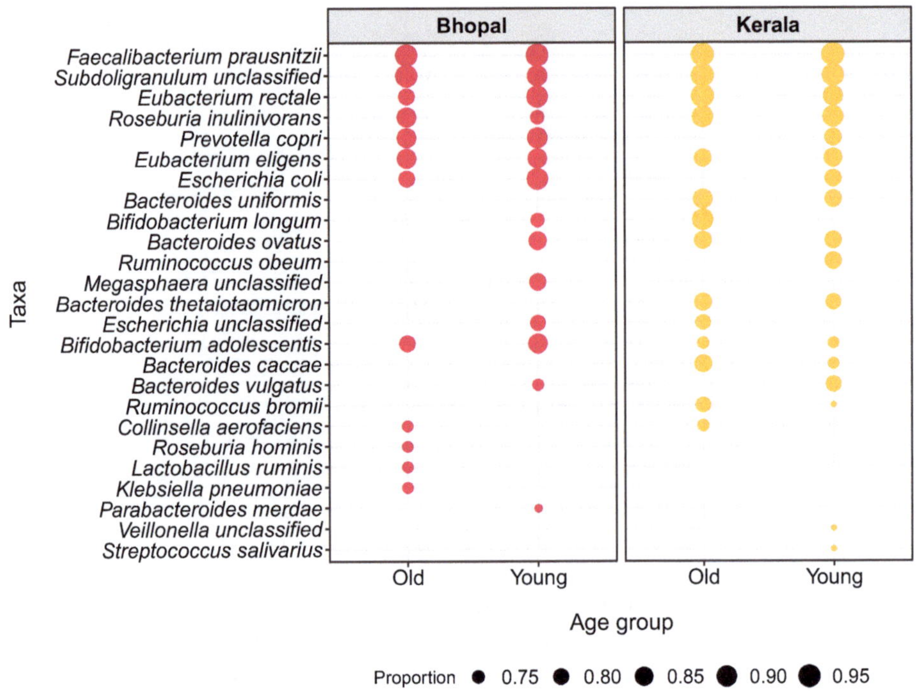

Fig. 7 Core taxa in both groups across two locations: Dot plot showing the core taxa in each group across two locations. Core taxa is defined as the taxa that is present in at least 70% of the samples. Size of the dot indicates the proportion of samples where the taxa is present

```
metadata <- metadata %>% column_to_rownames('sample_id')
maaslin2_out <- Maaslin2(input_data = species, normaliza-
tion = "TSS",
 transform = "log",analysis_method = "LM",
 min_prevalence = 0.7, min_abundance = 0.1,
 input_metadata = metadata, output = 'maaslin_res',
 fixed_effects = c('location', 'age_g'),
 correction = 'BH', standardize = FALSE,
 max_significance = 0.05)
significant_results <- maaslin2_out$results %>% filter(qval
< 0.05)
taxa_needed <-
 species %>%
 rownames_to_column('taxa') %>%
 dplyr::filter(taxa %in% significant_results$feature) %>%
 column_to_rownames('taxa') %>% t() %>% data.frame() %>%
 rownames_to_column('sample_id') %>%
 left_join(metadata %>% rownames_to_column('sample_id') %
>%
 dplyr::select('sample_id', 'location'), by='sample_id')
%>%
```

```
  dplyr::select(-c('sample_id')) %>%
  melt(variable.name = 'taxa', value.name = 'abundance') %
>%
  mutate(taxa = str_replace_all(taxa, '_', ' '))
diff_taxa <-
  ggplot(taxa_needed, aes(x=location, y=abundance, fill=lo-
cation)) +
  geom_boxplot() +
  scale_fill_manual(values=sampletypecol) +
  facet_wrap(~taxa, scales='free') +
  theme(legend.position = "none",
  strip.text = element_text(size=10, face='italic')) +
  xlab('')+ ylab('Relative abundance') +
  labs(fill='Location')
print(diff_taxa)
```

From the results (Fig. 8), we observe that there are three species that are differentially abundant between both the groups. It would be interesting to assess the robustness of these differentially abundant species by checking with other methods also (*see* **Note 5**).

5 Notes

1. Even after the decontamination step, residual human reads may remain in the sample. Depending on the classification tools used, it is crucial to ensure that these remaining human reads do not compromise the generated data.

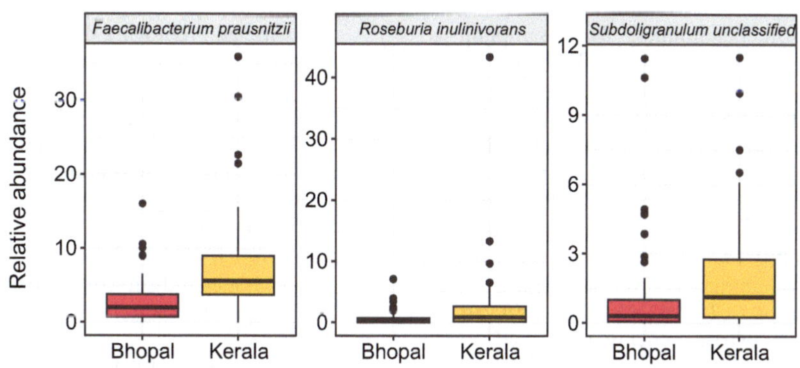

Fig. 8 Differentially abundant taxa between two groups across locations: Boxplot showing the relative abundances of differentially abundant taxa across two locations. Median is represented by the line at the center, upper and lower quartiles are represented by the limits of the box, and the minimum and maximum values are represented by whiskers. Outliers are represented as dots outside the whiskers

2. Many software tools used in upstream analyses are available as singularity images, facilitating ease of use in high-performance computing environments. Singularity usually runs with non-root privileges and hence is compatible with HPC resources.
3. For data analyses, it is beneficial to utilize docker images to avoid version inconsistencies that could affect the results. Further, docker has a large community support and can be used with cloud services, thereby making it ideal for ensuring consistency and reproducibility
4. Upstream analyses are typically conducted using `snakemake` or `nextflow` pipelines, ensuring uniform processing of high-throughput data. These pipelines provide reproducibility and are scalable across large datasets. Due to the streamlined data processing, using such pipelines helps in maintaining consistent workflows.
5. For downstream analyses, it is recommended to verify results across different methods and techniques to account for potential discrepancies (as discussed in [32, 33]). A more conservative approach involves investigating the conserved group of results from all methods for further hypothesis generation and gathering biological insights.

Conflict of Interest Statement

The authors have no conflicts of interest to declare that are relevant to the content of this chapter.

References

1. Quince C, Walker AW, Simpson JT et al (2017) Shotgun metagenomics, from sampling to analysis. Nat Biotechnol 35:833–844
2. Wesolowska-Andersen A, Bahl MI, Carvalho V et al (2014) Choice of bacterial DNA extraction method from fecal material influences community structure as evaluated by metagenomic analysis. Microbiome 2:19
3. Grahn N, Olofsson M, Ellnebo-Svedlund K et al (2003) Identification of mixed bacterial DNA contamination in broad-range PCR amplification of 16S rDNA V1 and V3 variable regions by pyrosequencing of cloned amplicons. FEMS Microbiol Lett 219:87–91
4. Yuan S, Cohen DB, Ravel J et al (2012) Evaluation of methods for the extraction and purification of DNA from the human microbiome. PLoS One 7:e33865
5. Fachrul M, Meric G, Inouye M et al (2022) Assessing and removing the effect of unwanted technical variations in microbiome data. Sci Rep 12:22236
6. Fernandez-Pato A, Sinha T, Gacesa R et al (2024) Choice of DNA extraction method affects stool microbiome recovery and subsequent phenotypic association analyses. Sci Rep 14:3911
7. Hornung BVH, Zwittink RD, Kuijper EJ (2019) Issues and current standards of controls in microbiome research. FEMS Microbiol Ecol 95(5):fiz045
8. Gweon HS, Shaw LP, Swann J et al (2019) The impact of sequencing depth on the inferred taxonomic composition and AMR gene content of metagenomic samples. Environ Microbiome 14:7

9. Beghini F, McIver LJ, Blanco-Miguez A et al (2021) Integrating taxonomic, functional, and strain-level profiling of diverse microbial communities with bioBakery 3. elife 10:e65088
10. Pasolli E, Schiffer L, Manghi P et al (2017) Accessible, curated metagenomic data through ExperimentHub. Nat Methods 14:1023–1024
11. Dhakan DB, Maji A, Sharma AK et al (2019) The unique composition of indian gut microbiome, gene catalogue, and associated fecal metabolome deciphered using multi-omics approaches. Gigascience 8(3):giz004
12. Chen S, Zhou Y, Chen Y, Gu J (2018) Fastp: an ultra-fast all-in-one FASTQ preprocessor. Bioinformatics 34:i884–i890
13. Li H (2013) Aligning sequence reads, clone sequences and assembly contigs with BWA-MEM. arXiv:1303.3997
14. Danecek P, Bonfield JK, Liddle J et al (2021) Twelve years of SAMtools and BCFtools. GigaScience 10. https://doi.org/10.1093/gigascience/giab008
15. Blanco-Miguez A, Beghini F, Cumbo F et al (2023) Extending and improving metagenomic taxonomic profiling with uncharacterized species using MetaPhlAn 4. Nat Biotechnol 41:1633–1644
16. Wood DE, Lu J, Langmead B (2019) Improved metagenomic analysis with kraken 2. Genome Biol 20:257
17. Lu J, Breitwieser FP, Thielen P, Salzberg SL (2017) Bracken: estimating species abundance in metagenomics data. PeerJ Comput Sci 3: e104. https://doi.org/10.7717/peerj-cs.104
18. Wright RJ, Comeau AM, Langille MGI (2023) From defaults to databases: parameter and database choice dramatically impact the performance of metagenomic taxonomic classification tools. Microb Genom 9(3):000949
19. Lu J, Rincon N, Wood DE et al (2022) Metagenome analysis using the kraken software suite. Nat Protoc 17:2815–2839
20. Inouye M, Dashnow H, Raven LA et al (2014) SRST2: Rapid genomic surveillance for public health and hospital microbiology labs. Genome Med 6:90
21. Salter SJ, Cox MJ, Turek EM et al (2014) Reagent and laboratory contamination can critically impact sequence-based microbiome analyses. BMC Biol 12:87
22. Paulson JN, Stine OC, Bravo HC, Pop M (2013) Differential abundance analysis for microbial marker-gene surveys. Nat Methods 10:1200–1202
23. Ma S, Shungin D, Mallick H et al (2022) Population structure discovery in meta-analyzed microbial communities and inflammatory bowel disease using MMUPHin. Genome Biol 23:208
24. Zhang Y, Parmigiani G, Johnson WE (2020) ComBat-seq: Batch effect adjustment for RNA-seq count data. NAR Genom Bioinform 2:lqaa078
25. Mallick H, Rahnavard A, McIver LJ et al (2021) Multivariable association discovery in population-scale meta-omics studies. PLoS Comput Biol 17:e1009442
26. Segata N, Izard J, Waldron L et al (2011) Metagenomic biomarker discovery and explanation. Genome Biol 12:R60
27. Lin H, Peddada SD (2020) Analysis of compositions of microbiomes with bias correction. Nat Commun 11:3514
28. Topçuoğlu BD, Lapp Z, Sovacool KL et al (2021) mikropml: User-friendly r package for supervised machine learning pipelines. J Open Source Softw 6:3073
29. Mölder F, Jablonski K, Letcher B et al (2021) Sustainable data analysis with snakemake [version 1; peer review: 1 approved, 1 approved with reservations]. F1000Res 10:33
30. Di Tommaso P, Chatzou M, Floden EW et al (2017) Nextflow enables reproducible computational workflows. Nat Biotechnol 35:316–319
31. Sui HY, Weil AA, Nuwagira E et al (2020) Impact of DNA extraction method on variation in human and built environment microbial community and functional profiles assessed by shotgun metagenomics sequencing. Front Microbiol 11:953
32. Nearing JT, Douglas GM, Hayes MG et al (2022) Microbiome differential abundance methods produce different results across 38 datasets. Nat Commun 13:342
33. Lin H, Peddada SD (2020) Analysis of microbial compositions: a review of normalization and differential abundance analysis. NPJ Biofilms Microbiomes 6:60

Chapter 2

Targeted Metagenomics Using Next-Generation Sequencing Methods

B. S. Yugandhar Reddy, S. Sripradha, and Achyut Kumar

Abstract

Metagenomics allows the discovery of the full diversity of all microbes present in a given niche. The technique is very powerful and has allowed very significant advances delineating the role of the microbiome in several disciplines including health, agriculture, ecology, industry, etc. Here, we describe the method required for processing of samples for metagenomic analysis using Next-Gen sequencing.

Key words Metagenomics, Metataxonomics, 16S sequencing, ITS sequencing, NGS, Microbiomics

1 Introduction

Metagenomics using either targeted or amplicon-based methods or shotgun-based metagenomics has revolutionized microbiology in the twenty-first century. The method involves extracting DNA or RNA and using it for sequencing to assess the full diversity of microbes in any given sample or niche [1]. This requires standardized protocols for wet lab handling of samples, extraction of nucleic acids, library preparation, sequencing, and further downstream processing. Samples for sequencing are varied and pose a significant challenge for optimal extraction of nucleic acids, failing which it can skew the diversity and composition determined [2]. For example, the inability to break complex cell walls or spores leads to underrepresentation or nondetection of such microbes. Optimization of extraction methods, kits, reagents, and primers is to be done, especially if novel samples are being sequenced for the very first time. In our experience with uncommon samples such as clothes, dust, and hard surfaces from indoor environments, we have had to iterate significantly to arrive at optimal methods while learning from the experience of other investigators [3, 4]. Sample

Karthik Raman and Gayathri Sambamoorthy (eds.), *Modelling the Microbiome*, Methods in Molecular Biology, vol. 3006,
https://doi.org/10.1007/978-1-0716-5080-6_2,

preparation methods have been reported to add sequencing bias, and therefore due care is to be exercised in these steps before any sequencing study is undertaken [5].

2 Materials

Sample collection (swabbing): Sterile swabs, sterile phosphate-buffered saline, sterile 15 mL tubes.

Sample collection (home dust): Vacuum with dust collection filters.

DNA extraction: Lysozyme, Tris-HCl, proteinase K, QIAamp BiOstic Bacteremia DNA Kit (cat# 12240-50).

DNA Quality Check: Qubit DNA BR Assay (Invitrogen, cat# Q32853), agarose, TAE buffer, GelRed.

Amplicon preparation: High-Fidelity 2X PCR Master Mix 250 rxns (cat# M0541L).

Library Preparation and Sequencing kits to be chosen as per the sequencing platform used.

For Illumina: NEBNext Ultra DNA Library Prep Kit for Illumina (cat# E7370L); NEBNext Multiplex Oligos for Illumina (96 Index Primers) (cat# E6609S); MiSeq Reagent Kit v2 500 cycle (cat# MS1022003).

3 Methods

All reagents and plasticware used during the experiment are to be nuclease-free for best results. Nuclease-free water is to be used for the preparation of reagents. Enzyme-containing solutions are to be thawed and used once, and repeated freeze–thawing should be avoided for optimal efficiency.

3.1 DNA Extraction from Skin Swabs/Skin Buffer Scrubs/Dust/Swatch/Swab Samples

3.1.1 Sample Collection

Microbiomes of any surface can be analyzed by either wet swabbing or doing a buffer scrub. Dry vacuuming hard surfaces and collecting the dust onto a filter accessory engineered to fit onto the front of the vacuuming accessory is ideal. Soft surfaces such as clothing can be cut into small pieces and directly used for extraction (*see* **Note 1**).

3.1.2 Wet Swabbing of Hard Surfaces

- Use sterile 0.01% Triton PBS as the wetting agent for the fresh sterile cotton swab (individually packed and low residue- or particulate-containing).
- Wet the sterile swab fully in the 0.01% Triton PBS and swab it onto the floor surface or surface of interest in an ordered manner, ensuring all corners and grouts are covered, if any (*see* **Notes 2** and **3**).

- Once swabbed, return the swab back into a sterile tube and store it at −20 °C for sequencing studies. In case of live bacteria culturing as a downstream process, store the samples at 4 °C.

3.1.3 Dry Vacuuming

- Ensure the presence of the dust collection contraption from Indoor Biotechnologies(cat#DU-ST-1, store.inbio.com), which snugly fits into the vacuum collection nozzle.
- Vacuum the floors or the desired hard surfaces until dust collects into filters completely.
- Use separate sterile filters for each new surface to be sampled for the study to avoid cross-contamination.
- Dust samples, if completely dry, can be stored at RT until further processing.

3.1.4 Extraction of DNA from Samples

Environmental samples are particularly challenging during DNA extraction. Various tweaks to protocol may be required for optimal yield and quality depending on the source of the sample (*see* **Notes 4–8**).

3.1.5 Preparation of Lysozyme Working Stock Solutions

- Dissolve 25 mg of lysozyme in 1 mL of 10 mM Tris-Cl (pH 8.0) to prepare the working stock.
- The working stock is aliquoted into 100 μL aliquots and stored at 4 °C.

3.1.6 Preparation of Proteinase K Working Stock Solution

- Prepare Proteinase K stock at a concentration of 22 mg/mL in nuclease-free water. Make working aliquots of 100 μL.
- Store all aliquots in a −20 °C deep freezer.

3.1.7 Procedure for Extraction When the Sample Is in a Buffer

- Centrifuge sample at 10,000 rpm for 10 min at 4 °C and carefully remove all the supernatant except the bottom 100 μL. Do not disturb any pellet that might have collected at the bottom. If pooling dilute samples, add buffer from the second tube and centrifuge again at 10,000 rpm for 10 min at 4 °C to collect all material into the same sample. Carefully remove the supernatant, except the bottom 100 μL.
- Vortex the sample to resuspend the pellet into the 100 μL supernatant.
- Add 20 μL of lysozyme (from 25 mg/mL stock of lysozyme). Vortex briefly and incubate at 37 °C for 15 min.
- Add 20 μL of proteinase K to the tube, vortex briefly, and incubate at 56 °C for 15 min. Observe for clearing of the solution post-incubation.
- Follow the next steps as per the QIAamp BiOstic Bacteremia DNA Kit (cat# 12240-50).

3.2 DNA Extraction from Dust Samples

3.2.1 Pre-processing Step

- Weigh 250 mg of dust sample.
- Add 450 μL of MBL lysis buffer provided by the QIAamp BiOstic Bacteremia DNA Kit.
- Mix the contents in the tube thoroughly and leave it for overnight lysis.
- After overnight lysis, add metallic beads in the tube for mechanical bead beating with 40 s of bead beating with 40 oscillations and 20 s of rest in between in ice to cool the samples. Repeat the process for each sample three times.
- After completing mechanical lysis, follow the QIAamp BiOstic Bacteremia DNA Kit protocol (cat# 12240-50).

3.3 DNA Extraction from Fabric Samples

3.3.1 Pre-processing Step

- Cut the fabric from the area of interest [e.g., handcuffs, collars, armpit, etc.] up to 250–300 mg of sample weight. Cut the sample further into smaller pieces to fit into a 25 mL centrifuge tube.
- Add 450 μL of MBL lysis buffer provided by the QIAamp BiOstic Bacteremia DNA Kit to the cloth sample.
- Mix the contents in the tube thoroughly and leave it for overnight lysis.
- After overnight lysis, add metallic beads in the tube for mechanical bead beating with 40 s of bead beating and 40 oscillations with 20 s of rest in between each set in ice. Repeat the process for each sample three times.
- After completing mechanical lysis, follow the QIAamp BiOstic Bacteremia DNA Kit protocol (cat# 12240-50).

3.4 DNA Sample Quality Check

The extracted DNA samples are quantified using Qubit DNA BR Assay (Invitrogen, Cat# Q32853). DNA purity and DNA integrity can be checked by running 2 μL on a 1% Agarose gel, or alternately, where available, analysis on Agilent TapeStation or an equivalent platform can also be conducted (*see* **Notes 9** and **10**).

3.5 Amplicon Preparation

3.5.1 Targeted Amplicon-Based Library Preparation

The first round of PCR is to amplify the complete 16S rRNA gene (~1.5 kb) from the sample using primer 16S F and 16S R with sequence, as described in Table 1 (*see* **Note 11**). This is followed by nested PCR using primers for the variable regions V1-V2 or V3-V4 as desired and the respective primers to be used for amplification as desired. The PCR conditions are as in Table 2 for full 16S, Table 3 for V3-V4, and Table 4 for V1-V2 regions, respectively. Following PCR amplification, samples are quantified using the Qubit DNA HS Assay (Invitrogen, Cat#Q32854).

Table 1
Primer details for 16S and variable regions

No	Primer name	Primer position	Forward/ Reverse	Primer sequence 5′–3′	Amplicon length
1	16S F	8F	Forward	AGAGTTTGATCCTGGCTCAG	~1500 bp
2	16S R	1492R	Reverse	GGTTACCTTGTTACGACTT	
3	16S V3V4 F	341F	Forward	CCTACGGGNGGCWGCAG	~460 bp
4	16S V3V4 R	785R	Reverse	GACTACHVGGGTATCTAA TCC	
5	16S V1F	8F	Forward	AGAGTTTGATCCTGGCTCAG	~250 bp
6	16S V2R	336R	Reverse	TGCTGCCTCCCGTAGGAGT	

Table 2
PCR conditions used for amplification of the complete 16S region

Temperature	Duration	Step
95 °C	3 min	1 cycle
95 °C	15 s	7 cycles
48 °C–54 °C	20 s	
72 °C	1 min	
95 °C	15 s	35 cycles
55 °C	20 s	
72 °C	1 min	
72 °C	7 min	1 cycle
4 °C	∞	Hold

Table 3
PCR profile for the amplification of 16S V3-V4 regions

Temperature	Duration	Step
95 °C	3 min	1 cycle
95 °C	30 s	
50 °C–54 °C	30 s	
72 °C	30 s	5 cycles
95 °C	30 s	
55 °C	30 s	
72 °C	30 s	35 cycles
72 °C	7 min	1 cycle
4 °C	∞	Hold

Table 4
PCR profile for the amplification of 16S V1-V2 regions

Temperature	Duration	Step
95 °C	3 min	1 cycle
95 °C 48 °C [Delta +1 °C/cycle] 72 °C	30 s 30 s 30 s	7 cycles
95 °C 55 °C 72 °C	30 s 30 s 30 s	35 cycles
72 °C	7 min	1 cycle
4 °C	∞	Hold

- The DNA samples are diluted to 5 ng/μL, and 1 μL is used for PCR amplification. For samples where DNA is too low, 5 μL of the sample can be taken for PCR and is amplified for the 16S rDNA region.
- The PCR product is then taken, followed by V1-V2/V3-V4 primers (*see* Table 1 for sequences) in a nested PCR approach and 1 μL of the full-length PCR product is used for the V1-V2/V3-V4 PCR reaction.
- Positive and no template controls are to be included in the two steps of PCR.

- PCR products are then analyzed on 1% agarose gel to check the amplicon presence and size.
- The PCR products are further taken for DNA library preparation depending on the sequencing platform to be used.

4 Notes

1. If sampling with buffers or swab materials, appropriate controls must be included to understand if any nucleic acid or microbial contamination exists in these materials. It is preferable to use nucleic acid-free water for all buffer preparations and also use nucleic acid-free plasticware for sample collection.
2. Sample quality is paramount for a good end result. Sample collection and storage play an important role in good-quality nucleic acid extraction. It is better to process samples at the earliest; if not, the right storage conditions, carrier medium, nuclease and/or protease inhibitors, etc. must be determined,

especially if they are to be transported to a processing facility. Optimal storage or preservation conditions must be determined *a priori.*

3. For swabbing hard surfaces such as floors, wetting a swab in an appropriate buffer, performing swabbing, and then storing it in a fresh tube have been shown to preserve the most bacterial counts (confirmed by plating on culture media). Returning the swab into buffer for storage showed reduced counts, so using the buffer only for wetting the swab prior to collection is recommended.
4. While yield is the major concern, quality and integrity of the nucleic acid are equally important. Pure, high-molecular-weight nucleic acids are good starting materials. If needed, the DNA can always be fragmented as required by some protocols by mechanical or enzymatic methods.
5. Extraction method is another major optimization step, and there is a plethora of choices; traditional phenol–chloroform extraction methods are robust and perform well, whereas others fail at times but require careful handling of toxic reagents and their disposal.
6. Kit-based methods are popular, and there are many vendors with several options, and choosing the right kit for the samples needs to be considered carefully by comparing the performance of a few kits on the samples. The kits mentioned in the above protocol are those that worked for our samples and not prescriptive.
7. Spore formers are a tougher nut to crack when it comes to cell lysis for nucleic acid extraction. A mixture of enzymatic lysis along with bead beating has yielded success. Enzymes such as lysozyme, mutanolysin, and lysostaphin have been used, along with proteinase K, followed by bead beating.
8. Samples which have undergone fixation with solvents undergo damage, such as Formalin-Fixed Paraffin-Embedded (FFPE), and typically yield low-quality nucleic acids. There are commercially available kits like NEBNext® FFPE DNA Repair v2 Module, SureSeq FFPE DNA Repair Mix, KAPA NGS FFPE DNA Polishing Kit, and others, which are designed to improve the quality and outcome of sequencing these samples.
9. When faced with low 260:280 ratios, further processing to remove protein or phenol contaminants will improve the purity of the samples.
10. Measurements like DIN or RIN values from Tapestation/ Bioanalyser or other such platforms are decent indicators of the quality of the extracted nucleic acid. While these are not foolproof, they offer a good idea on whether to include or reject samples for further processing.

11. Amplification is an optional step and must be carefully considered depending on the yield of nucleic acid in the extraction protocol. If samples are limited and the yield is low, there may be no choice but to first amplify before downstream sequencing.

Conflict of Interest Statement

The authors have no conflicts of interest to declare that are relevant to the content of this chapter.

References

1. Chaudhari HG, Prajapati S, Wardah ZH, Raol G, Prajapati V, Patel R, Shati AA, Alfaifi MY, Elbehairi SEI, Sayyed RZ (2023) Decoding the microbial universe with metagenomics: a brief insight. Front Genet 14:1119740
2. Felczykowska A, Krajewska A, Zielińska S, Łoś JM (2015) Sampling, metadata, and DNA extraction - important steps in metagenomic studies. Acta Biochim Pol 62(1):151–160. https://doi.org/10.18388/abp.2014_916
3. Ruiz-Calderon JF, Cavallin H, Song SJ, Novoselac A, Pericchi LR, Hernandez JN, Rios R, Branch OH, Pereira H, Paulino LC et al (2016) Walls talk: microbial biogeography of homes spanning urbanization. Sci Adv. 2(2): e1501061
4. Callewaert C, Van Nevel S, Kerckhof FM, Granitsiotis MS, Boon N (2015) Bacterial exchange in household washing machines. Front Microbiol 6:1381
5. Lear G, Dickie I, Banks J, Boyer S, Buckley HL, Buckley TR, Cruickshank R, Dopheide A, Handley KM, Hermans S, Kamke J, Lee CK, MacDiarmid R, Morales SE, Orlovich DA, Smissen R, Wood J, Holdaway R (2018) Methods for the extraction, storage, amplification and sequencing of DNA from environmental samples. N Z J Ecol 42:1–50. https://doi.org/10.20417/nzjecol.42.9

Chapter 3

Exploring the Ocean's Microbial World: Techniques and Protocols for Microbiome Research

Vijaya Raghavan Rangamaran, T. J. Sushmitha, Karthikeyan Kathan Tamilmani, Hinduja Murugesan, and Dharani Gopal

Abstract

Marine microbiomes play a crucial role in oceanic ecosystems, influencing biogeochemical cycles, climate regulation, and marine biodiversity. Accurate characterization of these microbial communities requires standardized protocols for sample collection, processing, sequencing and data analysis. This chapter provides a comprehensive guide to essential methodologies for marine microbiome research including field sampling strategies, DNA and RNA extraction techniques, high-throughput sequencing approaches (such as 16S rRNA amplicon sequencing and metagenomics) and bioinformatics pipelines for data interpretation. Additionally, we discuss quality control measures, best practices for reproducibility, and challenges associated with marine microbiome profiling. By adopting standardized methodologies, researchers can generate reliable, comparable datasets that enhance our understanding of marine microbial ecology and its broader environmental implications.

Key words Marine microbiome, Sampling protocols, Whole metagenome shotgun sequencing, Bioinformatics analysis

1 Introduction

The ocean is an abode of an incredibly diverse array of microorganisms that play pivotal roles in biogeochemical cycles, climate regulation and ecosystem functioning. From the surface to the deepest trenches, microbial communities dominate marine ecosystems, influencing everything from nutrient cycling to primary production [1]. Recent advancements in microbiome research, especially through high-throughput sequencing technologies have uncovered the vast diversity of microbial life that thrives in the ocean, most of which remain underexplored [2]. Understanding the structure, function and dynamics of these microbial communities is essential for comprehending the complex interactions that sustain

Karthik Raman and Gayathri Sambamoorthy (eds.), *Modelling the Microbiome*, Methods in Molecular Biology, vol. 3006,
https://doi.org/10.1007/978-1-0716-5080-6_3,

marine ecosystems and for addressing critical environmental challenges such as climate change and ocean health.

Marine microbiomes constitute a wide range of microorganisms including bacteria, archaea, viruses, which vary in composition and function across different marine habitats [3]. These microbes are pivotal to oceanic processes, including nutrient cycling, organic matter decomposition, and regulation of greenhouse gases, particularly methane and nitrogen compounds [4]. Exploration of marine microbiomes offers valuable insights into the ecology, evolution, and their responses to environmental changes, making them a key area of research in marine sciences.

To explore and analyze the microbial diversity within marine environments, several advanced techniques and protocols have been developed. Traditional culturing methods have been largely replaced by culture-independent approaches, such as 16S rRNA gene sequencing, whole metagenome sequencing, and metatranscriptomics. These methods allow for a comprehensive analysis of microbial communities without the need for cultivation [5]. High-throughput sequencing technologies have revolutionized the ability to identify and characterize previously unculturable microorganisms, thereby providing a more accurate picture of the microbial world beneath the waves.

This chapter aims to provide an overview of the key techniques and protocols used in marine microbiome research, starting from sample collection and preservation to DNA extraction, sequencing, and data analysis. Methods that have been developed to tackle the unique challenges of marine microbiology have been discussed as well. The rapid technological advancements in sequencing technologies combined with improved bioinformatics tools offer unprecedented opportunities for the study of the ocean's microbial inhabitants, and this research holds the potential to unlock new insights into the functioning and resilience of marine ecosystems.

2 Materials

2.1 Deep- Sea Sample Collection

1. Niskin Sampler for water collection.
2. Rosette Frame with CTD for real-time monitoring of physicochemical parameters.
3. Piston/Grab/Gravity corers for sediment sampling.

2.2 DNA Extraction

2.2.1 Phenol–Chloroform-Based Extraction

1. Water/Sediment sample.
2. Eppendorf tubes (2 mL and 1.5 mL) and stands.
3. Micropipette (1 mL, 20–200 μL) and tips.
4. Ethanol.
5. STE buffer (1×, pH 8).

6. TE buffer (1×, pH 8).
7. Lysozyme (conc. 10 mg/mL).
8. Proteinase K (conc. 10 mg/mL).
9. 10% SDS.
10. Tris-saturated phenol (pH 8).
11. Chloroform.
12. Water bath.
13. Ice box.
14. Centrifuge.
15. Freezer.
16. Vortexer.
17. Qubit/Nanodrop for quantification.

2.3 Bioinformatics Analysis

The list of bioinformatics tools and pipelines that are required to be installed toward the analysis of sequencing data is presented below.

2.3.1 16S rRNA Amplicon Sequencing for Microbial Profiling

1. **FastQC** (http://www.bioinformatics.babraham.ac.uk/projects/fastqc) for quality control.
2. **Cutadapt** [6] or **Trimmomatic** [7] for adapter trimming.
3. **DADA2** [8] and **QIIME2** [9] for sequence processing and analysis.
4. **R** [10] or **Python** for statistical analysis and visualization.

2.3.2 Whole Metagenome Shotgun Sequence Analysis and Generation of Metagenome-Assembled Genomes (MAGs)

1. **Trimmomatic** for quality trimming.
2. **FastQC** for quality assessment.
3. **SPAdes** or **MetaSPAdes** for specialized metagenomic assembly [11].
4. **QUAST** for assembly statistics [12].
5. **MetaBAT2** for binning metagenomic contigs [13].
6. **CheckM** for assessing MAG quality [14].
7. **Prodigal** for gene prediction [15].
8. **Anvi'o** for genome binning and visualization [16].

3 Methods

3.1 Sampling Design and Site Selection

Marine microbiome research begins with meticulous planning of sampling strategies. The choice of sampling sites should reflect the research objectives, whether focused on coastal ecosystems, open ocean environments, or extreme habitats like hydrothermal vents. Factors such as depth, temperature, salinity, and nutrient availability must be considered, since they significantly influence the microbial community structure and composition [17].

3.2 Sample Collection Techniques

3.2.1 Deep-Sea Water Sampling

1. **Deploy the CTD/water sampler**:
   - Position the CTD-equipped Rosette sampler on the research vessel.
   - Lower the sampler using the winch to the desired depth for water sampling.
   - Profile sampling: If conducting a vertical profile, record temperature, salinity, and oxygen levels at different depths (surface, mid-water, and bottom layers).
2. **Sampling**:
   - Once the Rosette sampler reaches the designated depth, activate the Niskin bottles to collect water samples from the specific depth.
   - For discrete sampling, collect water samples in sterile containers, ensuring no air or contamination enters the sample bottles.
   - For microbial or chemical analyses, filter water immediately through 0.22 μm filters or use appropriate collection methods for assays (*see* **Note 1**).
   - **Note**: If sampling from multiple depths, ensure each water sample is collected in a clean container without cross-contamination from other depths.
3. **Temperature-sensitive samples**:
   - For samples that need to be kept at low temperatures (e.g., microbial community analysis, enzymatic activity), place the water samples in insulated containers and return to the ship as quickly as possible for proper preservation.
4. **Label and record**:
   - Label each water sample container with the corresponding depth, date, time, and location (latitude/longitude).
   - Record sampling details (e.g., exact time of collection, CTD profiles) on a log sheet for traceability.

3.3 Deep-Sea Sediment Sampling

3.3.1 Deploy Sediment Sampler

- Depending on the study goals, choose the appropriate sediment corer (e.g., gravity corer, piston corer, or box corer) based on the depth and sediment composition.
- Lower the coring device to the seafloor using the vessel winch system. The exact depth and location are crucial for reproducibility and correlation with other data collected (e.g., water profiles).

3.3.2 Core Retrieval

- Once the sediment corer reaches the seafloor, initiate the coring mechanism.

- If using a gravity corer, the device will penetrate the sediment by its own weight. For piston corers, the device will use additional force to penetrate deeper sediments.
- Retrieve the core carefully to avoid disturbing the sediment layers.

3.3.3 Sediment Handling

- Once the core is retrieved, handle it with care to avoid contamination.
- Place the sediment core into sterile trays or containers immediately for preservation.
- For further analysis, cores can be subsampled by slicing into specific layers (e.g., surface, 10 cm, 20 cm, etc.) depending on the study goals.
- Store the sediment cores in cold storage or at −80 °C if they need to be preserved for microbiological analysis.

3.3.4 Sediment Grab Sampling

- For smaller, targeted sampling, a Van Veen Grab sampler can be used to obtain surface sediment.
- This is useful for shallow sampling or when using a box corer is not feasible.
- Record the exact location and depth of the grab sample, and label containers appropriately.

3.3.5 Sediment Description

- Record the physical characteristics of the sediment, such as color, texture, and smell (indicating potential microbial activity).
- Collect a small sample of sediment for the analysis of physical and chemical properties, such as grain size or pH.

3.3.6 Label and Record

- Label all sediment samples with their unique identifiers and record the depth, date, time, and coordinates of the sample collection.
- Maintain detailed log sheets for all sediment sampling activities.

3.4 Post-Sampling Procedures

3.4.1 Sample Preservation

- Store water samples in appropriate conditions.
- For microbiological analysis, filter or transfer to sterile vials and refrigerate.
- For chemical analysis, refrigerate or preserve with chemical preservatives (e.g., formaldehyde or EDTA) as needed.
- For sediment samples, refrigerate or freeze according to the type of analysis required (microbial, chemical, geochemical).

3.4.2 Data Documentation

- Ensure that all field and sample metadata (location, depth, water/sediment type) are accurately documented in a digital or physical logbook.
- Data should also include conditions such as temperature, salinity, and environmental parameters.

3.4.3 Sample Transport

- Upon return to the laboratory, safely transport water and sediment samples to avoid temperature fluctuations or contamination.
- Ensure samples are stored in clean, labeled containers for further analysis.

3.4.4 Quality Control

- Double-check all equipment and procedures to minimize cross-contamination during handling.
- Ensure that all samples are tracked by their unique IDs to maintain a proper chain of custody.

3.5 DNA and RNA Extraction Techniques

3.5.1 Challenges in Marine Samples

One of the major constraints in marine samples is that they often contain low biomass and high levels of inhibitors (e.g., salts and humic acids), complicating nucleic acid extraction. Robust protocols are required to ensure high-quality DNA and RNA for downstream analyses.

3.5.2 DNA Extraction

- Commercial kits: Kits such as the DNeasy PowerSoil Kit, DNeasy PowerWater Kit (Qiagen) are widely used for marine samples due to their efficiency in removing inhibitors and providing high DNA yield suitable for downstream sequencing (*see* **Note 2**).
- Phenol–chloroform extraction: This traditional method is effective for high-biomass samples but requires careful handling of hazardous chemicals. The detailed protocol is presented below:

  1. Take 1–2 g of sediment/filter paper for water samples and add 500 μL TE buffer and vortex well.
  2. Transfer 250 μL to a fresh autoclaved 2 mL tube and add 1 mL of STE buffer, vortex short, centrifuge at 8000 rpm for 5 min at 4 °C, and discard the supernatant.
  3. To the pellet, add 500 μL of TE buffer and do a quick vortex.
  4. Add 50 μL lysozyme of 10 mg/mL concentration and incubate for 1 h at 37 °C in a water bath and mix at regular intervals, for example, every 15 min.

5. After incubation, add 25 μL of 10% SDS and 20 μL of proteinase K (concentration 10 mg/mL), mix well, or vortex briefly.
6. Incubate at 37 °C for 2 h and mix well at regular intervals (every 15 min).
7. Now, add equal volume of Tris-saturated phenol (here, 500 μL) and mix gently for 1 min.
8. Centrifuge at 13,000 rpm for 10 min at 4 °C and transfer the upper aqueous phase into a fresh labeled 2 mL tube.
9. Add TE buffer to make up the final volume to 500 μL.
10. Add equal volume of chloroform (here, 500 μL) and mix gently by inversion for 1 min.
11. Centrifuge at 13,000 rpm for 5 to 10 min at 4 °C and transfer the upper aqueous phase carefully without disturbing the lower phase into a fresh labeled 2 mL tube.
12. Make up the final volume to 500 μL with TE (here, we will recover around 400 μL of the upper phase, so add 100 μL of TE).
13. Repeat steps 11 and 12, and transfer the upper aqueous phase to a fresh new labeled 1.5 mL tube.
14. Store the tube containing DNA at −40 °C for long term or at 4 °C for further recent use.
15. Confirm the presence of genomic DNA after the extraction procedure through agarose gel electrophoresis and quantify using a Qubit fluorometer or a Nanodrop spectrophotometer.

3.5.3 RNA Extraction

- RNA preservation: RNAlater or similar reagents are used to stabilize RNA during sampling.
- Extraction protocols: Kits like the RNeasy Kit (Qiagen) are optimized for microbial RNA extraction from marine samples.

3.5.4 Quality Assessment

The quality and quantity of extracted nucleic acids should be assessed using spectrophotometry (e.g., Nanodrop) or fluorometry (e.g., Qubit). Gel electrophoresis or bioanalyzer systems can further evaluate the integrity.

3.6 High-Throughput Sequencing Approaches

3.6.1 16S rRNA Amplicon Sequencing

Microbial profiling using 16S rRNA gene sequencing is a cornerstone of modern microbial ecology studies. It involves PCR amplification of hypervariable regions (e.g., V3–V4, V5–V6) from the metagenomic DNA, followed by sequencing on platforms like Illumina or Nanopore [18, 19]. This approach provides insights into microbial diversity and community structure but is limited by primer bias and resolution at the species level [20, 21]. With the advent of third-generation sequencing technologies like Nanopore, near-complete 16S genes could be sequenced, thereby providing greater resolution at the species level [22].

3.6.2 Metagenomics

Shotgun metagenomics involves sequencing the entire genomic content of a sample enabling taxonomic and functional profiling. This approach is particularly valuable for studying uncultured microorganisms and their associated metabolic pathways. Recent advances in long-read sequencing (e.g., PacBio and Oxford Nanopore; *see* **Note 3**) have improved the assembly of complex marine metagenomes [23].

3.6.3 Metatranscriptomics

Metatranscriptomics focuses on sequencing RNA to study gene expression and functional activity within microbial communities. This approach provides dynamic insights into microbial responses to environmental changes.

3.7 Bioinformatics Pipelines for Data Analysis

Bioinformatics analysis is a critical step in interpreting the vast amounts of data generated by high-throughput sequencing technologies. This section details the bioinformatics pipelines for 16S rRNA amplicon sequencing and whole-genome shotgun (WGS) metagenomics, including the commands for executing these analyses.

3.7.1 16S rRNA Amplicon Sequencing for Microbial Profiling

16S rRNA amplicon sequencing is widely used to profile microbial communities. The analysis involves several steps, including quality control, sequence processing, taxonomic classification, and diversity analysis.

Step 1: Quality Control and Preprocessing

Raw sequencing reads often contain adapters, low-quality bases, and primer sequences that need to be removed. Tools like FastQC and Trimmomatic are commonly used for quality control. Sample code snippets for executing this analysis are provided below. The examples consider paired-end datasets from Illumina.

Before starting, ensure having the necessary tools installed. Install the tools using conda (recommended):

```
conda create -n 16s_analysis -c bioconda fastqc cutadapt
qiime2 dada2 r-ggplot2 r-phyloseq
conda activate 16s_analysis
Bash Commands:
# Run FastQC to assess raw read quality
fastqc raw_reads_R1.fastq raw_reads_R2.fastq
```

```
# Trim adapters and low-quality bases using Trimmomatic
java -jar trimmomatic-0.39.jar PE -phred33 raw_reads_R1.
fastq raw_reads_R2.fastq \
   paired_R1.fastq   unpaired_R1.fastq   paired_R2.fastq
unpaired_R2.fastq
 illuminaclip:truseq3-pe.fa:2:30:10 leading:3 trailing:3
slidingwindow:4:15 minlen:36
```

Step 2: Sequence Processing and OTU/ASV Clustering
After preprocessing, sequences are clustered into Operational Taxonomic Units (OTUs) or Amplicon Sequence Variants (ASVs). QIIME 2 and DADA2 are popular tools for this step. The output of Trimmomatic can be used as the input in this step. The same files can also be used for other downstream analyses as required.

```
# Using DADA2 in QIIME 2 for ASV clustering
qiime dada2 denoise-paired --i-demultiplexed-seqs demux.
qza --p-trim-left-f 20 --p-trim-left-r 20 --p-trunc-len-f
250 --p-trunc-len-r 250 --o-table table.qza --o-represen-
tative-sequences   rep-seqs.qza   --o-denoising-stats
denoising-stats.qza
```

Step 3: Taxonomic Classification
Taxonomic classification assigns sequences to taxonomic groups using reference databases like Greengenes[24], SILVA [25], etc.

```
# Assign taxonomy using a pre-trained classifier in QIIME
2
qiime  feature-classifier  classify-sklearn  --i-classifier
gg-13-8-99-515-806-nb-classifier.qza \
 --i-reads rep-seqs.qza --o-classification taxonomy.qza
```

Step 4: Diversity Analysis
Alpha- and beta-diversity metrics are calculated to assess the microbial community structure and composition using MAFFT [26] and FastTree 2 [27].

Generate a phylogenetic tree for diversity analysis

```
qiime phylogeny align-to-tree-mafft-fasttree --i-
sequences  rep-seqs.qza  --o-alignment  aligned-rep-seqs.
qza  --o-masked-alignment  masked-aligned-rep-seqs.qza  --
o-tree unrooted-tree.qza --o-rooted-tree rooted-tree.qza
# Calculate alpha and beta diversity
qiime  diversity  core-metrics-phylogenetic  --i-phylogeny
```

```
rooted-tree.qza --i-table table.qza
 --p-sampling-depth 1000 --m-metadata-file metadata.tsv --
output-dir diversity_results
# Generate Visualizations
qiime metadata tabulate --m-input-file taxonomy.qza --o-
visualization taxonomy.qzv
```

3.7.2 Whole Metagenome Shotgun Sequence Data from Illumina and Generation of Metagenome-Assembled Genomes (MAGs)

This protocol aims to describe the process of analyzing shotgun whole-metagenome sequencing data from Illumina platforms, focusing on quality control, assembly, binning, and generation of metagenome-assembled genomes (MAGs). The efficiency of data analysis is also proportional to the hardware capabilities of the workstations in which these pipelines are executed. The CPUs and GPUs directly impact the basecalling efficiency of the sequencing machine and the underlying execution of bioinformatics pipelines (*see* **Note 4**).

The quality control steps mentioned in the previous section can be used for whole-metagenome shotgun datasets as well.

Assemble the cleaned sequence into contigs. For metagenomic data, **SPAdes** or **MetaSPAdes** is typically used.

```
spades.py -1 cleaned_1.fastq -2 cleaned_2.fastq -o spa-
des_assembly/ --meta --careful
The --meta option tells SPAdes to use metagenomic-specific
parameters.
```

3.7.3 Inspect Assembly Quality

Use Quast to assess the quality of the assembly.

```
quast.py spades_assembly/contigs.fasta -o quast_report/
```

This will generate an assembly report including statistics like N50, total length, and number of contigs.

3.7.4 Contig Binning (MAG Generation)

Bin the assembled contigs into separate metagenome-assembled genomes (MAGs). MetaBAT2 is a popular tool for binning metagenomic contigs. Tools like Maxbin2 and Concoct can also be used for binning metagenome sequence datasets.

```
conda install -c bioconda metabat2
metaBAT2 -i spades_assembly/contigs.fasta -d bins/ -o
bin_output -m 1500 --maxBin 100
```

```
* -m 1500 specifies the minimum contig length for binning.
After binning, assess the quality of the generated MAGs
using CheckM.
conda install -c bioconda checkm-genome
checkm lineage_wf -x fa bins/ checkm_output/
```

This will provide metrics like completeness, contamination, and strain heterogeneity of each MAG.

3.7.5 Gene Prediction

For each MAG, perform gene prediction using Prodigal.

```
conda install -c bioconda prodigal
prodigal -i bin_file.fasta -o bin_file.gff -a bin_file.faa
```

This will generate gene predictions (in GFF format) and amino acid sequences (FASTA format).

3.7.6 Functional Annotation

To assign functions to the genes identified in the MAGs, tools including DIAMOND [28] or BLAST (https://blast.ncbi.nlm.nih.gov/Blast.cgi) can be used.

```
conda install -c bioconda diamond
diamond blastp -d nr.dmnd -q bin_file.faa -o bin_file.daa
Convert DIAMOND results to a human-readable format:
diamond view -a bin_file.daa -o bin_file_annotated.tsv
```

3.7.7 Data Visualization and Further Analysis

```
Anvi'o can be used to visualize and analyze the binned
MAGs.
conda install -c bioconda anvi'o
anvi-gen-genomes-storage -g bin_file.fasta -o anvi'o_sto-
rage
anvi-profile -i spades_assembly/contigs.bam -c anvi'o_con-
fig -o anvi'o_profiles
Use Anvi'o to explore and visualize the genome bins, their
quality, and functional content.
bash
Copy
anvi-interactive -c anvi'o_config -p anvi'o_profiles
```

This will launch a web interface for interactive analysis of the MAGs.

3.8 Quality Control Measures and Reproducibility

3.8.1 Contamination Control

Contamination is a major concern in low-biomass marine samples. Negative controls should be included at every stage, from sampling to sequencing, to identify and mitigate contamination.

3.8.2 Standardization of Protocols

Adopting standardized protocols, such as those recommended by the Earth Microbiome Project, enhances reproducibility and comparability across studies [29].

3.8.3 Data Sharing and Metadata Reporting

Public repositories like the Sequence Read Archive (SRA) and the European Nucleotide Archive (ENA) facilitate data sharing. Comprehensive metadata reporting, following standards like MIxS (Minimum Information about any (x) Sequence), is essential for data reuse and integration.

3.9 Challenges in Marine Microbiome Profiling

3.9.1 Technical Challenges

- Low biomass: Marine samples often yield limited DNA/RNA, requiring sensitive extraction and amplification methods.
- Inhibitors: High salt and organic matter content can interfere with molecular analyses.
- Sequencing depth: Achieving sufficient sequencing depth to capture rare taxa remains a challenge.

3.9.2 Biological Complexity

Marine microbial communities are highly dynamic and influenced by environmental gradients. Understanding these complexities requires integrated multi-omics approaches and long-term monitoring.

3.9.3 Data Interpretation

The vast amount of data generated by high-throughput sequencing poses computational challenges: Availability of databases for specific annotations of large-scale sequence datasets (*see* **Note 5**). Advanced machine learning and artificial intelligence tools are increasingly being employed to uncover patterns and predict ecological interactions [30].

4 Notes

1. Microbiome analysis from water samples might require filtration through a series of filter papers depending on the turbidity of the sample. This would also help in reducing the clog that happens while directly filtering on 0.22 μm filters.
2. Enzymatic pretreatment of the water and sediment samples will help in improving the profiling of Gram+ve microbial communities. The pre treatment can be followed by any manual of kit-based extraction protocols.

3. Library preparation for NGS (especially for Nanopore sequencing) can be optimized by varying the end-repair incubation time and magnetic bead-based wash steps to retain more DNA for the downstream processes.
4. Basecalling efficiency in latest Nanopore sequencers has been directly impacted with the graphic card available. We have observed that the basecalling in Nanopore Promethion2 solo is more efficient and faster with a 24 GB graphic card than in lower-capacity workstations.
5. Creation of customized 16S rDNA reference database for profiling of microbial communities increases the classification efficiency. Also, HMM models can be generated with specific functional gene targets to improve the annotation from whole-metagenome sequence datasets. The custom sequence database can be used as a training set for AI/ML-based microbial classification.

Acknowledgments

The authors are thankful to the Director, National Institute of Ocean Technology, and the Ministry of Earth Sciences, Government of India.

Conflict of Interest Statement The authors have no conflicts of interest to declare that are relevant to the content of this chapter.

References

1. Fuhrman JA (2009) Microbial community structure and its functional implications. Nature 459:193–199. https://doi.org/10.1038/nature08058
2. Hug LA, Coelho LP, Aylward FO (2016) A new view of the tree of life. Nat Microbiol 1: 16048. https://doi.org/10.1038/nmicrobiol.2016.48
3. Zinger L, Bonin A, Jean P (2011) Global patterns of marine microbial diversity. Nat Rev Microbiol 9:779–789. https://doi.org/10.1038/nrmicro2644
4. Bernhard AE, Field KG, Lipp EK (2005) Ecology of sulfur-cycling prokaryotes in the ocean. Curr Opin Microbiol 8:281–288. https://doi.org/10.1016/j.mib.2005.04.006
5. Comeau AM, Douglas GM, Langille MG (2017) Microbial communities in the ocean: a tool for understanding the microbial ecology of marine environments. Environ Microbiol Rep 9:167–174. https://doi.org/10.1111/1758-2229.12462
6. Martin M (2011) Cutadapt removes adapter sequences from high-throughput sequencing reads. EMBnet J 17:10–12. https://doi.org/10.14806/ej.17.1.200
7. Bolger AM, Lohse M, Usadel B (2014) Trimmomatic: a flexible trimmer for Illumina sequence data. Bioinformatics 30:2114–2120. https://doi.org/10.1093/bioinformatics/btu170
8. Callahan BJ, McMurdie PJ, Rosen MJ, Han AW, Johnson AJA, Holmes SP (2016) DADA2: high-resolution sample inference from Illumina amplicon data. Nat Methods 13:581–583. https://doi.org/10.1038/nmeth.3869
9. Bolyen E, Rideout JR, Dillon MR, Bokulich NA, Abnet CC, Al-Ghalith GA et al (2019) Reproducible, interactive, scalable and extensible microbiome data science using QIIME 2. Nat Biotechnol 37:852–857. https://doi.org/10.1038/s41587-019-0209-9

10. Ihaka R, Gentleman R (1996) R: a language for data analysis and graphics. J Comput Graph Stat 5:299–314. https://doi.org/10.1080/10618600.1996.10474713
11. Nurk S, Meleshko D, Korobeynikov A, Pevzner PA (2017) metaSPAdes: a new versatile metagenomic assembler. Genome Res 27: 824–834. https://doi.org/10.1101/gr.213959.116
12. Gurevich A, Saveliev V, Vyahhi N, Tesler G (2013) QUAST: quality assessment tool for genome assemblies. Bioinformatics 29:1072–1075. https://doi.org/10.1093/bioinformatics/btt086
13. Kang DD, Li F, Kirton E, Thomas A, Egan R, An H, Wang Z (2019) MetaBAT 2: an adaptive binning algorithm for robust and efficient genome reconstruction from metagenome assemblies. PeerJ 7:e7359. https://doi.org/10.7717/peerj.7359
14. Parks DH, Imelfort M, Skennerton CT, Hugenholtz P, Tyson GW (2015) CheckM: assessing the quality of microbial genomes recovered from isolates, single cells, and metagenomes. Genome Res 25:1043–1055. https://doi.org/10.1101/gr.186072.114
15. Hyatt D, Chen GL, LoCascio PF, Land ML, Larimer FW, Hauser LJ (2010) Prodigal: prokaryotic gene recognition and translation initiation site identification. BMC Bioinformatics 11:119. https://doi.org/10.1186/1471-2105-11-119
16. Eren AM, Esen ÖC, Quince C, Vineis JH, Morrison HG, Sogin ML, Delmont TO (2015) Anvi'o: an advanced analysis and visualization platform for 'omics data'. PeerJ 3: e1319. https://doi.org/10.7717/peerj.1319
17. Sunagawa S, Acinas SG, Bork P, Bowler C, Tara Oceans Coordinators (2020) The Tara Oceans voyage: a global study of marine microbial ecosystems. Sci Adv 6:eabc2797. https://doi.org/10.1126/sciadv.abc2797
18. Rangamaran VR, Sankara Subramanian SH, Balachandran KRS, Gopal D (2023) Vertical microbial profiling of Arabian Sea oxygen minimal zone reveals complex bacterial communities and distinct functional implications. Microb Ecol 85:357–371. https://doi.org/10.1007/s00248-021-01952-z
19. Sushmitha TJ, Rajeev M, Murthy PS, Rao TS, Pandian SK (2023) Planktonic and early-stage biofilm microbiota respond contrastingly to thermal discharge-created seawater warming. Ecotoxicol Environ Saf 264:115433. https://doi.org/10.1016/j.ecoenv.2023.115433
20. Rajeev M, Sushmitha TJ, Toleti SR, Pandian SK (2020) Sediment-associated bacterial community and predictive functionalities are influenced by choice of 16S ribosomal RNA hypervariable region(s): an amplicon-based diversity study. Genomics 112:4968–4979. https://doi.org/10.1016/j.ygeno.2020.09.006
21. Apprill A, McNally S, Parsons R, Weber L (2021) Minor revision to V4 region SSU rRNA 806R gene primer greatly increases detection of SAR11 bacterioplankton. Aquat Microb Ecol 85:129–137. https://doi.org/10.3354/ame01948
22. Sushmitha TJ, Rajeev M, Pandian SK (2020) Next generation sequencing advances and applications in the world of bacterial diversity. Microb Syst:178–209. https://doi.org/10.1201/9780429053535-9
23. Tyson GW, Lo CC, Hugenholtz P (2022) Metagenomic insights into the uncultured microbial majority. Annu Rev Microbiol 76: 431–450. https://doi.org/10.1146/annurev-micro-041320-111021
24. DeSantis TZ, Hugenholtz P, Larsen N, Rojas M, Brodie EL, Keller K et al (2006) Greengenes, a chimera-checked 16S rRNA gene database and workbench compatible with ARB. Appl Environ Microbiol 72:5069–5072. https://doi.org/10.1128/AEM.03006-05
25. Quast C, Pruesse E, Yilmaz P, Gerken J, Schweer T, Yarza P et al (2012) The SILVA ribosomal RNA gene database project: improved data processing and web-based tools. Nucleic Acids Res 41:D590–D596. https://doi.org/10.1093/nar/gks1219
26. Katoh K, Misawa K, Kuma KI, Miyata T (2002) MAFFT: a novel method for rapid multiple sequence alignment based on fast Fourier transform. Nucleic Acids Res 30:3059–3066. https://doi.org/10.1093/nar/gkf436
27. Price MN, Dehal PS, Arkin AP (2010) FastTree 2–approximately maximum-likelihood trees for large alignments. PLoS One 5: e9490. https://doi.org/10.1371/journal.pone.0009490
28. Buchfink B, Xie C, Huson DH (2015) Fast and sensitive protein alignment using DIAMOND. Nat Methods 12:59–60. https://doi.org/10.1038/nmeth.3176
29. Thompson LR, Sanders JG, McDonald D, Earth Microbiome Project Consortium (2021) A communal catalogue reveals Earth's multiscale microbial diversity. Nature 551: 457–463. https://doi.org/10.1038/nature24621
30. Ruff SE, Felden J, Hinrichs KU (2023) Machine learning approaches for analyzing microbial community structure and function. Nat Rev Microbiol 21:123–135. https://doi.org/10.1038/s41579-022-00820-2

Chapter 4

Shotgun Metagenomic Analysis of Microbial Community Dynamics in Wastewater Treatment Through Constructed Wetlands

Georgios Miliotis and Anna Tumeo

Abstract

Constructed wetlands (CWs) offer a sustainable, nature-based solution to wastewater treatment, supporting diverse and dynamic microbial communities that drive nutrient cycling, pollutant degradation, and pathogen removal. This chapter presents an end-to-end methodology for performing shotgun metagenomic analyses on microbial populations from CW influent and effluent. We detail approaches for site selection, sample collection, filtration, DNA extraction, and the incorporation of positive and negative controls to ensure reproducibility and data quality. Two modular bioinformatic workflows encompassing quality control, assembly, taxonomic/functional annotation, and metagenome-assembled genome recovery are described alongside options for detecting antimicrobial resistance genes, pathogens, toxins, and plasmids. In addition, an example workflow for the calculation of alpha and beta diversity is provided. Guidelines for data standardization, replication, and compliance with community-driven reporting standards (MIMS, MIMAG) are also included. Incorporating this protocol will facilitate standardized, reproducible insights into CW microbial dynamics, thereby informing ecological understanding and guiding practical interventions that enhance wastewater treatment efficacy and improve public health outcomes.

Key words Constructed wetlands, Shotgun metagenomics, Microbial communities, Wastewater treatment, Antimicrobial resistance, Metagenome-assembled genomes, Taxonomic and functional annotation, Pathogen detection

1 Introduction

Protecting public health and ensuring sustainable water resources are global challenges as urbanization and industrial activities intensify. Increasing volumes and complexity of wastewater demand effective treatment approaches. Conventional methods such as activated sludge processes and membrane filtration are energy-

Supplementary Information The online version contains supplementary material available at https://doi.org/10.1007/978-1-0716-5080-6_4.

Karthik Raman and Gayathri Sambamoorthy (eds.), *Modelling the Microbiome*, Methods in Molecular Biology, vol. 3006, https://doi.org/10.1007/978-1-0716-5080-6_4,

intensive, costly, and often fail to fully remove emerging contaminants, including pharmaceuticals, antibiotics, antimicrobial-resistant organisms (AROs), and antimicrobial resistance genes (ARGs) [1]. In response, nature-based solutions such as constructed wetlands (CWs) have gained attention for their ability to leverage ecological processes and diverse microbial communities for effective and more sustainable wastewater management [2].

CWs mimic natural wetlands by using vegetation, soil, and associated microbial communities to degrade, adsorb, and transform pollutants [3, 4]. CWs have been shown to remove micropollutants, nutrients, and suspended solids, while potentially mitigating the spread of antimicrobial resistance (AMR) [5]. Integrated Constructed Wetlands (ICWs), an Irish adaptation of the CW design, while originally developed for improved ecological performance and landscape integration, also proved to enhance removal of pathogens and ARGs [6]. By combining ecological principles with engineering design, CWs and ICWs could offer a sustainable alternative to conventional wastewater treatment, reducing pollutant loads and minimizing public health risks.

Wastewater environments are rich in pollutants and nutrients, fostering complex and dynamic microbial communities that vary across wastewater types and geographic regions [7]. Studies of urban and livestock wastewater commonly identify *Proteobacteria* as the dominant phylum, owing to its versatile metabolic roles in nutrient cycling and contaminant degradation. In addition, *Actinobacteria*, *Bacteroidetes*, and *Firmicutes* are frequently abundant, likely due to their ability to tolerate environmental stresses and degrade polysaccharide and other complex organic compounds [8–10].

At the genus level, *Arcobacter*, *Acinetobacter*, *Pseudomonas*, *Bacteroides*, and *Prevotella* are usually predominant. These genera are commonly linked to fecal contamination and pollutant degradation, reflecting both anthropogenic influence and the diverse functional roles that microorganisms fulfill in wastewater-impacted ecosystems [11–13]. Nonetheless, a growing number of studies have flagged the presence in wastewater of opportunistic pathogenic species of public health concern, including from the genera *Aeromonas*, *Enterobacter*, *Escherichia-Shigella*, and *Klebsiella* [14–16]. Overall, the diversity of wastewater microbial communities reflects the complexity of these ecosystems and underscores the need for future research to comprehensively characterize their composition and functional potential, as well as to determine the associated public and environmental health risks.

Wastewater is increasingly recognized as a reservoir of AROs and ARGs. Horizontal gene transfer facilitates their spread between microbial species, while agricultural and industrial activities can amplify dispersion into downstream environments, elevating the risk of transmission to humans, animals, and the food chain

[17]. The fate of AROs and ARGs during wastewater treatment is complex and depends on numerous factors, including treatment efficiency, environmental conditions, and the mobility of genetic elements. Although conventional treatment processes can reduce some resistance determinants, residual AROs and ARGs may still enter natural water bodies, promoting the dissemination of resistance within environmental microbial populations and posing ongoing threats to public health and ecosystem stability [18].

Traditional approaches to water quality assessment often focus on known fecal indicators and opportunistic pathogens (e.g., *Escherichia coli*, coliforms, *Enterococcus* spp.) using culture-based methods or targeted PCR assays [19, 20]. Although valuable, these techniques are inherently limited by their dependence on cultivation and prior knowledge of target DNA sequences. As a result, such methodologies may fail to capture the full complexity of wastewater-associated microbial communities, overlooking both understudied opportunistic pathogens and uncultured or novel taxa that harbor AMR determinants and can serve as ARG reservoirs.

Shotgun metagenomic sequencing (SMS) provides a cultivation-independent approach to comprehensively characterize microbial communities [4]. Unlike targeted methods, SMS analyzes all extracted DNA, revealing both known and novel taxa, their functional capabilities, and the distribution of ARGs and other genetic features. However, applying SMS to CWs and ICWs presents specific challenges. These systems exhibit varying microbial biomass levels, complex biofilm formation, and dynamic physicochemical conditions, including fluctuating nutrient concentrations, pH, and redox states [21]. Additionally, in contrast to conventional wastewater treatment plants, which typically employ mechanical and energy-intensive processes such as aeration and chemical treatments, they rely on passive biological processes and natural attenuation [22]. With these unique characteristics, CWs and ICWs require specialized, end-to-end protocols that ensure high-quality, unbiased, and reproducible results.

A robust SMS experimental design should account for the possible introduction of contamination during any step of the workflow. Contamination from reagents, consumables (referred to as the "kitome"), and laboratory handling (the "splashome") can lead to false positives and distort microbial community profiles [23, 24]. To address this, best practices include the use of negative controls, aseptic sample handling, and rigorous laboratory techniques [25–27]. Similarly, efficient environmental DNA (eDNA) extraction, often aided by enzymatic treatments (i.e., MetaPolyzyme, ExoPolyzyme), is beneficial for lysing diverse environmental taxa, including those in biofilms or with thick cell walls which might be missed by traditional DNA extraction protocols [28].

Biological replication is essential for distinguishing genuine ecosystem variability from artifacts introduced during sampling and processing [29, 30]. Utilizing standardized metadata reporting frameworks, such as the Minimum Information about any (x) Sequence (MIxS) and its metagenome extension (MIMS) for environmental and methodological metadata, together with the Minimum Information about Metagenome-Assembled Genomes (MIMAG) defined by the Genomic Standards Consortium (GSC), enhances data comparability and reproducibility across studies [31, 32]. Adhering to these guidelines, alongside the use of standardized modular bioinformatics pipelines, ensures accurate and reproducible taxonomic and functional assignments [33, 34].

Moreover, the advent of long-read sequencing technologies and complementary multi-omics approaches, such as metatranscriptomics and metaproteomics, can deepen our understanding of microbial dynamics and ecosystem functions within CWs and ICWs [35, 36]. Integrating these methodologies with rigorous best practices enables the optimization of wastewater treatment strategies, the containment of AMR dissemination, and the enhancement of public and environmental health outcomes.

This chapter presents a comprehensive, standardized end-to-end methodology encompassing site selection, sampling, eDNA extraction, SMS, and bioinformatics to characterize microbial community composition, functionality, and the presence of opportunistic pathogens and AROs, that applies to both CWs and ICWs (Fig. 1). Through practical examples of the integrated use of

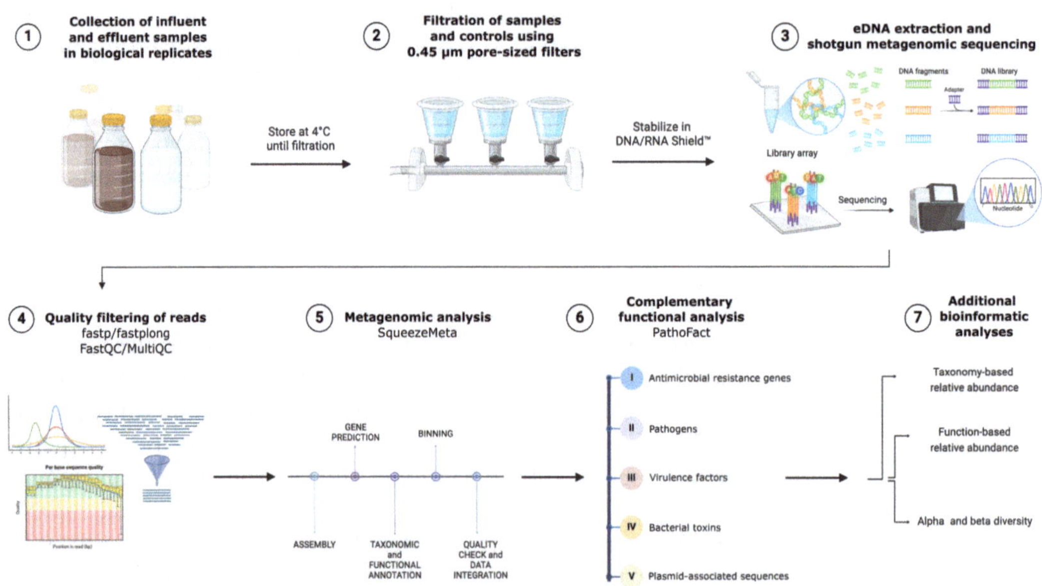

Fig. 1 Overview of the shotgun metagenomic workflow. Samples are collected, filtered, and stabilized (1–2); the eDNA is extracted and sequenced (3); raw reads undergo quality control (4) and integrated metagenomic analysis with SqueezeMeta (5); PathoFact identifies ARGs, pathogens, toxins, and plasmids (6); downstream analyses visualize and quantify microbial composition, function, and diversity (7)

bioinformatics pipelines such as SqueezeMeta [37] and PathoFact [38], this chapter provides a powerful complementary framework for metagenomic analyses of wetland environments. While SqueezeMeta delivers a comprehensive view of microbial community composition, diversity, and functional potential, PathoFact focuses on detecting and characterizing pathogenic and resistance-related genetic elements. Together, these tools enable a more complete understanding of both microbial ecology and associated public health risks, offering a robust platform for assessing the efficacy of wastewater treatment interventions. By informing the refinement of wetland design and operational strategies, this end-to-end protocol with a dual-pipeline approach ultimately aims to support improved contaminant removal, mitigate the spread of antimicrobial resistance, and contribute to the protection of environmental and human health.

2 Materials

All procedures should be performed following strict contamination control measures and standard laboratory safety protocols.

2.1 Laboratory Equipment and Consumables

1. Microsart® Manifold with 500 mL Stainless-Steel Funnels (168 M3-SS500, Sartorius).
2. Biological Safety Cabinet (Class II, BSL-2 Hood).
3. Microsart® e.jet Laboratory Liquid Transfer Pump (166MP-4, Sartorius).
4. Sterile tweezers.
5. Eppendorf 5 mL screw-cap DNA LoBind tubes (10,686,313, Fisher).
6. Autoclave.

2.2 Personnel Health and Safety

1. Personnel handling human and animal wastewater should follow institutional biosafety guidelines and, where applicable, ensure that vaccinations recommended by regional health authorities are up to date (e.g., Hepatitis A, Hepatitis B, Tetanus, Measles, Mumps and Rubella [MMR], Varicella).

2.3 Sample Collection

1. 1000 mL Polypropylene Sterile Bottles with Sodium Thiosulfate.
2. Ice packs and insulated cooler boxes.

2.4 Sample Filtration

1. Sartorius Microsart® ADDfilter 250 mL, 0.45 μm (FIL5614, Sartorius).
2. Sterile Phosphate-Buffered Saline (PBS).
3. Rely+On™ Virkon® tablets (115-0020, VWR).

2.5 Sample Stabilization Chemicals

1. Zymo Research DNA/RNA Shield™ stabilization reagent (R1100-250, Zymo Research).

2.6 Cell Lysis and DNA Extraction Chemicals

1. Magen Biotechnology MagPure Stool DNA Kit.
2. MetaPolyzyme, DNA free (MAC4LDF, Merck).

2.7 Positive Controls (n = 2)

1. Filtration Positive Control: ATCC® MSA-1000 (10 Strain Even Mix Genomic Material) (ATCC-MSA-1000, LGC standards).
2. Sequencing Positive Control: ATCC® MSA-2003 (10 Strain Even Mix Whole Cell Material) (ATCC-MSA-2003, LGC standards).

2.8 Negative Controls (n = 5)

1. PBS Negative Control: Pure Sterile PBS.
2. Filtration Negative Control: Filtered Sterile PBS.
3. DNA Extraction Negative Control: DNA-free water.
4. Library Preparation Negative Control: No-Template Control (NTC).
5. Stabilization Reagent Negative Control: DNA/RNA Shield™.

2.9 Shotgun Metagenomic Sequencing

1. Sequencing reagents, kits, and platform as required. Second-generation sequencing options include Illumina and MGI Tech. Third-generation sequencing options include PacBio and Oxford Nanopore Technologies (ONT).

2.10 Software

1. *Fastp v0.24.0 / fastplong v0.2.2* [39]:
   For trimming and filtering raw reads by quality and length, ensuring high-quality input for assembly.
2. *FastQC v0.12.1* [40]:
   For quality assessment of raw SMS data.
3. *MultiQC v1.21* [41]:
   For synthesis of quality metrics from multiple samples and pipelines into a single, comprehensive report.
4. *SqueezeMeta v1.6.3* [37]:
   For end-to-end metagenomic analysis including assembly, gene prediction, taxonomic and functional annotation, binning, and data integration.
5. *Anvi'o v8* [42]:
   For interactive data visualization and refinement of genome bins.
6. *PathoFact 1.0* [38]:
   For ARGs and virulence factors prediction.

7. *GTDB-Tk v2.4.0* [43]:
   For classification of recovered bins into a standardized, phylogeny-based taxonomy.
8. *CheckM v1.2.4* [34]:
   For assessment of completeness and contamination of recovered bins.
9. *Decontam v1.12* [44]:
   For identification and removal of contaminant sequences detected in negative controls.
10. *NCBI SRA Toolkit and BioProject/BioSample Submission Portal*
    For submission of raw SMS data and metadata to public repositories, ensuring data accessibility and reproducibility.

2.11 Guidelines and Standards

1. Minimum Information about Metagenomic Sequences (MIMS) [32]: Ensures that essential metadata such as sampling location, environmental conditions, and methodological details accompany sequence data submissions. This enhances data reusability and integration with other studies.
2. Minimum Information about Metagenome-Assembled Genomes (MIMAG) [31]: Provides guidelines and quality metrics for reporting genome bins (i.e., metagenome-assembled genomes). Following MIMAG standards ensures transparency and facilitates meaningful comparisons across different studies and datasets.

3 Methods

3.1 Experimental Design and Sample Collection

Establishing a robust experimental design begins with the careful selection of CW sites that collectively represent the full range of systems relevant to the study's objectives. Explicit selection criteria need to be considered: (i) wetland configuration (e.g., surface flow, subsurface flow, vertical flow); (ii) operational scale, ranging from pilot to full-scale; (iii) influent quality and complexity, including pollutant load, wastewater source, and seasonal variability; and (iv) geographical context, including climatic region, catchment land use, and environmental setting. Employing a structured, criteria-based selection process ensures that the study captures diversity in design, function, and environmental influence, thereby enhancing the relevance and transferability of the findings.

Once candidate sites are identified, a detailed description of each CW should be recorded, including design features (layout, vegetation, substrate materials), operational parameters (hydraulic retention time, influent and effluent flow rates), and any pretreatment steps. The nature and origin of the influent wastewater (industrial, municipal, agricultural), the surrounding land-use patterns, climate

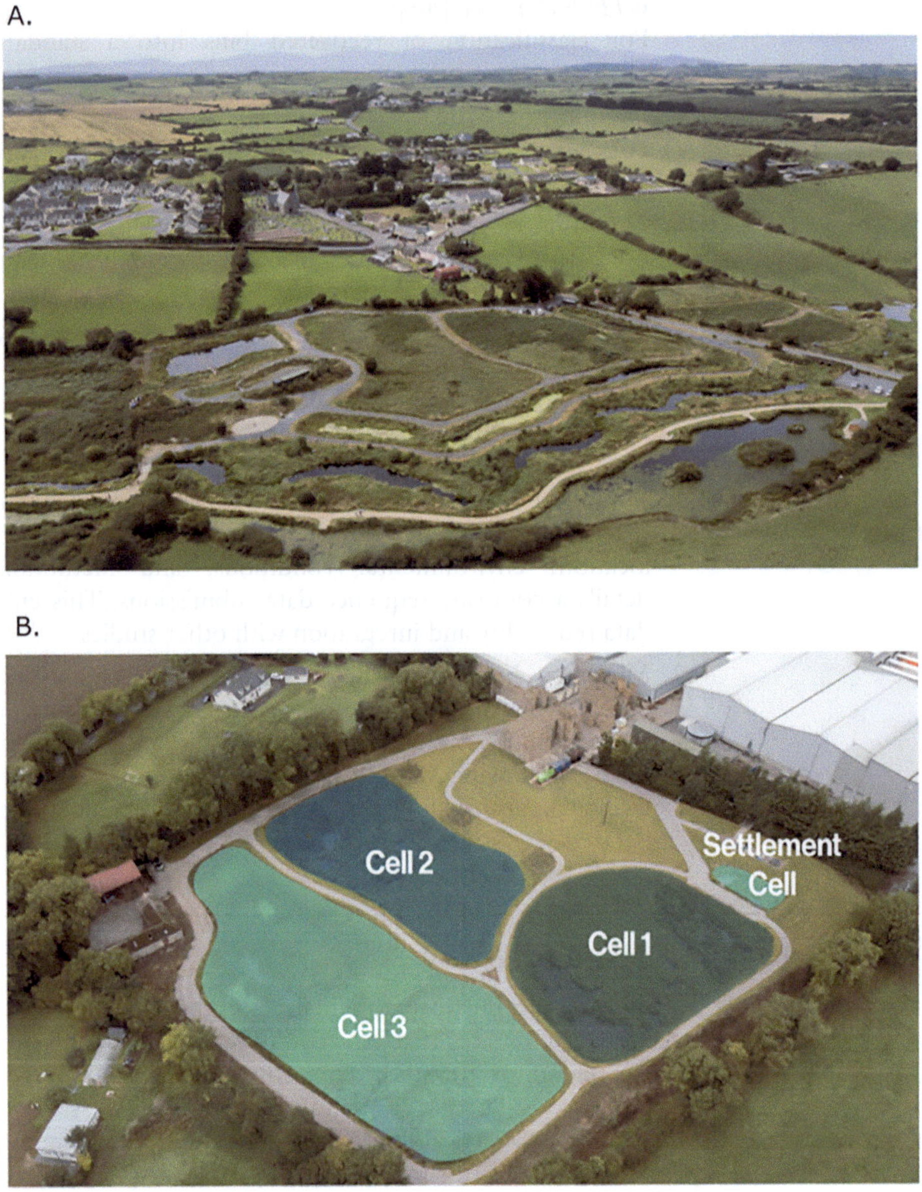

Fig. 2 Integrated Constructed Wetlands (ICWs). (**a**) ICW receiving and treating storm and wastewater from Dunhill village (Ireland). (**b**) A closer view of an ICW treating industrial wastewater. Settlement and treatment cells are arranged to enhance pollutant removal and microbial activity, improving water quality before discharge. Photo credits: VESI Environmental

zone, and other ecological attributes that may affect microbial community dynamics should also be documented (Fig. 2a, b).

Once sites have been designated, collect water samples (1L) from both the influent and effluent streams, representing pre- and post-treatment conditions, using sterile collection bottles or Whirl-Pak bags (*see* **Notes 1** and **2**). Acquire a minimum of three independent biological replicates per sampling point to ensure statistical robustness and help discern natural variability from sampling artifacts. Sample transport to the laboratory should occur immediately after collection. Maintain samples at approximately 4 °C in insulated cooler boxes provided with ice packs or ice to minimize microbial community shifts.

In compliance with the MIMS/MIMAG reporting standards, ensure that all relevant sample metadata are consistently captured. This includes recording both environmental and water quality parameters at the time of sampling, such as on-site weather conditions (temperature, precipitation, humidity, wind), as well as water chemistry parameters (pH, conductivity, total dissolved solids).

3.2 Sample Filtration

Sample temperature should be measured and recorded prior to filtration. If the temperature differs substantially (i.e., more than a few degrees), flag the sample as potentially altered and interpreted with caution. To maximize eDNA yield and enhance the representation of microbial taxa, concentrate microbial cells by vacuum filtration. Filter up to 1 L of each sample through a 0.45 μm pore-sized cellulose nitrate membrane using a vacuum manifold under controlled pressure (e.g., ~600 mbar). Should filter clogging occur before 1 L is processed, record the exact filtered volume for normalization during downstream analyses (*see* **Note 3**).

Using sterile, DNA/RNA-free forceps, carefully transfer the membrane filter into a sterile 5 mL DNA LoBind Eppendorf tube. Immediately immerse the filter in 1.5 mL of DNA/RNA Shield™ stabilization solution to halt microbial activity, preserve genetic material, and ensure safer handling of potentially infectious agents. The stabilized sample can be stored at 4 °C for short-term preservation or at room temperature for limited periods (e.g., days to weeks), with DNA/RNA Shield™ helping to minimize nucleic acid degradation during storage and potential freeze–thaw cycles. For long-term storage (e.g., months to years), samples should be kept at −20 °C or below, according to the manufacturer's recommendations.

3.3 Use of Positive and Negative Controls

Incorporate positive and negative controls at each critical processing stage, including filtration, DNA extraction, and library preparation to detect contamination, monitor process efficiency, and ensure reliable results. In total, two positive and five negative controls are recommended. Subject controls to the same storage conditions and, as appropriate for the type of control, to the same extraction and/or library preparation and SMS steps as the other samples, as detailed below.

3.3.1 Positive Controls for Filtration and Sequencing

The use of at least two positive controls is recommended to verify assay consistency and efficiency:

(i) *Filtration Positive Control*: use ATCC® 10 Strain Even Mix Whole Cell Material (MSA-2003) to assess filtration efficiency. MSA-2003 consists of ten well-characterized bacterial strains in equal proportions. Reconstitute MSA-2003 according to the certificate of analysis and the manufacturer's instructions, and prepare a dedicated positive control sample by spiking the reconstituted material into sterile buffer or into a separate aliquot reserved exclusively for quality control, rather than into environmental samples directly. Process this positive control sample as described in Subheading 3.2 and subject it to the same storage, extraction, library preparation, and SMS steps as the other samples. This approach enables evaluation of filtration performance and detection of any retention biases.

(ii) *Sequencing Positive Control*: for sequencing, ATCC® 10 Strain Even Mix Genomic Material (MSA-1000) is used as a benchmark to verify eDNA extraction efficiency and SMS accuracy. Reconstitute MSA-1000 according to the manufacturer's instructions. Mix equal volumes of the MSA-1000 solution and DNA/RNA Shield™ (e.g., 10 μL MSA-1000 in 10 μL DNA/RNA Shield™) to achieve a final concentration suitable for sequencing. Subject this control to library preparation and SMS in parallel with the environmental samples. By comparing the expected and observed microbial profiles, this control helps confirm that the sequencing process accurately represents the underlying eDNA composition.

3.3.2 Negative Controls for Filtration and Sequencing

Five negative controls are recommended to detect and minimize the introduction of contamination bias at various stages of the analysis:

(i) *PBS Negative Control*: prepare a negative control using only sterile PBS (e.g., 1 mL) without sample DNA. Subject this control to the same storage, extraction, library preparation, and SMS steps as the other samples. Sequencing PBS alone enables detection and quantification of background "kitome" associated with the reagent, and helps distinguish this signal from contamination originating from the filtration equipment.

(ii) *Filtration Negative Control*: filter a defined volume (e.g., 200 mL) of sterile PBS through the system alongside the environmental samples. Subject this control to the same storage, extraction, library preparation, and SMS steps as the other samples. Because sterile PBS should not contain microbial DNA, the presence of any reads in this control suggests procedural contamination introduced during filtration.

(iii) *DNA Extraction Negative Control*: subject a known volume (e.g., 200 μL) of DNA-free water to the same DNA extraction, library preparation, and SMS steps as the other samples. If microbial DNA is detected, it may originate from reagents, consumables, or other contamination sources in the extraction workflow.

(iv) *Library Preparation Negative Control (NTC)*: include an NTC containing all library preparation reagents but no DNA template in the SMS round alongside the environmental samples. The detection of microbial sequences in the NTC indicates contamination introduced during library preparation or sequencing, enabling such reads to be excluded from downstream metagenomic analyses [26].

(v) *Stabilization Reagent Negative Control*: prepare an additional negative control using only DNA/RNA Shield™ (e.g., 200 μL) without sample DNA. Subject this control to the same storage, extraction, library preparation, and SMS steps as the other samples. Sequencing the stabilization reagent alone allows one to detect and quantify any background "kitome" associated with the reagent, thereby improving the accuracy of downstream metagenomic analyses.

3.4 DNA Extraction and Improving Total DNA Yield

Following stabilization in DNA/RNA Shield™, eDNA extraction should be performed as soon as possible. Preferably use a lytic enzyme blend such as MetaPolyzyme, DNA Free to enhance cell lysis, particularly for Gram-positive bacteria and biofilm-associated organisms [28]. A pilot analysis in which representative samples (e.g., one influent and one effluent sample per CW site) are processed with and without lytic enzymes is recommended to determine whether enzymatic lysis significantly increases yield or alters the detected community structure, enabling selection of a consistent extraction strategy.

Extract the eDNA using a standard commercial kit suitable for wastewater samples (e.g., Magen MagPure Stool DNA Kit) according to the manufacturer's instructions, ensuring consistency across all samples and controls. Quantify the resulting eDNA (e.g., using Qubit or a similar fluorometric assay) to confirm sufficient yield and quality. Store purified DNA at −20 °C until downstream library preparation and SMS.

3.5 Shotgun Metagenomic Library Preparation and Normalization

Before initiating library construction, quantify DNA concentrations for all samples using, for instance, a fluorometric assay such as Qubit and adjust them to a uniform concentration as dictated by the selected library preparation kit (e.g., Illumina DNA Prep or NEBNext Ultra II DNA Library Prep). Normalizing DNA ensures that differences in SMS output reflect genuine differences in microbial communities rather than sample loading discrepancies.

For library preparation, follow the manufacturer's recommended protocols, ensuring consistent reagent volumes, reaction conditions, and fragmentation methods. Short-read sequencing platforms (e.g., Illumina NovaSeq or NextSeq) are widely employed due to their high throughput, low error rates, and extensive bioinformatic support. These second-generation sequencing platforms produce relatively short reads (~150–300 bp) that, when assembled, can yield robust metagenomic profiles. Long-read platforms (e.g., Oxford Nanopore, PacBio) may be considered for improved contiguity and enhanced recovery of MAGs. However, long-read data often require additional polishing and error-correction steps due to higher intrinsic error rates.

The optimal sequencing depth should be determined using pilot sequencing and preliminary rarefaction analyses (*see* Subheading 3.7) to assess sequencing sufficiency and compare sampling effort, helping to balance cost against community resolution. For complex wastewater and CW microbiomes, short-read sequencing depths on the order of 10–20 million reads per sample are common, although optimal coverage may vary with community complexity and study objectives.

3.6 Bioinformatics Analysis

3.6.1 Quality Control and Removal of Contaminants

Quality control includes trimming low-quality bases, removing adapters, and discarding reads of inappropriate length using tools such as fastp or fastplong [39]. Per-sample quality checks are usually performed with FastQC [40], and the results are aggregated into a single report with MultiQC [41]. After this initial cleanup, negative control data and, where available, DNA concentration data should be integrated to identify reagent-derived or "kitome" contaminants. Statistical evaluation of sequence frequencies in negative controls and environmental samples can be conducted with the decontam R package [45], which classifies potential contaminants based on their increased prevalence in the negative controls and/or an inverse relationship between feature frequency and sample DNA concentration. Where host-associated reads are expected (e.g., in human or animal wastewater), additional filtering against appropriate host reference genomes (e.g., via Bowtie2) should be performed before downstream taxonomic and functional analysis. This approach ensures that identified taxa and genes represent authentic environmental signals rather than laboratory artifacts.

3.6.2 Assembly and Annotation

SqueezeMeta serves as an integrated metagenomic analysis pipeline for assembly, gene prediction, taxonomic assignment, functional annotation, binning, and data integration. Gene prediction is performed with Prodigal [46], while taxonomic and functional annotations are obtained as homology searches using DIAMOND [47] against reference databases (eggNOG, KEGG, GenBank nr) [48–50].

3.6.3 Assembly Strategy Consideration

One of the key differences among metagenomic pipelines is whether an assembly step is included. Assembly combines metagenomic reads into longer contiguous sequences (contigs), whereas assembly-free approaches rely on direct annotation of raw reads via homology searches against reference databases. Assembly-free approaches can be useful for samples with low microbial biomass and high proportions of host DNA, where assembly may be limited by insufficient coverage. However, accurate taxonomic and functional annotation of short reads requires large reference databases and substantial computational resources [51], and short reads often lack sufficient phylogenetic signal for precise assignment [52, 53]. Assembly is therefore generally advisable for more reliable taxonomic and functional gene annotation and recovery of individual MAGs through binning, which advances the study of the uncultivated microbial majority [54]. Although assembly artifacts can occur (e.g., chimeric contigs), and reads from low-abundance taxa may remain unassembled, co-assembly strategies can partly mitigate these issues.

SqueezeMeta is a modular pipeline that allows the choice of different assembly modes (i.e., sequential, co-assembly, merged, and seqmerge) and assemblers (i.e., MEGAHIT, SPAdes, Canu, and Flye). The choice of assembly mode depends on the experimental design, type and number of samples to analyze, and computational resources available (Table 1). While sequential mode assembles samples sequentially and independently, co-assembly allows pooling of reads from related metagenomes, such as biological replicates of the same sample type, into one comprehensive assembly (i.e., one metagenome). Thereby, co-assembly can improve recovery of shared genomic content, reduce the percentage of non-assembled reads, and maximize the possibility to retrieve individual MAGs with different levels of completion through binning strategies that leverage co-variation across samples. Binning can also be performed on sequential assemblies, but typically with reduced performance compared with co-assembly. However, co-assembly has extensive memory requirements, possibly making it unsuitable when there is a high number of samples.

If co-assembly cannot be supported, the merged and seqmerge modes constitute viable alternatives, although less preferable when sufficient memory is available. In merged and seqmerge modes, samples are assembled individually to reduce computational demands, and resulting contigs are subsequently merged to let the analysis proceed as in the co-assembly mode. Merging can be performed in a single step (merged mode) or multiple steps (seqmerge mode), progressively combining the two most similar assemblies based on amino acid identity, thereby further reducing memory requirements. Although computationally less intensive than co-assembly, merged and seqmerge modes can still be slow on large datasets and increase the risk of creating chimeric

Table 1
Summary of characteristics, applicability, pros, and cons of each assembly strategy available in SqueezeMeta

Assembly strategy	Characteristics	Suitable for	Pros	Cons
Sequential	Samples are treated and assembled sequentially into separate metagenomes	Independent analysis of samples	Lower memory requirements	Higher percentage of chimeras and unassembled reads, reduced binning performance
Co-assembly	Reads from multiple samples of the same type are pulled together into a single co-assembly	Replicates of the same sample type	Low percentage of non-assembled reads, easier retrieval of MAGs	Extensive memory requirements
Merged and seqmerge	Reads from different samples are assembled individually, and resulting contigs merged all together (merged) or stepwise (seqmerge) into one co-assembly	Replicates of the same sample type	Co-assembly analysis at lower computational cost	Time-consuming, increased risk of chimeric contigs

contigs. In addition, SqueezeMeta provides alternative assembly-free analysis modes to perform direct taxonomic and functional assignments on either short or long raw metagenomic reads.

Likewise, the choice of assembler depends on the computational resources available and on the nature of the metagenomic reads. MEGAHIT and SPAdes are the recommended assemblers with short Illumina reads, while Canu and Flye support long reads from PacBio or ONT-Minion. While MEGAHIT is less computationally demanding than SPAdes, SPAdes can often produce high-quality assemblies, although its advantages over MEGAHIT depend on dataset characteristics and computational resources.

3.6.4 Example SqueezeMeta Run on CW Samples

For an optimal analysis and comprehensive representation of the microbial community, run SqueezeMeta in co-assembly mode on biological replicates of the same sample type (e.g., the effluent of one selected CW site) (*see* **Note 4**).

An example command run has the following syntax:

```
SqueezeMeta.pl -m coassembly -p effluent -s effluent_reps.txt
-f /path/to/reads -t 64
```

1. -**m**: mode.
2. -**p**: project name.
3. -**s**: a sample file that specifies sample names, the names of their corresponding raw read files, and the sequencing pair represented in those files, separated by tabulators. In this example run, the samples file would have the following structure:

```
effluent_replicate_1 effluent_replicate_1_1.fq.gz pair1
effluent_replicate_1 effluent_replicate_1_2.fq.gz pair2
effluent_replicate_2 effluent_replicate_2_1.fq.gz pair1
effluent_replicate_2 effluent_replicate_2_2.fq.gz pair2
effluent_replicate_3 effluent_replicate_3_1.fq.gz pair1
effluent_replicate_3 effluent_replicate_3_2.fq.gz pair2
```

4. -**f**: path to the directory where the read files specified in the sample file are stored.
5. -**t**: threads. A high number of threads (i.e. ≥32) is preferable due to the high computational requirements of the co-assembly mode (*see* **Note 5**).

3.7 Rarefaction Curves

Rarefaction curves can be generated in R using functions such as rarecurve() in the vegan (https://github.com/vegandevs/vegan) package. By plotting the number of detected taxa, as defined by the selected taxonomic profiler or annotation method, against the number of sequences sampled, rarefaction curves help assess whether additional sequencing is likely to recover substantial additional taxa or whether the current depth provides a reasonable representation of the detectable diversity. In the context of SMS, rarefaction is most useful for evaluating sequencing sufficiency and comparing sampling effort across samples, rather than as a direct estimator of true community richness. Interpreting rarefaction curves before large-scale sequencing helps ensure that sequencing efforts are adequate without being excessively redundant. Example rarefaction curves and corresponding interpretation are shown in Fig. 3.

3.8 Diversity Analyses

3.8.1 Alpha Diversity

Alpha diversity indices evaluate the richness and evenness of taxa within individual samples, assessing the complexity of microbial communities. Key alpha diversity indices include the Shannon Index, which accounts for both taxa abundance and evenness, and the Simpson Index, which measures the probability that two individuals randomly selected from a sample will belong to the same taxon (Fig. 4a). For alpha diversity analysis, QIIME 2 [64] and the vegan (https://github.com/vegandevs/vegan) package in R are commonly used, as shown below (Subheading 3.8.3), with taxonomic or functional abundance tables (e.g., read counts or normalized abundances per taxon or gene family).

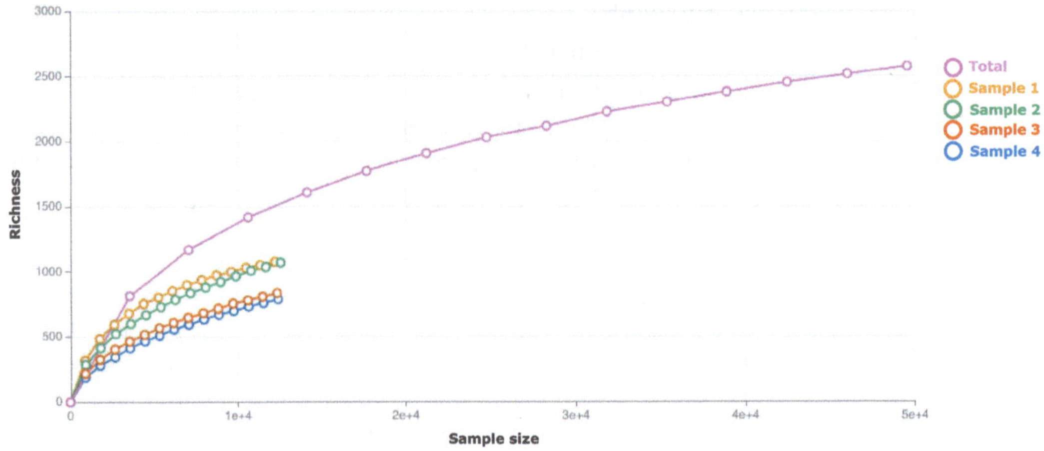

Fig. 3 Example rarefaction curves. Sample size (*X*-axis) shows the number of reads subsampled during real-time analysis. Richness (*Y*-axis) indicates the number of unique taxonomic units observed in subsampled reads. Sample curves approaching a plateau suggest that most of the detectable diversity in a given sample has already been captured. Steeply rising curves otherwise indicate that additional sequencing would likely recover additional taxa. The total curve, representing pooled richness, reflects the extent of unshared diversity across the dataset. Comparing curve shapes and slopes across samples helps identify those that may remain under-sequenced or display unusually high or low diversity. Overlapping or similarly shaped curves suggest comparable community complexity, whereas diverging curves indicate differences in richness that may reflect either biological (e.g., community heterogeneity) or technical factors (e.g., uneven sequencing depth)

3.8.2 Beta Diversity

Beta diversity analysis assesses differences in microbial community composition between samples, enabling comparisons of microbial ecosystems across treatment conditions. Key beta diversity metrics include the Bray–Curtis dissimilarity index, which quantifies compositional differences based on species or feature abundance, and UniFrac distances, which incorporate phylogenetic information by using a tree derived from the observed taxa (e.g., based on marker genes or concatenated protein alignments). Note that, therefore, UniFrac distances cannot be calculated directly from a taxonomic abundance table without first constructing an appropriate phylogenetic tree. Software tools such as QIIME 2 [64] and R packages including Phyloseq and vegan (https://github.com/vegandevs/vegan) are commonly used for beta diversity analyses. For example, QIIME 2 provides methods for calculating Bray–Curtis and UniFrac distances, along with visualization techniques like Principal Coordinates Analysis (PCoA) and Non-metric Multidimensional Scaling (NMDS) to aid interpretation of results (Fig. 4b). The Phyloseq package also facilitates the integration of phylogenetic and compositional data, supporting robust beta diversity assessments. In any attempted visualization of beta diversity, tridimensional NMDS plotting is beneficial for capturing more of the full complexity and underlying variation in the data.

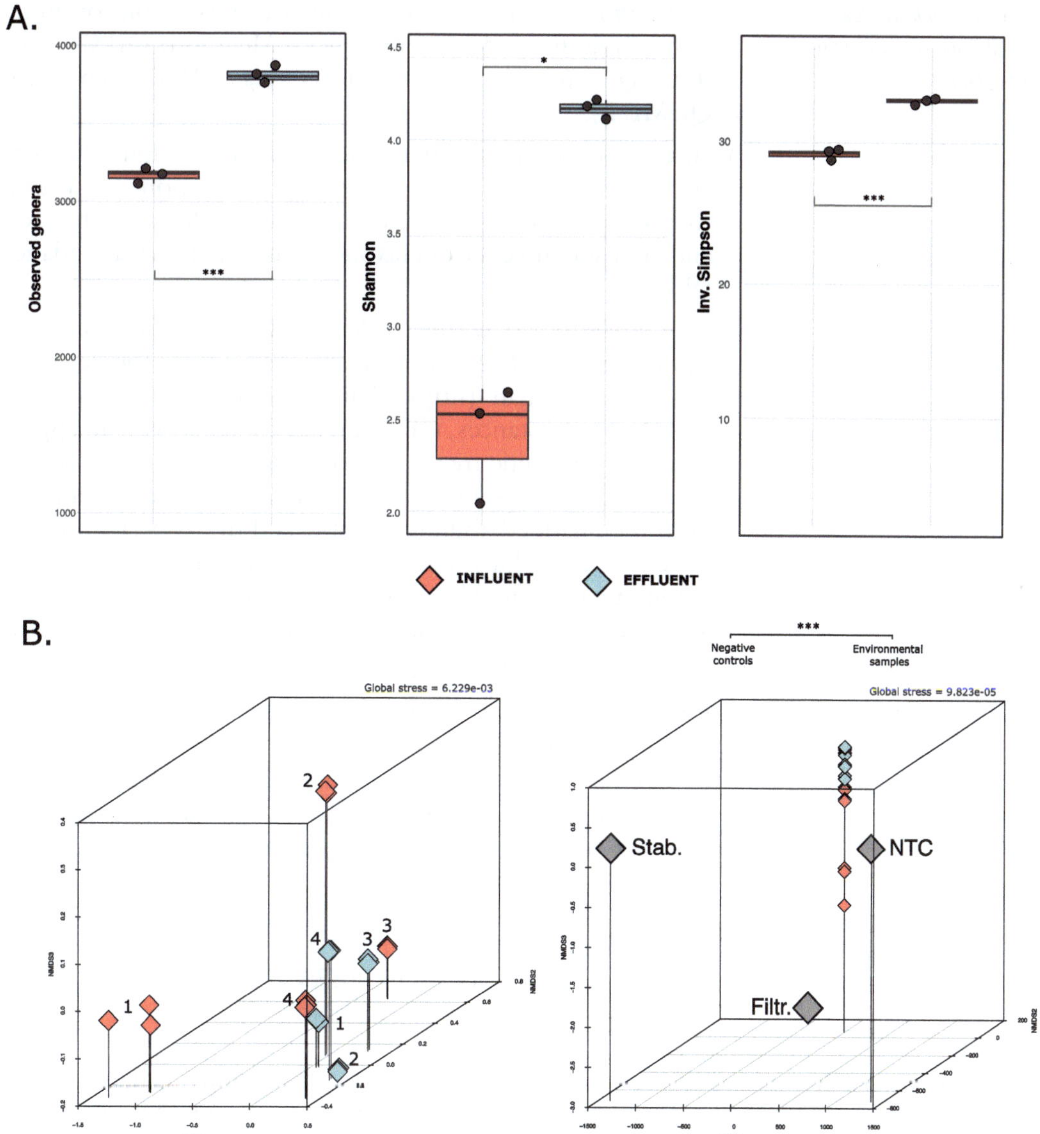

Fig. 4 Example alpha and beta diversity analyses. (**a**) Boxplots showing statistically significant (*p*-value <0.01) differences in key alpha diversity indices (number of observed genera, Shannon and Simpson's Reciprocal Index) between influent and effluent samples collected from an Integrated Constructed Wetland (ICW). Statistical significance of the comparison between paired influent and effluent groups, as assessed using a paired *t*-test, is also displayed. Values are shown for all biological replicates. (**b**) Non-metric multidimensional scaling ordination displaying beta diversity of wastewater microbial communities from multiple ICWs (1–4). Clustering of environmental samples ($n = 24$) is shown respectively (left) excluding and (right) together with the negative controls ($n = 3$). These include one Filtration Negative Control (Filtr.), one Library Preparation Negative Control (NTC), and one Stabilization Reagent Negative Control (Stab.). Statistical significance of the differences in clustering between environmental samples and negative controls, as assessed using PERMANOVA, is also shown

3.8.3 Example Workflow for Alpha and Beta Diversity Calculation

(i) *Data import and preparation*: import taxonomic or functional assignment data from SqueezeMeta into QIIME 2 or R using appropriate data formats (e.g., BIOM tables for QIIME 2).

(ii) *Alpha diversity calculation*: using QIIME 2 alpha diversity analysis methods or the diversity() function in vegan, (https://github.com/vegandevs/vegan) calculate Shannon and Simpson indices from taxonomic or functional abundance tables.

(iii) *Beta diversity calculation*: calculate Bray–Curtis dissimilarities using QIIME 2 beta diversity methods or the vegdist() function in vegan (https://github.com/vegandevs/vegan). For UniFrac distances, use QIIME 2 together with an appropriate phylogenetic tree (e.g., constructed from the observed taxa based on marker genes or MAGs), as UniFrac depends on both the provided abundance table and the associated phylogenetic relationships. Example alpha and beta diversity calculations are presented in Supplementary File 1.

(iv) *Visualization*: Generate diversity plots using QIIME 2 visualization tools or the ggplot2 package in R to interpret and present the diversity metrics.

(v) *Statistics: (Alpha diversity)* Perform statistical comparison between paired groups (e.g., influent/effluent from the same ICW) using a paired *t*-test or, if normality assumptions are not met, the Wilcoxon signed-rank test. *(Beta diversity)* Assess statistical significance of differences in community composition between groups using PERMANOVA (Permutational Multivariate Analysis of Variance), and the homogeneity of multivariate dispersion using PERMDISP (Permutation Test for Multivariate Dispersion) (*see* **Note 6**).

3.9 Binning and MAG Recovery

Assembly strategies support recovery of individual, complete MAGs (i.e., bins) as the starting point for deeper analysis of the microbial community. SqueezeMeta supports different binning programs, including CONCOCT and MetaBAT [55, 56], to group contigs from the same organism into bins based on similar oligonucleotide composition and abundance and co-variation patterns across different samples [37]. For each bin, SqueezeMeta assigns a consensus taxonomy based on the predominant taxonomic annotation among its constituent contigs. Furthermore, SqueezeMeta uses CheckM [34] to estimate bin completeness, contamination, and strain heterogeneity according, among others, to presence and copy number of lineage-specific marker genes in each bin.

3.9.1 Handling of MAGs

For each metagenome obtained using any assembly strategy, SqueezeMeta outputs a list of identified MAGs and related completeness, contamination, and strain heterogeneity. To report MAGs in publications or deposit them in public databases, it is best practice to follow GSC's MIMAG and report standard quality metrics, including assembly statistics and estimates of genome completeness and contamination [31], both conveniently produced by SqueezeMeta. Accordingly, each MAG can be classified as finished, high-quality draft, medium-quality draft, or low-quality draft [31].

3.9.2 Taxonomically Unclassified High-Quality MAGs

SqueezeMeta can recover high-quality MAGs that fail to be taxonomically classified at the species level or above. To further assess taxonomic novelty, it is good practice to examine individual MAGs with tools such as GTDB-Tk [43], which can provide objective taxonomic placement of bins. Note that GTDB-Tk can optionally be run within the SqueezeMeta workflow using the "--gtdbtk" flag. In either case, ensure that the most recent version of GTDB-Tk reference database is used. MAGs may warrant further investigation if GTDB-Tk hints at taxonomic novelty and if all minimum MIMAG standards for high quality are met. These include: (i)$\geq$ 90% completeness; (ii) $\leq$5% contamination; (iii) presence of the 23S, 16S, and 5S rRNA genes; (iv) presence of $\geq$18 tRNA genes representing $\geq$18 amino acids (Fig. 5a, b). MAGs and associated metadata can be deposited in online databases such as GOLD [57] or Seqcode [65] and, where appropriate, proposed as *Candidatus* taxa. However, full characterization and formal definition of novel prokaryotic taxa under the International Code of Nomenclature of Prokaryotes generally requires cultivation of the type strain.

3.10 ARGs, Pathogens, and Plasmid Detection with PathoFact

To complement the microbial community and functional analyses performed by SqueezeMeta, PathoFact [38] can be used as a complementary modular pipeline to identify and characterize ARGs, virulence factors (VFs), bacterial toxins, and plasmid-associated sequences from assembled metagenomic contigs and their predicted protein-coding genes. PathoFact integrates the use of curated databases with advanced machine-learning-based prediction approaches to detect and quantify pathogen-associated genomic features, thereby providing data essential for understanding the distribution of AMR Antimicrobial resistance (AMR), pathogenicity, and horizontal gene transfer events in complex microbial communities (Fig. 6a–c).

3.10.1 Databases and Tools Used by PathoFact

(i) *Comprehensive Antibiotic Resistance Database (CARD)* [58]: for identification and classification of ARGs.

(ii) *Virulence Factor Database (VFDB)* [59]: for detection of virulence factors associated with potential pathogens.

(iii) *PlasmidFinder or PlasFlow* [60, 61]: for identification of plasmid-associated sequences.

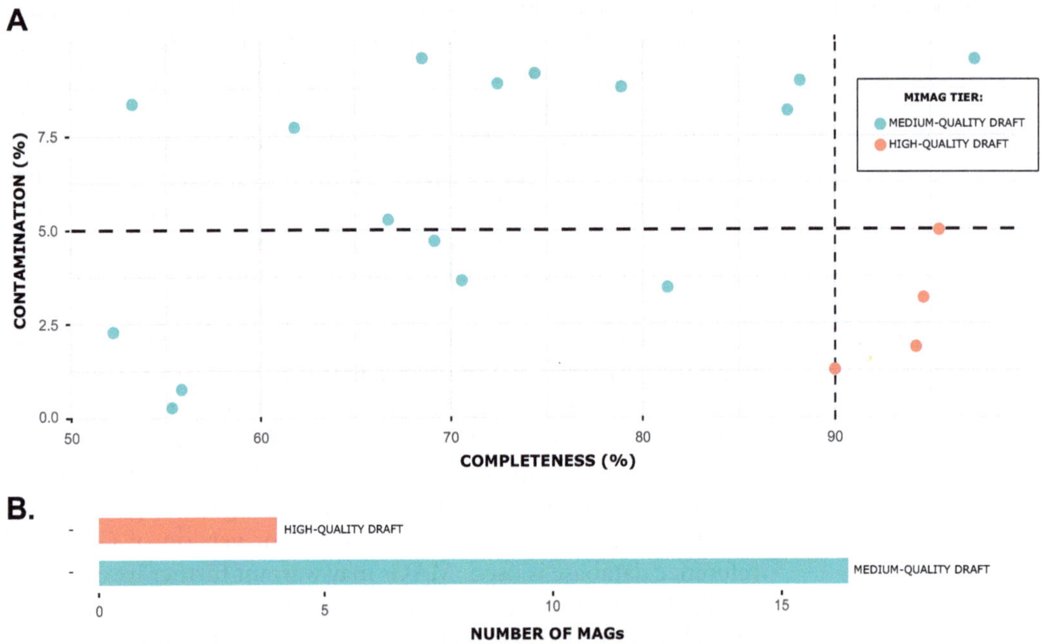

Fig. 5 Completeness and contamination of synthetic metagenome-assembled genomes (MAGs). (**a**) Scatterplot showing completeness versus contamination for 20 synthetic MAGs, illustrating the application of MIMAG standards for genome quality assessment. Dashed lines indicate commonly used thresholds for high-quality draft genomes (≥90% completeness and ≤5% contamination) and medium-quality draft genomes (≥50% completeness and ≤10% contamination). (**b**) Bar plot showing the number of MAGs falling into the high- and medium-quality draft categories according to MIMAG criteria. Synthetic data are used solely for demonstration of typical CheckM-like outputs in metagenomic workflows

(iv) *Toxin-Antitoxin Database (TADB)* [62]: for identification of toxins that contribute to bacterial pathogenicity.

(v) *DIAMOND* [47]: for high-speed, sensitive sequence alignment against the aforementioned databases, enabling efficient processing of large metagenomic datasets.

(vi) *deepARG* [63]: an advanced deep learning tool specifically designed for the detection and classification of ARGs with high accuracy.

3.10.2 Outputs Generated by PathoFact

(i) *Annotated Tables:*

- *ARGs*: detailed listings of identified ARGs, including resistance mechanisms and associated bacterial taxa.
- *VFs*: detailed listings of detected VFs, their functions, secreted/non-secreted status, and corresponding microbial hosts.
- *Plasmid-associated sequences*: information on the genomic context of predicted ARGs, VFs, and toxins, including their localization on plasmid-associated sequences, which facilitates the understanding of potential horizontal gene transfer events.

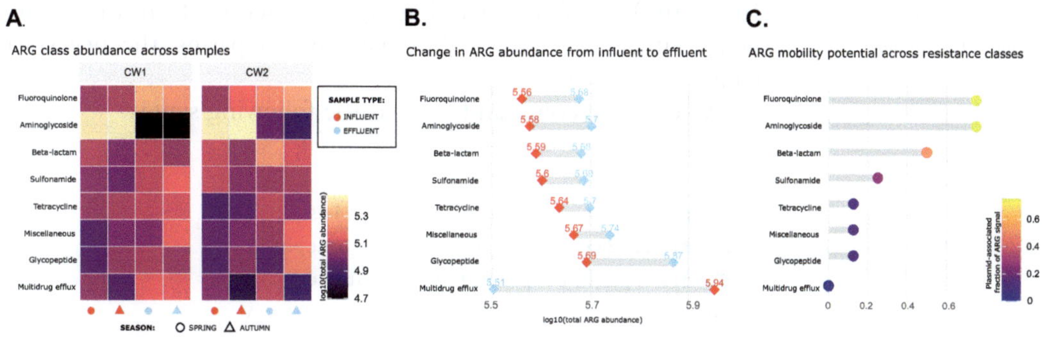

Fig. 6 PathoFact-style outputs integrating abundance of antimicrobial resistance gene (ARG) abundance and sample metadata. (**a**) *ARG class abundance heatmap.* Relative abundance of eight ARG classes across influent and effluent samples from two constructed wetlands (CW1 and CW2). Values represent log10-transformed relative abundances derived from synthetic PathoFact-style outputs. Sample metadata (site, sample type, and season) are encoded below the heatmap. The plot illustrates typical influent–effluent shifts, including strong enrichment of beta-lactam and multidrug efflux determinants in influent samples. (**b**) *Influent–effluent comparison of total ARG signal by resistance class.* A dumbbell plot shows paired log10-scaled total ARG abundances for each resistance class in influent and effluent samples. Lines connect matched values for each class, highlighting direction and magnitude of treatment-related changes. Beta-lactam, fluoroquinolone, and aminoglycoside resistance classes show the largest decreases following treatment. (**c**) *Plasmid-associated fraction of ARGs.* Proportion of ARG signal located on plasmid-associated contigs for each resistance class. This panel summarizes the potential mobility of resistance determinants, with beta-lactam, fluoroquinolone, and aminoglycoside classes showing the highest plasmid linkage, consistent with their elevated horizontal gene transfer potential

- *Toxins*: listings of bacterial toxins identified, along with their functional annotations and microbial sources.

(ii) *Quantitative and Taxonomic Information:*

- *Abundance Estimates*: counts or relative abundances of identified ARGs, VFs, toxins, and plasmid-associated sequences across samples.
- *Taxonomic Assignments*: integration of taxonomic data to link detected genes with specific microbial taxa.

3.10.3 Visualization and Interpretation of PathoFact Outputs

PathoFact provides rich tabular data and annotations but does not inherently generate advanced visualizations (e.g., heatmaps, bar charts, network diagrams) or interpretative risk assessments. Users should employ external tools and statistical frameworks to visualize these outputs and relate them to ecological or public health questions. For example, data from PathoFact can be integrated with R or Python visualization libraries to create heatmaps or bar plots. Network analysis tools can reveal co-occurrence patterns, and the user's own interpretation or additional modeling can translate these findings into broader ecological or public health contexts.

3.10.4 Example PathoFact Command

Run PathoFact after generating gene predictions from the assembled contigs (e.g., using Prodigal) and preparing the necessary databases (*see* **Note 7**).

An example command run has the following syntax:

```
python pathofact.py run \
--contigs /path/to/assembled_contigs.fasta \
--prodigal_dir /path/to/prodigal_output \
--db_dir /path/to/pathofact_databases \
--output_dir /path/to/pathofact_results \
--sample_id effluent_sample \
--run_name effluent_pathofact_run \
--threads 46
```

1. --**contigs**: path to the FASTA file containing assembled contigs. These contigs have been generated in previous steps, such as those performed by SqueezeMeta.
2. --**prodigal_dir**: directory containing gene prediction results (e.g.,.faa files) generated by Prodigal. PathoFact requires these predicted protein sequences to perform ARG, virulence factor, and plasmid-associated gene identification.
3. --**db_dir**: directory containing all databases and reference files that PathoFact uses, such as CARD and VFDB.
4. --**output_dir**: directory where PathoFact will write results. If it does not exist, PathoFact will create it. The output will typically include tables listing detected ARGs, virulence factors, plasmid-associated sequences, and toxin genes.
5. --**sample_id**: label used to identify the sample being analyzed. This is useful when processing multiple samples.
6. --**run_name**: custom prefix or identifier for this particular analysis run, aiding in keeping track of multiple executions or datasets.
7. --**threads**: number of threads to use. Increasing the number of threads can substantially reduce the run time for large datasets.

4 Notes

1. Because sediments can affect downstream sample processing and analysis, collection of clear samples is, when possible, desirable. Collection of samples from custom buckets designed to exclude sediments is therefore recommended.
2. The performance of CWs and ICWs may sometimes be constrained by environmental conditions and seasonal fluctuations in temperature and rainfall, which can determine nearly

complete absence of flow through the wetland. Therefore, while 1 L is the recommended sampling volume, consider collecting smaller volumes when flow rate is low. Likewise, sampling from inner cells prior to discharge can represent a viable alternative in the absence of outflow. Any resulting deviations in the microbial community must be interpreted accordingly.

3. The volume of filtrate obtained during a typical filtration period, which may extend up to ~6 h for high-biomass or sediment-rich samples, can range from 100 mL to 1 L depending on biomass density, sediment load, and sample type. Because extended filtration times may influence microbial activity and community composition, total filtration duration should be recorded and considered during data interpretation. Statistical analyses have shown that differences in filtrate volume, when accounted for during downstream normalization, do not significantly impact the estimated composition of microbial communities.

4. The approach of co-assembling raw reads from replicates of the same sample type collected at one specific time point is known as "conservative" co-assembly. Where data from multiple time points and sufficient computational resources are available, a "longitudinal" approach co-assembling all replicates of the same sample type collected at different time points (e.g., over a 12-month period) can allow a more comprehensive representation of the microbial diversity.

5. Due to its high memory requirements, access to at least 90 GB of memory is desirable to prevent co-assembly from crashing. Additional tips for running SqueezeMeta in low-memory environments are available at: jtamames/SqueezeMeta.

6. Very small sample group sizes (e.g., $n < 5$) limit the statistical power of both the Wilcoxon test (used for alpha diversity) and PERMANOVA (beta diversity), reducing the stability and interpretability of p-values. These constraints should be explicitly considered when interpreting results. In those cases, effect sizes and, where applicable, confidence intervals, can provide additional context for comparison of small sample sizes.

7. Before running PathoFact, ensure that all prerequisite steps (e.g., genome assemble and gene prediction) have been completed successfully. The required databases for PathoFact should also be correctly installed and accessible. PathoFact outputs can be readily integrated with the broader metagenomic analyses performed with SqueezeMeta. However, additional data processing and separate visualization tools will be required for producing visualizations such as heatmaps, bar charts, or network diagrams, and in-depth ecological and/or risk assessments.

References

1. Liu Y, Blowes DW, Ptacek CJ, Groza LG (2019) Removal of pharmaceutical compounds, artificial sweeteners, and perfluoroalkyl substances from water using a passive treatment system containing zero-valent iron and biochar. Sci Total Environ 691:165–177. https://doi.org/10.1016/j.scitotenv.2019.06.450
2. Wu H, Zhang J, Ngo HH et al (2015) A review on the sustainability of constructed wetlands for wastewater treatment: design and operation. Bioresour Technol 175:594–601. https://doi.org/10.1016/j.biortech.2014.10.068
3. Usharani B (2019) Metagenomics study of the microbes in constructed wetland system treating sewage. Int Lett Nat Sci 74:26–48. https://doi.org/10.18052/www.scipress.com/ILNS.74.26
4. Bai Y, Liang J, Liu R et al (2014) Metagenomic analysis reveals microbial diversity and function in the rhizosphere soil of a constructed wetland. Environ Technol 35:2521–2527. https://doi.org/10.1080/09593330.2014.911361
5. Lavrnić S, Nan X, Blasioli S et al (2020) Performance of a full scale constructed wetland as ecological practice for agricultural drainage water treatment in Northern Italy. Ecol Eng 154:105927. https://doi.org/10.1016/j.ecoleng.2020.105927
6. Scholz M, Harrington R, Carroll P, Mustafa A (2007) The integrated constructed wetlands (ICW) concept. Wetlands 27:337
7. Sanderson H, Brown RS, Hania P et al (2019) Antimicrobial resistant genes and organisms as environmental contaminants of emerging concern: addressing global public health risks. In: Management of emerging public health issues and risks. Elsevier, pp 147–187
8. Buelow E, Bayjanov JR, Majoor E et al (2018) Limited influence of hospital wastewater on the microbiome and resistome of wastewater in a community sewerage system. FEMS Microbiol Ecol 94. https://doi.org/10.1093/femsec/fiy087
9. Tang J, Bu Y, Zhang X-X et al (2016) Metagenomic analysis of bacterial community composition and antibiotic resistance genes in a wastewater treatment plant and its receiving surface water. Ecotoxicol Environ Saf 132:260–269. https://doi.org/10.1016/j.ecoenv.2016.06.016
10. Xue J, Schmitz BW, Caton K et al (2019) Assessing the spatial and temporal variability of bacterial communities in two Bardenpho wastewater treatment systems via Illumina MiSeq sequencing. Sci Total Environ 657:1543–1552. https://doi.org/10.1016/j.scitotenv.2018.12.141
11. Jia S, Zhang X-X, Miao Y et al (2017) Fate of antibiotic resistance genes and their associations with bacterial community in livestock breeding wastewater and its receiving river water. Water Res 124:259–268. https://doi.org/10.1016/j.watres.2017.07.061
12. Szekeres E, Baricz A, Chiriac CM et al (2017) Abundance of antibiotics, antibiotic resistance genes and bacterial community composition in wastewater effluents from different Romanian hospitals. Environ Pollut 225:304–315. https://doi.org/10.1016/j.envpol.2017.01.054
13. Yang Y, Li B, Zou S et al (2014) Fate of antibiotic resistance genes in sewage treatment plant revealed by metagenomic approach. Water Res 62:97–106. https://doi.org/10.1016/j.watres.2014.05.019
14. Lu X, Zhang X-X, Wang Z et al (2015) Bacterial pathogens and community composition in advanced sewage treatment systems revealed by metagenomics analysis based on high-throughput sequencing. PLoS One 10:e0125549. https://doi.org/10.1371/journal.pone.0125549
15. Numberger D, Ganzert L, Zoccarato L et al (2019) Characterization of bacterial communities in wastewater with enhanced taxonomic resolution by full-length 16S rRNA sequencing. Sci Rep 9:9673. https://doi.org/10.1038/s41598-019-46015-z
16. Yao S, Ye J, Yang Q et al (2021) Occurrence and removal of antibiotics, antibiotic resistance genes, and bacterial communities in hospital wastewater. Environ Sci Pollut Res 28:57321–57333. https://doi.org/10.1007/s11356-021-14735-3
17. Huijbers PMC, Blaak H, de Jong MCM et al (2015) Role of the environment in the transmission of antimicrobial resistance to humans: a review. Environ Sci Technol 49:11993–12004. https://doi.org/10.1021/acs.est.5b02566
18. Bürgmann H, Frigon D, Gaze WH et al (2018) Water and sanitation: an essential battlefront in the war on antimicrobial resistance. FEMS Microbiol Ecol 94. https://doi.org/10.1093/femsec/fiy101
19. Lamori JG, Xue J, Rachmadi AT et al (2019) Removal of fecal indicator bacteria and

antibiotic resistant genes in constructed wetlands. Environ Sci Pollut Res 26:10188–10197. https://doi.org/10.1007/s11356-019-04468-9

20. Prendergast DM, O'Doherty Á, Burgess CM et al (2022) Critically important antimicrobial resistant Enterobacteriaceae in Irish farm effluent and their removal in integrated constructed wetlands. Sci Total Environ 806:151269. https://doi.org/10.1016/j.scitotenv.2021.151269
21. Truu M, Juhanson J, Truu J (2009) Microbial biomass, activity and community composition in constructed wetlands. Sci Total Environ 407:3958–3971. https://doi.org/10.1016/j.scitotenv.2008.11.036
22. Dotro G, Langergraber G, Molle P et al (2017) Treatment wetlands. Biological wastewater treatment series, vol 7. IWA Publishing, London
23. Salter SJ, Cox MJ, Turek EM et al (2014) Reagent and laboratory contamination can critically impact sequence-based microbiome analyses. BMC Biol 12:87. https://doi.org/10.1186/s12915-014-0087-z
24. Olomu IN, Pena-Cortes LC, Long RA et al (2020) Elimination of "kitome" and "splashome" contamination results in lack of detection of a unique placental microbiome. BMC Microbiol 20:157. https://doi.org/10.1186/s12866-020-01839-y
25. Eisenhofer R, Minich JJ, Marotz C et al (2019) Contamination in low microbial biomass microbiome studies: issues and recommendations. Trends Microbiol 27:105–117. https://doi.org/10.1016/j.tim.2018.11.003
26. Kim D, Hofstaedter CE, Zhao C et al (2017) Optimizing methods and dodging pitfalls in microbiome research. Microbiome 5:52. https://doi.org/10.1186/s40168-017-0267-5
27. Karstens L, Asquith M, Davin S et al (2019) Controlling for contaminants in low-biomass 16S rRNA gene sequencing experiments. mSystems 4. https://doi.org/10.1128/mSystems.00290-19
28. Tighe S, Afshinnekoo E, Rock TM et al (2017) Genomic methods and microbiological technologies for profiling novel and extreme environments for the extreme microbiome project (XMP). J Biomol Tech 28:31–39. https://doi.org/10.7171/jbt.17-2801-004
29. Prosser JI (2010) Replicate or lie. Environ Microbiol 12:1806–1810. https://doi.org/10.1111/j.1462-2920.2010.02201.x
30. Goodrich JK, Di Rienzi SC, Poole AC et al (2014) Conducting a microbiome study. Cell 158:250–262. https://doi.org/10.1016/j.cell.2014.06.037
31. Bowers RM, Kyrpides NC, Stepanauskas R et al (2017) Minimum information about a single amplified genome (MISAG) and a metagenome-assembled genome (MIMAG) of bacteria and archaea. Nat Biotechnol 35:725–731. https://doi.org/10.1038/nbt.3893
32. Yilmaz P, Kottmann R, Field D et al (2011) Minimum information about a marker gene sequence (MIMARKS) and minimum information about any (x) sequence (MIxS) specifications. Nat Biotechnol 29:415–420. https://doi.org/10.1038/nbt.1823
33. Quince C, Walker AW, Simpson JT et al (2017) Shotgun metagenomics, from sampling to analysis. Nat Biotechnol 35:833–844. https://doi.org/10.1038/nbt.3935
34. Parks DH, Imelfort M, Skennerton CT et al (2015) CheckM: assessing the quality of microbial genomes recovered from isolates, single cells, and metagenomes. Genome Res 25:1043–1055. https://doi.org/10.1101/gr.186072.114
35. Laver T, Harrison J, O'Neill PA et al (2015) Assessing the performance of the Oxford Nanopore technologies minion. Biomol Detect Quantif 3:1–8. https://doi.org/10.1016/j.bdq.2015.02.001
36. VerBerkmoes NC, Denef VJ, Hettich RL, Banfield JF (2009) Functional analysis of natural microbial consortia using community proteomics. Nat Rev Microbiol 7:196–205. https://doi.org/10.1038/nrmicro2080
37. Tamames J, Puente-Sánchez F (2019) SqueezeMeta, a highly portable, fully automatic metagenomic analysis pipeline. Front Microbiol 9. https://doi.org/10.3389/fmicb.2018.03349
38. de Nies L, Lopes S, Busi SB et al (2021) PathoFact: a pipeline for the prediction of virulence factors and antimicrobial resistance genes in metagenomic data. Microbiome 9:49. https://doi.org/10.1186/s40168-020-00993-9
39. Chen S (2023) Ultrafast one-pass FASTQ data preprocessing, quality control, and deduplication using fastp. iMeta 2. https://doi.org/10.1002/imt2.107
40. Andrews S (2010) FastQC: a quality control tool for high throughput sequence data [Online]. Available at: http://www.bioinformatics.babraham.ac.uk/projects/fastqc/
41. Ewels P, Magnusson M, Lundin S, Käller M (2016) MultiQC: summarize analysis results for multiple tools and samples in a single

report. Bioinformatics 32:3047–3048. https://doi.org/10.1093/bioinformatics/btw354

42. Eren AM, Kiefl E, Shaiber A et al (2020) Community-led, integrated, reproducible multi-omics with anvi'o. Nat Microbiol 6:3–6. https://doi.org/10.1038/s41564-020-00834-3
43. Chaumeil P-A, Mussig AJ, Hugenholtz P, Parks DH (2022) GTDB-Tk v2: memory friendly classification with the genome taxonomy database. Bioinformatics 38:5315–5316. https://doi.org/10.1093/bioinformatics/btac672
44. Davis NM, Proctor DM, Holmes SP et al (2018) Simple statistical identification and removal of contaminant sequences in marker-gene and metagenomics data. Microbiome 6:226. https://doi.org/10.1186/s40168-018-0605-2
45. Field D, Garrity G, Gray T et al (2008) The minimum information about a genome sequence (MIGS) specification. Nat Biotechnol 26:541–547. https://doi.org/10.1038/nbt1360
46. Hyatt D, Chen G-L, LoCascio PF et al (2010) Prodigal: prokaryotic gene recognition and translation initiation site identification. BMC Bioinformatics 11:119. https://doi.org/10.1186/1471-2105-11-119
47. Buchfink B, Xie C, Huson DH (2015) Fast and sensitive protein alignment using DIAMOND. Nat Methods 12:59–60. https://doi.org/10.1038/nmeth.3176
48. Huerta-Cepas J, Szklarczyk D, Forslund K et al (2016) eggNOG 4.5: a hierarchical orthology framework with improved functional annotations for eukaryotic, prokaryotic and viral sequences. Nucleic Acids Res 44:D286–D293. https://doi.org/10.1093/nar/gkv1248
49. Kanehisa M (2000) KEGG: Kyoto encyclopedia of genes and genomes. Nucleic Acids Res 28:27–30. https://doi.org/10.1093/nar/28.1.27
50. Benson DA, Cavanaugh M, Clark K et al (2012) GenBank. Nucleic Acids Res 41:D36–D42. https://doi.org/10.1093/nar/gks1195
51. Pignatelli M, Aparicio G, Blanquer I et al (2008) Metagenomics reveals our incomplete knowledge of global diversity. Bioinformatics 24:2124–2125. https://doi.org/10.1093/bioinformatics/btn355
52. Carr R, Borenstein E (2014) Comparative analysis of functional metagenomic annotation and the mappability of short reads. PLoS One 9:e105776. https://doi.org/10.1371/journal.pone.0105776
53. Wommack KE, Bhavsar J, Ravel J (2008) Metagenomics: read length matters. Appl Environ Microbiol 74:1453–1463. https://doi.org/10.1128/AEM.02181-07
54. Tyson GW, Chapman J, Hugenholtz P et al (2004) Community structure and metabolism through reconstruction of microbial genomes from the environment. Nature 428:37–43. https://doi.org/10.1038/nature02340
55. Alneberg J, Bjarnason BS, de Bruijn I et al (2014) Binning metagenomic contigs by coverage and composition. Nat Methods 11:1144–1146. https://doi.org/10.1038/nmeth.3103
56. Kang DD, Li F, Kirton E et al (2019) MetaBAT 2: an adaptive binning algorithm for robust and efficient genome reconstruction from metagenome assemblies. PeerJ 7:e7359. https://doi.org/10.7717/peerj.7359
57. Reddy TBK, Thomas AD, Stamatis D et al (2015) The Genomes OnLine Database (GOLD) v.5: a metadata management system based on a four level (meta)genome project classification. Nucleic Acids Res 43:D1099–D1106. https://doi.org/10.1093/nar/gku950
58. Alcock BP, Huynh W, Chalil R et al (2023) CARD 2023: expanded curation, support for machine learning, and resistome prediction at the comprehensive antibiotic resistance database. Nucleic Acids Res 51:D690–D699. https://doi.org/10.1093/nar/gkac920
59. Liu B, Zheng D, Jin Q et al (2019) VFDB 2019: a comparative pathogenomic platform with an interactive web interface. Nucleic Acids Res 47:D687–D692. https://doi.org/10.1093/nar/gky1080
60. Carattoli A, Zankari E, García-Fernández A et al (2014) *In Silico* detection and typing of plasmids using PlasmidFinder and plasmid multilocus sequence typing. Antimicrob Agents Chemother 58:3895–3903. https://doi.org/10.1128/AAC.02412-14
61. Krawczyk PS, Lipinski L, Dziembowski A (2018) PlasFlow: predicting plasmid sequences in metagenomic data using genome signatures. Nucleic Acids Res 46:e35–e35. https://doi.org/10.1093/nar/gkx1321
62. Guan J, Chen Y, Goh Y-X et al (2024) TADB 3.0: an updated database of bacterial toxin–antitoxin loci and associated mobile genetic elements. Nucleic Acids Res 52:D784–D790. https://doi.org/10.1093/nar/gkad962
63. Arango-Argoty G, Garner E, Pruden A et al (2018) DeepARG: a deep learning approach

for predicting antibiotic resistance genes from metagenomic data. Microbiome 6:23. https://doi.org/10.1186/s40168-018-0401-z

64. Bolyen E, Rideout JR, Dillon MR, Bokulich NA, Abnet CC, Al-Ghalith GA & Caporaso JG et al (2019) Reproducible, interactive, scalable and extensible microbiome data science using QIIME 2. Nature biotechnology, 37(8):852–857. https://doi.org/10.1038/s41587-019-0209-9

65. Hedlund BP, Chuvochina M, Hugenholtz P, Konstantinidis KT, Murray AE, Palmer M & Whitman WB (2022) SeqCode: a nomenclatural code for prokaryotes described from sequence data. Nature Microbiology, 7(10): 1702-1708. https://doi.org/10.1038/s41564-022-01214-9

Chapter 5

Dilution-to-Stimulation: A Method for Selecting Polymer-Transforming Microbial Consortia

Diego Javier Jiménez, Laura Díaz-García, Lila Aldakheel, and Alexandre Soares Rosado

Abstract

The utilization of liquid enrichment cultures for selecting microbial communities has been employed to recover and increase the abundance of desired microbes capable of thriving on specific carbon sources. This strategy facilitates the development of microbial consortia designed to serve as models for studying the eco-enzymology of particular metabolic processes. In this context, we introduce a top–down method referred to as "dilution-to-stimulation" to artificially select polymer-transforming microbial consortia from soil samples.

Key words Enrichment cultures, Soil, Microbiome engineering, Plastics, Lignocellulose

1 Introduction

Currently, the engineering of microbial consortia is a relevant topic in applied microbiology and microbial ecology [1]. Microbial consortia can be defined as collections of species that interact (both positively and negatively) in a particular environment where they perform a specific function. Microbial consortia can be designed using two different strategies (i.e., bottom–up and top–down) [1–3]. In the bottom–up strategy, the design focuses on the mixing of axenic microbial populations [4, 5]. However, its successful implementation depends on the prior knowledge of each community member [6]. Conversely, the top–down strategy relies on the selection of stable microbial consortia along consecutive growth cycles. In this approach, natural microbial communities are stimulated to survive in specific culture conditions, allowing the progressive selection of the fittest microbial species [1]. Commonly, the selection of these microbial consortia is driven by the carbon source. This enrichment method, termed "dilution-to-stimulation" [7], has frequently been used to design lignocellulose-degrading

Karthik Raman and Gayathri Sambamoorthy (eds.), *Modelling the Microbiome*, Methods in Molecular Biology, vol. 3006, https://doi.org/10.1007/978-1-0716-5080-6_5,

microbial consortia [2, 3, 8]. Notably, various factors can shape the structure and function of the resulting microbial consortia, including the microbial inoculum source, inclusion of specific substrates as the sole carbon source (e.g., different agricultural residues), culture conditions (e.g., oxygen availability, temperature, and pH), and the dilution factor between growth cycles [9–11]. This enrichment method has also been used to develop soil-derived keratinolytic microbial consortia [12], and it is an excellent strategy for artificially selecting plastic-transforming microbial communities from different ecosystems [13]. During the enrichment process, the accurate measurement of microbial growth (e.g., by quantifying the cell concentration or optical density [OD]) and consumption and/or modification of carbon sources are imperative. Additionally, microbial diversity analyses (e.g., 16S rRNA gene surveys) can help determine community stability [2, 3] during the selection process. Once the consortia are stable, the interactions and functional roles of each member can be assessed using different meta-omics approaches [14, 15]. The resulting polymer-transforming microbial consortia are excellent sources of enzymes with utility in many biotechnological applications. In this chapter, we describe "dilution-to-stimulation" as a top–down method for selecting/enriching plant biomass (i.e., lignocellulose) and polyethylene terephthalate (PET)-transforming microbial consortia from soil samples.

2 Materials

2.1 Solutions

Prepare all solutions using ultrapure water (prepared by purifying deionized water) and analytical grade reagents (*see* **Note 1**).

1. Buffer stock solution (10×): 0.493 M of anhydrous Na_2HPO_4 and 0.114 M of anhydrous K_2HPO_4. Add 100 mL of water to a 1 L glass beaker. Then, transfer 70 g of Na_2HPO_4 and 20 g of K_2HPO_4 to the glass beaker. Add water to a volume of 900 mL. Mix the components using a magnetic stir bar until the salts are completely dissolved. Increase the volume to 1 L with water and store the solution at room temperature.
2. Mineral salt stock solution (10×): Add approximately 100 mL of water to a 1 L glass beaker. Then, transfer 10 g of $(NH_4)_2SO_4$, 1 g of $Ca(NO_3)_2$, and 2 g of $MgCl_2$ to the glass beaker. Add water to a volume of 900 mL. Mix the components using a magnetic stir bar until the salts are completely dissolved. Increase the volume to 1 L with water and store the solution at room temperature.
3. Trace element stock solution (TESS; 1000×): Add approximately 20 mL of water to a 250 mL glass beaker. Add 500 mg of EDTA, 300 mg of $FeSO_4 \cdot 7\,H_2O$, 5 mg of $CoCl_2 \cdot 6\,H_2O$, 5 mg

of $ZnSO_4{\cdot}7\ H_2O$, 3 mg of $MnCl_2{\cdot}4\ H_2O$, 3 mg of $NaMoO_4{\cdot}2\ H_2O$, 2 mg of $NiCl_2{\cdot}6\ H_2O$, 2 mg of H_3BO_3, and 1 mg of $CuCl_2{\cdot}2\ H_2O$ into the water and mix the components using a magnetic stir bar until they are completely dissolved. Increase the volume to 200 mL with water. Sterilize the solution via filtration through a 0.22 μm filter. Prepare 10 mL aliquots and store them at 4 °C or −20 °C (for long-term use).

4. Vitamin stock solution (VSS; 1000×): Add approximately 10 mL of water to a 100 mL glass beaker. Add 10 mg of Ca-pantothenate, 10 mg of cyanocobalamin (B_{12}), 10 mg of nicotinic acid, 10 mg of pyridoxal-HCl (B_6), 10 mg of riboflavin, 10 mg of thiamine-HCl (B_1), 1 mg of biotin, 1 mg of folic acid, and 1 mg of *p*-aminobenzoic acid into the water and mix the components using a magnetic stir bar until they are completely dissolved. Increase the volume to 100 mL with water. Sterilize the solution via filtration through a 0.22 μm filter. Prepare 10 mL aliquots and store them at 4 °C or −20 °C (for long-term use).
5. Mineral salt medium (MSM): Add 100 mL of buffer stock solution (10×) and 100 mL of mineral salt stock solution (10×) to a 1 L glass bottle and increase the volume to 1 L with water. Autoclave the solution at 121 °C and 15 psi for 15 min. After cooling the solution to 40 °C, add 1 mL of sterile filtered TESS (1000×) and 1 mL of sterile VSS (1000×) (*see* **Note 2**). Adjust the pH to 7.0.

3 Substrates

Select an appropriate polymer substrate (e.g., corn stover or amorphous PET). Reduce the particle size of the substrates using a milled knife through a 1 mm screen (*see* **Note 3**). Wash the substrates twice with sterile water and 70% (vol/vol) ethanol (*see* **Note 4**). Dry the substrates in an oven at 50 °C for 24–48 h and store them in 50 mL Falcon tubes at room temperature until use. In the case of PET, irradiate the polymer with short-wave ultraviolet (UV) light (100–280 nm) for 10–20 min (*see* **Note 5**).

4 Microbial Source

Retrieve a homogeneous soil sample from nature (e.g., approximately 500 g). Sieve the sample to remove leaves, roots, and rocks. Mix 10 g of the soil sample and 10 g of sterile glass beads (1–5 mm) with 90 mL of MSM in a 250 mL bottle (Fig. 1). Shake for 20 min at 250 rpm and allow it to settle for 2 min. Immediately inoculate the enrichment flasks with an aliquot of the supernatant (see below).

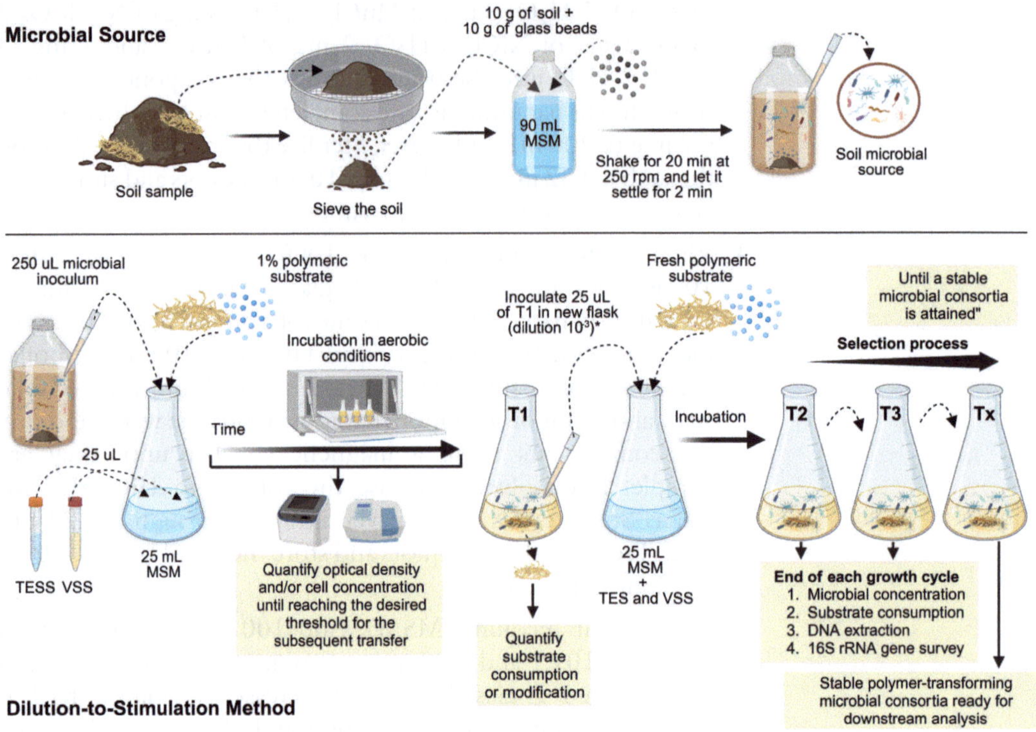

Fig. 1 Schematic representation of the dilution-to-stimulation method, including the preparation of the soil inoculum (upper part). MSM, mineral salt medium; TESS, trace element stock solution; VSS, vitamin stock solution. Created using BioRender

5 Methods

Perform all the following procedures at room temperature and under a laminar flow cabinet unless otherwise specified.

1. Add 1% of the polymeric substrates (*see* **Note 6**) to the MSM (*see* **Note 1**).
2. Inoculate 250 μL of the microbial source in a sterilized 100 mL Erlenmeyer flask containing 25 mL of MSM and 1% polymeric substrate to achieve a 1×10^{-2} dilution (Fig. 1).
3. Create two negative controls: (i) Inoculate 25 mL of MSM and 1% polymeric substrate (without microbial source) in a 100 mL Erlenmeyer flask and (ii) inoculate 250 μL of the microbial source in a 100 mL Erlenmeyer flask containing 25 mL of MSM (without carbon source).
4. Before incubation, retrieve a 200 μL sample from each flask and quantify the number of cells per milliliter (*see* **Note 7**).
5. Incubate the flasks at 25 °C and 130 rpm (aerobic conditions) for 5–15 days (*see* **Note 8**).

6. During and at the end of batch culture, gently shake the flasks and allow the solids to settle. Then, take a sample (e.g., 200 μL) to determine OD at 600 nm (*see* **Note 9**), number of cells per milliliter, and consumption/modification of the substrate (*see* **Note 10**).
7. Once growth is achieved (e.g., up to 1 × 10^7 cells/mL), transfer 25 μL (*see* **Note 11**) of the microbial suspension (after **step 5**) to a new 100 mL Erlenmeyer flask containing fresh 25 mL of MSM + 1% polymeric substrate.
8. Incubate the flasks at 25 °C and 130 rpm (aerobic conditions) for 5–15 days (*see* **Note 8**).
9. Repeat these growth cycles (**steps** 7 and **8**) 6–10 times or until stability is determined (*see* **Note 12**).
10. After each growth cycle, the remaining cultures can be used to extract DNA, assess microbial diversity, and perform enzymological analysis.

6 Notes

1. The final volume of the solutions can be adjusted according to the number of flasks, treatments, and culture conditions to be used in the "dilution-to-stimulation" method.
2. MSM can be prepared directly in the flasks used for the "dilution-to-stimulation" method as follows: Add 2.5 mL of buffer stock solution (10×) and 2.5 mL of MSM solution (10×) to a 100 mL Erlenmeyer flask. Add water to a volume of 25 mL. Autoclave the mixture at 121 °C and 15 psi for 15 min. After cooling to 40 °C, add 25 μL of sterile filtered TESS (1000×) and 25 μL of sterile VSS (1000×). Subsequently, add substrates and the microbial source to the flasks.
3. Particle size is a parameter to be defined by the researcher. This is critical in determining the contact surface between microbes and substrates, which has a direct effect on their growth and performance. In the case of PET-active consortia, polymer nanoparticles can be prepared following the method reported in [16].
4. This process aims to sterilize the surface of the polymer. However, in the case of lignocellulose, it also removes soluble oligosaccharides.
5. Excessive UV exposure can affect the chemical structure of PET.
6. The percentage of the polymers can range between 0.1% and 10% depending on the research aims and types of substrates. A mixture of different substrates (e.g., 1:1) can be used [3].

7. Ideally, before incubation, the flasks should contain approximately 1×10^4–1×10^5 cells/mL, and the cell density should reach approximately 1×10^7–1×10^9 cells/mL after incubation. Cell quantification can be performed using a Neubauer chamber or a flow cytometer.
8. Culture conditions (e.g., temperature, pH, and oxygen) are determined by the researcher. The duration of incubation depends on microbial growth and the type of polymer. This can also vary between growth cycles.
9. OD at 600 nm is used as a proxy for cell density. When enrichment cultures reach an OD of >0.5, they may be ready for the next growth cycle. However, this value must be compared with the negative controls.
10. At the end of batch culture, assessing the consumption of the substrate can help determine the effectiveness of the community in transforming the polymer. Because of the recalcitrance of PET, this measurement is only possible with lignocellulose using the following protocol: Collect the substrate and dry it in an oven at 50 °C until a constant weight is achieved. Then, calculate the percentage of weight loss using the following formula: substrate weight loss (%) = $[(a - b)/c] \times 100$, where a denotes the residual control substrate weight, b denotes the residual substrate weight, and c denotes the total substrate weight. To evaluate significant differences in microbial growth and substrate weight loss along the transfers, analysis of variance and a post hoc Tukey–Kramer test can be performed, with a confidence interval of 99% ($\alpha = 0.01$). For PET, different thermogravimetric techniques can help determine the structural modification of the polymer [17].
11. In this case, the dilution factor is 1×10^{-3}; however, it can be modified depending on the cell concentration and the goals of the study.
12. Stability can be determined based on the growth rate, substrate consumption, or microbial composition of the consortia [3, 8].

Acknowledgments

We thank the editors for the invitation.

Conflict of Interest Statement This study was funded by baseline resources (BAS/1/1096-01-01) obtained from A.S.R.

References

1. Hu H et al (2022) Guided by the principles of microbiome engineering: accomplishments and perspectives for environmental use. mLife 1(4):382–398. https://doi.org/10.1002/mlf2.12043
2. Diaz-Garcia L et al (2021) Top-down enrichment strategy to co-cultivate lactic acid and lignocellulolytic bacteria from the Megathyrsus maximus phyllosphere. Front Microbiol 12: 744075. https://doi.org/10.3389/fmicb.2021.744075
3. Díaz-García L et al (2021) Dilution-to-stimulation/extinction method: a combination enrichment strategy to develop a minimal and versatile lignocellulolytic bacterial consortium. Appl Environ Microbiol 87(2):e02427-20. https://doi.org/10.1128/aem.02427-20
4. Jia Y et al (2020) A bottom-up approach to develop a synthetic microbial community model: application for efficient reduced-salt broad bean paste fermentation. Appl Environ Microbiol 86(12):e00306-20. https://doi.org/10.1128/AEM.00306-20
5. Jimenez DJ et al (2020) Defining the eco-enzymological role of the fungal strain Coniochaeta sp. 2T2.1 in a tripartite lignocellulolytic microbial consortium. FEMS Microbiol Ecol 96(1):fiz186. https://doi.org/10.1093/femsec/fiz186
6. San Leon D, Nogales J (2022) Toward merging bottom-up and top-down model based designing of synthetic microbial communities. Curr Opin Microbiol 69:102169. https://doi.org/10.1016/j.mib.2022.102169
7. Ho K-L, Lee D-J, Su A, Chang J-S (2012) Biohydrogen from cellulosic feedstock: dilution-to-stimulation approach. Int J Hydrog Energy 37(20):15582–15587. https://doi.org/10.1016/j.ijhydene.2012.01.093
8. Jiménez DJ, Dini-Andreote F, van Elsas JD (2014) Metataxonomic profiling and prediction of functional behaviour of wheat straw degrading microbial consortia. Biotechnol Biofuels 7:92. https://doi.org/10.1186/1754-6834-7-92
9. de Lima Brossi MJ, Jimenez DJ, Cortes-Tolalpa L, van Elsas JD (2016) Soil-derived microbial consortia enriched with different plant biomass reveal distinct players acting in lignocellulose degradation. Microb Ecol 71(3): 616–627. https://doi.org/10.1007/s00248-015-0683-7
10. Cortes-Tolalpa L et al (2016) Different inocula produce distinctive microbial consortia with similar lignocellulose degradation capacity. Appl Microbiol Biotechnol 100(17): 7713–7725. https://doi.org/10.1007/s00253-016-7516-6
11. Jimenez DJ et al (2017) Ecological insights into the dynamics of plant biomass-degrading microbial consortia. Trends Microbiol 25(10): 788–796. https://doi.org/10.1016/j.tim.2017.05.012
12. Kang D et al (2018) Enrichment and characterization of an environmental microbial consortium displaying efficient keratinolytic activity. Bioresour Technol 270:303–310. https://doi.org/10.1016/j.biortech.2018.09.006
13. Jimenez DJ, Sanchez A, Dini-Andreote F (2023) Engineering microbiomes to transform plastics. Trends Biotechnol 42(3):265–268. https://doi.org/10.1016/j.tibtech.2023.09.011
14. Jimenez DJ, Maruthamuthu M, van Elsas JD (2015) Metasecretome analysis of a lignocellulolytic microbial consortium grown on wheat straw, xylan and xylose. Biotechnol Biofuels 8: 199. https://doi.org/10.1186/s13068-015-0387-8
15. Jimenez DJ, Chaves-Moreno D, van Elsas JD (2015) Unveiling the metabolic potential of two soil-derived microbial consortia selected on wheat straw. Sci Rep 5:13845. https://doi.org/10.1038/srep13845
16. Rodríguez-Hernández AG, Munoz-Tabares JA, Aguilar-Guzmán JC, Vazquez-Duhalt R (2019) A novel and simple method for polyethylene terephthalate (PET) nanoparticle production. Environ Sci Nano 6(7):2031–2036. https://doi.org/10.1039/c9en00365g
17. Jiménez DJ et al (2022) Merging plastics, microbes, and enzymes: highlights from an international workshop. Appl Environ Microbiol 88(14):e00721-00722. https://doi.org/10.1128/aem.00721-22

Chapter 6

Computational Microbial and Viral Ecology Analysis

James C. Kosmopoulos and Karthik Anantharaman

Abstract

The explosion in known microbial diversity in the last two decades has made it abundantly clear that microbes in the environment do not exist in isolation; they are members of communities. Accordingly, omics approaches such as metagenomics have revealed that interactions between diverse groups of community members such as archaea, bacteria, and viruses (bacteriophages) are common and have significant impacts on entire microbiomes. Thus, to have a well-developed understanding of microbes as they naturally exist in the environment, biological entities of all kinds must be studied together. While numerous protocols for metagenome analysis exist, comprehensive published protocols for the simultaneous analysis of viruses and prokaryotes together are scarce. Further, as bioinformatic methods for microbiology rapidly advance, existing metagenomic tools and pipelines require frequent re-evaluation. This ensures the adherence to best practices for microbiome and metagenomic data analysis. Here, we offer an expansive approach for the joint analysis of bulk sequence data from a mixed microbial community (metagenomes) and viral-sized fraction communities (viromes). This chapter serves as a beginner's-level guide for researchers with limited bioinformatics expertise who wish to engage in multiscale metagenome and virome analyses. We cover steps from initial study design to sequence read processing, metagenome assembly, quality control, virus identification, microbial and viral genome binning, taxonomic characterization, species-level clustering, and host–virus predictions. We also provide the bioinformatic scripts used in our workflow for reuse in one's own computational methods. Lastly, we discuss additional approaches a researcher can take after processing data with this workflow.

Key words Microbial ecology, Viral ecology, Metagenomics, Viromics, Host-virus interactions, Bacteriophage

1 Introduction

In the current age of genomics, microbiology as a field is rapidly growing and evolving as DNA sequence data becomes more accessible and higher in quality. Accordingly, metagenomics is a powerful tool to study the genetic material of entire microbiomes directly from biological samples. Since these approaches can be cultivation-independent, the application of metagenomics and bioinformatics to the study of microbiomes enables a deeper understanding of microbial ecology, evolution, and the roles that microorganisms

Karthik Raman and Gayathri Sambamoorthy (eds.), *Modelling the Microbiome*, Methods in Molecular Biology, vol. 3006, https://doi.org/10.1007/978-1-0716-5080-6_6,

play in the environment and human health. While metagenomes of a mixed community of bacteria, archaea, and viruses offer a broad overview of the biological entities within an environment [1–4], focusing on virus-like particle-sized fraction metagenomes (viromes) can provide an additional layer of focus. Viruses (particularly bacteriophages) play crucial roles in regulating microbial populations [5–7], horizontal gene transfer [8, 9], and influencing microbial genetic diversity [10–12]. Viromes can uncover the diversity, abundance, and functions of viruses in microbiomes [4, 13, 14]. Therefore, analyzing viromes in conjunction with mixed-community metagenomes can greatly enhance a researcher's understanding of microbial and viral dynamics, interactions, and impacts on their hosts and environments.

To streamline the complex processes involved in metagenomic and virome analyses, several workflows and pipelines have been developed. Tools like MetaWRAP [15], ViWrap [16], the JGI Metagenome Workflow [17], and Anvi'o [18] offer comprehensive suites for processing, analyzing, and visualizing metagenomic data. These tools facilitate quality control, assembly, binning, annotation, and comparative genomics, allowing researchers to derive meaningful insights from their data efficiently. Despite the existence of these useful automated pipelines, a step-by-step manual that is presented in this chapter remains crucial. This chapter is intended to serve as a resource for understanding the intricacies of each analytical step in common metagenomics and viromics workflows. We also explain the rationale behind methodological choices and discuss why they may or may not apply to a given research context.

2 Materials

2.1 Hardware and Operating System

When working with one or two individual samples, performing in silico analyses of total microbial and viral communities can typically be done on a personal computer with at least 6 GB of RAM, a processor with at least 4 cores, and tens of GB of available storage. However, the incredible diversity of microbes and their viruses often requires biological replicates across temporal, spatial, and/or physiochemical scales to perform in-depth investigations (*see* Methods). Scaling bioinformatic analyses with multiple samples will often require the use of a high-performance computing server. These servers are sometimes available to researchers at individual research institutions, or access to one can be purchased from an online host. There are also web servers, such as Galaxy [19], that allow users to access resources for bioinformatics for free.

Moreover, the operating system of the machine performing bioinformatic analyses is a critical component. Many bioinformatic tools require a command-line environment (a *shell*) to operate and often have specific OS requirements. The vast majority of tools are compatible with Linux operating systems such as Ubuntu or Debian, and, to a slightly lesser extent Mac OS (Unix). Except for a handful of tools that exist within a graphical user interface or online, most tools will require either a Linux or Mac (Unix) operating system. Researchers running analyses locally (without a server) on a Windows machine can install a virtual machine for free that will allow them to access a Linux operating system, such as Windows Subsystem for Linux (learn.microsoft.com/en-us/windows/wsl).

The analyses performed here were conducted on a server with an Ubuntu 20.04.06 OS with access to 256 CPUs, 2 TB of RAM, and 180 TB of storage. However, the analyses to be run will be capped to use at most 16 CPUs, 100 GB of RAM, and 2 TB of disk space.

2.2 Software

The software used in the analyses in this chapter are listed in Table 1. The versions of the packages used here are also provided for reproducibility, but installing the same versions of these packages may not be required for all aspects of these analyses.

Table 1
List of tools used in Section 3

Tool	Version	Source	Citation	Section(s)
sra-toolkit	3.0.7	github.com/ncbi/sra-tools		3.2
BBTools	38.86	sourceforge.net/projects/bbmap	[32]	3.3
SPAdes	3.13.0	github.com/ablab/spades	[35]	3.4
Bowtie2	2.5.1	github.com/BenLangmead/bowtie2	[40]	3.5
SAMtools	1.16.1	htslib.org	[42]	3.5
metaQUAST	5.2.0	github.com/ablab/quast	[44]	3.5
bioawk	1.0	github.com/lh3/bioawk		3.5
MetaBAT2	2.15	bitbucket.org/berkeleylab/metabat	[55]	3.6
CheckM	1.2.2	ecogenomics.github.io/CheckM	[60]	3.6
GTDB-Tk	2.1.1	github.com/Ecogenomics/GTDBTk	[71]	3.7 and 3.12
geNomad	1.7.4	portal.nersc.gov/genomad	[80]	3.8
vRhyme	1.1.0	github.com/AnantharamanLab/vRhyme	[88]	3.9 and 3.12
CheckV	1.0.1	bitbucket.org/berkeleylab/checkv	[84]	3.9
dRep	3.5.0	github.com/MrOlm/drep	[52]	3.10 and 3.11
prodigal-gv	2.11.0-gv	github.com/apcamargo/prodigal-gv	[80]	3.11
vConTACT2	0.11.3	bitbucket.org/MAVERICLab/vcontact2	[97]	3.11
iPHoP	1.2.0	bitbucket.org/srouxjgi/iphop	[109]	3.12

When working with multiple software packages, as is typical in metagenomic analyses, we highly recommend the use of the package manager Conda, implemented as either Anaconda, Miniconda, or Mamba. It is not uncommon for one tool to require an exact version of a software package that conflicts with another tool. Conda allows a user to install packages to separate *environments* to avoid conflict between the dependencies of multiple tools. The installation of Conda or the individual tools used in the analyses here will not be covered, but instructions can be found by following the linked webpages for each within their cited publications.

Furthermore, workflow managers such as Snakemake [20] can play a significant role in managing complex analyses. Snakemake automates the execution of workflows, ensuring reproducibility and scalability. It allows researchers to define their analysis pipelines in a flexible, readable, and modular manner, handling dependencies and automating the execution across different computational environments. This not only saves time but also enhances the reliability of the results by reducing manual errors. The steps carried out in this chapter are treated as a "one-off" analysis of metagenomes and viromes. In the real world of bioinformatics and metagenomics, it is common to refine one's methods in an iterative approach when figuring out which tools, parameters, and data are necessary to test their hypotheses. We encourage the researcher to consider workflow managers in their computational research, as they can greatly aid in the organization of computational methods and promote reproducibility of one's results.

2.3 *Tara* Oceans Dataset

To analyze the metagenomes of microbial and viral communities in pairs, we will leverage the publicly available *Tara* Oceans dataset [21]. This dataset contains globally sampled metagenomes from 12 size fractions ranging from <0.2 μm to >680 μm [22]. The focus of this study will be on the <0.2 μm fraction (enrichment for viruses and other virus-like particles; hereafter viromes) and the 0.2–3 μm fraction (enrichment for bacteria and archaea; hereafter metagenomes). In total, there are 21 samples with both fractions available that are available to download from the NCBI Sequence Read Archive under BioProjects PRJEB1787 (metagenomes) and PRJEB4419 (viromes). This same subset of the *Tara* Oceans data was recently compared in depth to paired viromes and metagenomes from other environments [23]. Given the large size of the raw sequence read data for all 42 samples, this study will focus on only 10 sample pairs (20 samples total) representing the South Atlantic Gyre [21] (Table 2). Downloading the raw sequence data for these samples will be done with the NCBI SRA Toolkit (*see* Methods).

Table 2
Summary of metagenomic sequence data to be analyzed

Sample	Size fraction	TARA sample label	Sampling depth [m]	Environmental feature	NCBI SRA accession number
TARA_068_DCM	Metagenome	TARA_068_DCM_0.22-3	50	(DCM) deep chlorophyll maximum layer	ERR599017
TARA_068_DCM	Virome	TARA_068_DCM_<-0.22	50	(DCM) deep chlorophyll maximum layer	ERR594415
TARA_068_SRF	Metagenome	TARA_068_SRF_0.22-3	5	(SRF) surface water layer	ERR599174
TARA_068_SRF	Virome	TARA_068_SRF_<-0.22	5	(SRF) surface water layer	ERR594391
TARA_070_MES	Metagenome	TARA_070_MES_0.22-3	800	(MES) mesopelagic zone	ERR599044
TARA_070_MES	Virome	TARA_070_MES_<-0.22	800	(MES) mesopelagic zone	ERR594407
TARA_070_SRF	Metagenome	TARA_070_SRF_0.22-3	5	(SRF) surface water layer	ERR599165
TARA_070_SRF	Virome	TARA_070_SRF_<-0.22	5	(SRF) surface water layer	ERR594353
TARA_072_DCM	Metagenome	TARA_072_DCM_0.22-3	100	(DCM) deep chlorophyll maximum layer	ERR599133
TARA_072_DCM	Virome	TARA_072_DCM_<-0.22	100	(DCM) deep chlorophyll maximum layer	ERR594379
TARA_072_MES	Metagenome	TARA_072_MES_0.22-3	800	(MES) mesopelagic zone	ERR599005
TARA_072_MES	Virome	TARA_072_MES_<-0.22	800	(MES) mesopelagic zone	ERR594388
TARA_072_SRF	Metagenome	TARA_072_SRF_0.22-3	5	(SRF) surface water layer	ERR598984
TARA_072_SRF	Virome	TARA_072_SRF_<-0.22	5	(SRF) surface water layer	ERR594364
TARA_076_DCM	Metagenome	TARA_076_DCM_0.22-3	150	(DCM) deep chlorophyll maximum layer	ERR599148

(continued)

Table 2
(continued)

Sample	Size fraction	TARA sample label	Sampling depth [m]	Environmental feature	NCBI SRA accession number
TARA_076_DCM	Virome	TARA_076_DCM_<-0.22	150	(DCM) deep chlorophyll maximum layer	ERR594355
TARA_076_SRF	Metagenome	TARA_076_SRF_0.22-3	5	(SRF) surface water layer	ERR599126
TARA_076_SRF	Virome	TARA_076_SRF_<-0.22	5	(SRF) surface water layer	ERR594354
TARA_078_SRF	Metagenome	TARA_078_SRF_0.22-3	5	(SRF) surface water layer	ERR599006
TARA_078_SRF	Virome	TARA_078_SRF_<-0.22	5	(SRF) surface water layer	ERR594411

3 Methods

3.1 Study Design

There are various factors that one can consider when designing a metagenomics-based microbiome study. The nuances in the design of a project will depend on one's study system and the individual research questions being asked. Thus, study design and sampling methods can have a significant impact on one's interpretation of the ecology of microbiomes [23, 24]. For any investigation into microbial and viral metagenomics, we suggest three key factors for the researcher to consider when designing their study. These three factors will also dictate the hardware requirements to analyze the resulting sequence data (*see* Materials), so each factor should be carefully considered.

The first factor one should consider is the scale of the microbiome study, that is, the variation of biological and technical replicates. We recommend prioritizing biological replicates over technical replicates. This is because biological replicates can improve the power and accuracy of sequence-based analyses better than technical replicates [25]. Second, it should be considered whether to enrich for any biological groups. For example, any microbiome study with an interest in viruses should generate virus-sized fraction metagenomes (viromes) from their samples [14, 23]. We argue that any holistic investigation into a microbial community should consider viruses in addition to the microbes that they infect. Third, the target sequencing depth of each biological sample is crucial. The more DNA sequence data generated per sample, the greater the resolution will be for fine-level variation of genes and genomes [26]. Greater sequencing depth will also allow for the detection of rare members of a community [27, 28]. However, increasing the desired sequencing depth will also increase the cost of the study. This presents a trade-off between the financial costs of increasing sequencing depth and information able to be generated with the increased depth [29].

There are virtually endless microbiomes not yet investigated or sequenced on Earth with an equally long list of questions that one could ask about them. Yet, there is already a vast amount of publicly available microbiome data. These data are stored in digital repositories such as NCBI GenBank, JGI IMG, and EMBL-EBI ENA, among others. Before embarking on a metagenomic sequencing project, small or large, we encourage the researcher to ask themselves if their hypotheses can be tested using already publicly available sequencing data. In doing so, it is the responsibility of the researcher to obtain the consent of those who originally generated the data. One resource that was explicitly generated for the widespread use of the scientific public is the *Tara* Oceans dataset [21], which will be used for the analyses here (*see* Materials). This dataset involves a study design that spans time (sample collection date), spatial (geography and water column depth), and physiochemical (water column depth and environmental feature) scales. The dataset also includes the enrichment of multiple biological entities such as microbes and viruses. Thus, it is well suited for multiple in-depth investigations into whole microbial and viral communities of microbiomes in the environment.

3.2 Accessing Publicly Available, Raw Metagenomic Sequence Reads

The subset of *Tara* data used in this analysis (*see* Materials) will be downloaded using the `prefetch` command from the NCBI SRA Toolkit, followed by `fasterq-dump` to produce raw sequence read files in `.fastq` format. To download multiple `.fastq` files in bulk, a list of accession numbers can be provided. A simple text file can be created and edited using a command-line text editor like `vim` or `nano` which usually come pre-installed in many operating systems. A column of accession numbers such as that in Table 2 can be copied from a local spreadsheet and pasted into the text editor. Using the `cat` command, we show the contents of this file below.

```
cat acc_list.txt
ERR599017
ERR594415
ERR599174
ERR594391
ERR599044
ERR594407
ERR599165
ERR594353
ERR599133
ERR594379
ERR599005
ERR594388
ERR598984
ERR594364
ERR599148
ERR594355
ERR599126
ERR594354
ERR599006
ERR594411
```

The file `acc_list.txt` can now be provided to `prefetch`. The flag `--max-size 50g` is added because some of the files will be larger than the default maximum file size of 20 GB.

```
prefetch --option-file acc_list.txt --output-directory
prefetch_out --max-size 50g
```

The folder `prefetch_out` now contains one folder for each accession in the list. Within each folder are `.sra` files that will be read by `fasterq-dump`.

```
fasterq-dump prefetch_out/ERR* --outdir raw_reads
spots read: 47,220,407
reads read: 94,440,814
reads written: 94,440,814
# (Remaining output truncated)
```

"*" following `prefetch_out/ERR` is known as a *wildcard* character. Using it here returned all files or folders under `prefetch_out` that have the prefix ERR so that each individual folder did not have to be typed in the `fasterq-dump` command. There should now be a new folder named `raw_reads` that contains two `.fastq` files per sample (`SAMPLENAME_1.fastq`, `SAMPLENAME_2.fastq`). This is because these are *paired-end* sequence reads. Each sample has a forward (`_1`) and a reverse (`_2`) set of reads (a *mate*) that are used together as a pair. After the paired-end `.fastq` files have been downloaded into `raw_reads`, the `prefetch_out` folder can be removed to save on storage using the command `rm -r prefetch_out`.

To further save on disk space, each of the generated `.fastq` files will be compressed using the command `pigz`. The wildcard character "*" will be used again, this time to specify all files under `raw_reads` that end with the file extension `.fastq`. This will convert all `.fastq` files in the `raw_reads` folder to compressed files with the extension `.fastq.gz`. Many bioinformatics tools that handle sequence reads can conveniently handle `.fastq.gz` files instead of the large `.fastq` files.

```
pigz raw_reads/*.fastq
```

3.3 Sequence Read Quality Control (QC)

The most important step in any metagenomics project involves processing and QC of the raw sequenced reads obtained directly from a sequencing center or data repository. As the old saying, "*garbage in, garbage out*", suggests, even with highly accurate next-generation sequencing technologies, raw sequence read libraries contain poor quality and/or contaminating reads that need to be removed. There have been multiple tools developed to address this issue. Some of the more popular and robust tools include (but are not limited to) FastQC for initial read quality assessment [30], Trimmomatic for the removal of low-quality bases in Illumina paired-end reads [31], and BBDuk for decontamination [32]. One should consider using a tool like FastQC to assess the quality of reads before QC and again on reads after QC. Comparing the two reports can reveal how one's choice of QC methods improved (or perhaps did not improve) the sequence read quality.

The rqcfilter2 program from BBTools [32] is a package that will be used here to perform read adapter removal, read trimming, contaminant filtering, and read quality filtering, all in one execution. It requires an associated database that is 106 GB in size. A *for-loop* can be executed on the command-line using user-defined variables `DATADIR`, `THREADS`, `MEM`, and `OUT` to set the location of the RQCFilter data, maximum-allowable CPU usage, maximum-allowable memory usage, and the output folder for the filtered reads. Within the for-loop, the IDs from each mate in the pair of `.fastq.gz` files can be inferred using bash string manipulation and wildcards so that they can be dynamically provided to `rqcfilter2.sh`, instead of manually providing them one-by-one. Using `nano` or `vim` again, the code shown below can be written to a file named `qc_reads.sh`:

```
#!/bin/sh
# qc_reads.sh

DATADIR="RQCFilterData"
THREADS="16"
MEM="64"
OUT="filtered_reads"

for FASTQ in raw_reads/*_1.fastq.gz; do
 BASE=$(basename "$FASTQ")
# Extract the sample name
 SAMPLE=$(echo "$BASE" | cut -d'_' -f1)
 F="$FASTQ"
# Extract the reverse mate filename
 R=$(echo "$F" | sed 's/_1.fastq.gz/_2.fastq.gz/')

 echo "Running rqcfilter2.sh (BBTools) on $F $R"
 rqcfilter2.sh \
 in="$F" in2="$R" \
 rqcfilterdata="$DATADIR" \
 trimfragadapter=t \
 qtrim=r trimq=0 \
 maxns=3 maq=3 \
 minlen=51 mlf=0.33 \
 phix=t removehuman=t \
 removedog=t removecat=t \
 removemouse=t removemicrobes=f \
 clumpify=t tmpdir= \
```

```
    barcodefilter=f trimpolyg=5 \
    usejni=f chastityfilter=f \
    tmpdir=rqcfilter2_tmp \
    path="$OUT" \
    out="${SAMPLE}_1.fastq.gz" \
    out2="${SAMPLE}_2.fastq.gz" \
    threads="$THREADS" -Xmx"$MEM"g

  rm -r rqcfilter2_tmp
    echo "Done $SAMPLE."
  done
```

The flags `phix=t removehuman=t removedog=t removecat=t removemouse=t` removes contaminants by mapping the reads to phage Phi X (commonly added to sequencing runs), human, dog, cat, and mouse reference sequences in the `RQCFilterData` database, while the flag `removemicrobes=f` ensures that reads mapping to microbial reference sequences are retained. The remaining flags such as `trimfragadapter=t qtrim=r trimq=0 maxns=3 maq=3 minlen=51 mlf=0.33` are quality trimming parameters that the user can adjust. These values were chosen to mimic the parameters used in the JGI Metagenome Workflow [17]. The remaining flags `clumpify=t tmpdir= barcodefilter=f trimpolyg=5 usejni=f chastityfilter=f` are additional flags that the user may or may not wish to enable depending on the sequencing technology used and how the reads were obtained. There are many parameters that can be changed when running `rqcfilter2.sh`; we encourage the user to review and consider them with `rqcfilter2.sh --help`.

This may take some time to run, so to execute this script in the background, the `nohup` command can be used with the addition of "`&`". The text that would have been printed to the user's screen can be saved in a file using a "`>`" operator specifying an output file for the screen text:

```
nohup sh qc_reads.sh > qc_reads.nohup.out &
```

After the for-loop is completed, the contents of `qc_reads.nohup.out` can be viewed with the `less` command to see what is printed to the "screen" when running the script, if desired.

The folder `filtered_reads` now contains the filtered mates produced by `rqcfilter2.sh`. Although not required for genome assembly, error-correcting filtered reads before they are assembled into a metagenome can greatly improve the assembly quality [33, 34]. However, it should be noted that error correction may not always improve assembly quality. Error correction typically is not beneficial when the sequencing depth is low [33, 34], and results can vary depending on the error-correction software and/or assembler used [33, 34]. Judging by the file sizes alone, the sequence reads in the *Tara* dataset here were sequenced deeply. Additionally, SPAdes [35], the assembly workflow to be applied below (Section 3.4), performs optimally with the read error correction. There are several tools available for read error correction, and their applications vary by sequencing technology [33]. Below, we demonstrate how to error-correct filtered reads using yet another tool from the BBTools suite, `bbcms.sh`. Just as what was done to run `rqcfilter2.sh`, another bash script, `correct_reads.sh`, can execute this command for each pair of filtered reads.

```
#!/bin/sh
# correct_reads.sh

THREADS="16"
MEM="64"
OUT="corrected_reads"

for FASTQ in filtered_reads/*_1.fastq.gz; do
 BASE=$(basename "$FASTQ")
# Extract the sample name
 SAMPLE=$(echo "$BASE" | cut -d'_' -f1)
 F="$FASTQ"
# Extract the reverse mate filename
 R=$(echo "$F" | sed 's/_1.fastq.gz/_2.fastq.gz/')
# Set the output path for forward corrected reads
 OUTF="${OUT}/${SAMPLE}_1.fastq.gz"
# Set the output path for reverse corrected reads
 OUTR="${OUT}/${SAMPLE}_2.fastq.gz"

 echo "Running bbcms.sh (BBTools) on $F $R"
 bbcms.sh \
 -Xmx"$MEM"g \
 t="$THREADS" \
 mincount=2 \
 highcountfraction=0.6 \
 in="$F" \
 in2="$R" \
 out="$OUTF" \
 out2="$OUTR"

 echo "Done $SAMPLE."
done
```

The options `mincount=2 highcountfraction=0.6` are specified to discard reads largely composed of very rare k-mers, which can lead to issues in assembly downstream. These options were chosen to mimic the implementation of `bbcms.sh` in the JGI Metagenome Workflow [17] to adhere to standardization. However, the user should consider all options available with `bbcms.sh` and set the parameters that are most suited for their data.

The folder `corrected_reads` now exists and contains the error-corrected, filtered reads. However, during the QC and error-correction process, it is not uncommon for one mate in the pair to have been discarded while the other did not. This results in an unequal number between the forward and reverse reads for each sample. Genome and metagenome assemblers typically do not allow this, so the unpaired (singleton) reads will have to be separated into their own file. This can be done with another tool in the BBTools suite, `repair.sh`, which will be executed in another for-loop in the script below, named `repair_reads.sh`.

```
#!/bin/sh
# repair_reads.sh

DATADIR="RQCFilterData"
MEM="64"
OUT="corrected_reads"

for FASTQ in corrected_reads/*_1.fastq.gz; do
 BASE=$(basename "$FASTQ")
# Extract the sample name
 SAMPLE=$(echo "$BASE" | cut -d'_' -f1)
 F="$FASTQ"
# Extract the reverse mate filename
 R=$(echo "$F" | sed 's/_1.fastq.gz/_2.fastq.gz/')
# Set the output path for forward repaired reads
 FILTOUTF="${OUT}/${SAMPLE}_F.fastq.gz"
# Set the output path for reverse repaired reads
 FILTOUTR="${OUT}/${SAMPLE}_R.fastq.gz"
# Set the unpaired reads filename
 FILTOUTS="${OUT}/${SAMPLE}_U.fastq.gz"

# Separate filtered reads into F, R, and U reads
 echo "Running repair.sh (BBTools)"
 repair.sh \
 -Xmx"$MEM"g \
 in="$F" in2="$R" \
 out="$FILTOUTF" \
 out2="$FILTOUTR" \
 outs="$FILTOUTS" \
```

```
  repair

  echo "Done $SAMPLE."
done
```

Now, all of the raw reads have been properly QC'd and error-corrected. The non-"repaired" and the non-error-corrected reads can be removed with the following command:

```
rm -r corrected_reads/*_1.fastq.gz corrected_reads/*_2.
fastq.gz filtered_reads
```

Since the raw reads were obtained from an online archive, they can be removed as well to save on space with `rm -r raw_reads`. However, the original raw reads for every sequencing project should always be saved or archived. If reads were filtered from a collection of raw reads that have not yet been published or archived, the raw reads should *not* be removed.

3.4 Metagenome Assembly

The assembly of reads is the nucleus of any metagenomics project. Given the high complexity of metagenomic sequence data, even after QC and error correction, it is imperative to pick an assembler that is most suited for the data. Not all assemblers are able to assemble metagenomes, and several assemblers are designed for a particular sequencing technology [36–38]. Paired-end, next-generation (i.e., *Illumina*) sequencing is the most common technology used for metagenomes, so it may be unsurprising that most modern assemblers are designed for these reads. While several genome and metagenome assemblers have been developed over the years and continue to be developed to this day, two assemblers have proven to be particularly reliable. SPAdes [35] and MEGAHIT [38] are paired-end read assemblers designed for single genomes and metagenomes. They are both extensively documented and maintained which make them quite user-friendly. Neither tool consistently performs better than the other, and their performance varies from dataset to dataset [28, 39]. However, the metagenome assembly mode of SPAdes (metaSPAdes) requires substantially more RAM than MEGAHIT [39]. Since the performance of each tool varies on a case-by-case basis, we recommend that researchers assemble their metagenomic reads with multiple assemblers and then compare the results (*see* Section 3.5). For the purposes of this chapter, we will only demonstrate how to assemble the error-corrected metagenomic reads with metaSPAdes.

metaSPAdes can be run on every set of error-corrected reads using another shell script with a for-loop, `assemble.sh`.

```
#!/bin/sh
# assemble.sh

THREADS="16"
MEM="100"
OUT="assemblies"

mkdir -p $OUT

for FASTQ in corrected_reads/*_F.fastq.gz; do
 BASE=$(basename "$FASTQ")
# Extract the sample name
 SAMPLE=$(echo "$BASE" | cut -d'_' -f1)
 F="$FASTQ"
# Extract the reverse mate filename
 R=$(echo "$F" | sed 's/_F.fastq.gz/_R.fastq.gz/')
# Extract the unpaired reads filename
 U=$(echo "$F" | sed 's/_F.fastq.gz/_U.fastq.gz/')

 echo "Running metaSPAdes to assemble reads $F $R $U"
 spades.py \
 -1 "$F" -2 "$R" -s "$U" \
 -o "$OUT"/"$SAMPLE" \
 --meta \
 --only-assembler \
 --tmp-dir spades_tmp \
 -t "$THREADS" -m "$MEM"

rm -r spades_tmp
 echo "Done $SAMPLE."
done
```

Here, the `--meta` flag enables metagenome assemble mode. The flag `--only-assembler` disables error correction since this was already done with `bbcms.sh`, above. Assuming this script was run in the background and had the screen output directed to a text file with `nohup sh assemble.sh > assemble.nohup.out &`, one can quickly check how many SPAdes runs were successful by counting the occurrences of the phrase "`SPAdes pipeline finished`" in the screen output. This is because SPAdes will print this message to the screen after a successful assembly. Of course, counting the occurrences of this phrase will not apply to other assemblers, but the user can run `grep -c` another phrase that their assembler of choice may print only when the assembly has finished.

```
grep -c "SPAdes pipeline finished" assemble.nohup.out
20
```

If any of the runs failed to complete because the 100 GB memory limit was exceeded, and no more memory is available to use, attempting the assembly again with MEGAHIT will be a good option.

The folder `assemblies` should now contain one subfolder for each *Tara* sample assembled. Inside of each subfolder is a file `contigs.fasta` that contains reads that were assembled into *contigs*. A contig is a contiguous stretch of an assembled DNA sequence. The paired-end sequencing technology used allowed for the contigs to be joined by SPAdes into *scaffolds* in the file `scaffolds.fasta`. Scaffolds are essentially multiple contigs that were joined by a sequence of Ns. The Ns represent unknown nucleotides. Here, the number of Ns in a sequence corresponds to the number of base pairs between the stretches of DNA represented by the contigs. In essence, having a set of contigs that were joined into a scaffold means that the orientation of each contig relative to each other is known, the distance (base pairs) between the contigs is known, but the exact nucleotides that the Ns represent are unknown. The advantage of scaffolds is that they allow for a greater contiguity of an assembly, which can help retain the genomic context of encoded genes. If one prefers to analyze sequences where every base is known, then one can choose to continue their analysis with the contigs instead of scaffolds. Here, the assembled scaffolds will be used for further analysis.

A new folder can be made to store just the assembled scaffolds with `mkdir scaffolds`. Next, the scaffold files can be copied to the new folder, each file renamed with their original sample identifiers, with the following code:

```
find assemblies -mindepth 2 -maxdepth 2 -type f -name
"scaffolds.fasta" -exec sh -c 'cp {} scaffolds/$(basename
$(dirname {}))_scaffolds.fasta' \;
```

The contents of the scaffold folder can be inspected with ls scaffolds. The folder should now contain the following:

```
ERR594353_scaffolds.fasta
ERR594354_scaffolds.fasta
ERR594355_scaffolds.fasta
ERR594364_scaffolds.fasta
ERR594379_scaffolds.fasta
ERR594388_scaffolds.fasta
```

```
ERR594391_scaffolds.fasta
ERR594407_scaffolds.fasta
ERR594411_scaffolds.fasta
ERR594415_scaffolds.fasta
ERR598984_scaffolds.fasta
ERR599005_scaffolds.fasta
ERR599006_scaffolds.fasta
ERR599017_scaffolds.fasta
ERR599044_scaffolds.fasta
ERR599126_scaffolds.fasta
ERR599133_scaffolds.fasta
ERR599148_scaffolds.fasta
ERR599165_scaffolds.fasta
ERR599174_scaffolds.fasta
```

Having each scaffold file be named by their sample origin will be useful when comparing assemblies to each other. The remaining SPAdes output files will not be used hereafter. They will be archived using the tar command with compression using `pigz`, and the original folder will be removed to save on disk space:

```
tar --use-compress-program="pigz --best --recursive" -cf
assemblies.tar.gz assemblies && rm -r assemblies
```

The `&&` simply means that when the first part of the code finishes running without error, the second part `rm -r assemblies` will be executed.

3.5 Metagenome QC

Just as QC was performed on sequence reads before assembly, the assembled metagenome scaffolds will also need to undergo QC. SPAdes successfully ran for each of the 20 samples, but it is not clear at this point what fraction of sequence reads ended up in the final assembly, how many scaffolds were assembled in each sample, how long they are, or what fraction of the scaffolds is composed of Ns. These are the important indicators of the overall quality of a (meta)genome assembly.

An ideal metagenome will have been assembled from all input reads, but in reality, not all input reads will end up in the final assembly. The *read recruitment* of scaffolds is a good indicator of how much genomic information in samples went unobserved. Read recruitment is the percentage of reads that went into an assembler that ends up mapping back to the output scaffolds. Mapping the filtered, error-corrected reads to the scaffolds will allow for the read recruitment to be calculated. A higher read recruitment of a metagenome assembly is indicative of a higher overall metagenome quality. In other words, a metagenome with a read recruitment of 90% is a better representative of the actual sequenced microbial community than a metagenome with a read recruitment of 25%. Unless deliberate measures were taken to filter the scaffolds resulting from the assembly, the assemblies should recruit at least 50% of the reads.

The tool Bowtie2 [40] will be used to map the filtered, error-corrected reads to the scaffolds. In addition to calculating read recruitment, mapping reads this way will be useful for further analyses below. Bowtie2 is a great read-mapper for short-read data. For metagenomes assembled from long reads, minimap2 [41] is recommended. First, the scaffolds need to be *indexed* by Bowtie2. This will be done using a short shell script `bowtie_index.sh` with a for-loop, sending the output indices to a new folder bowtie2_`index`:

```
#!/bin/sh
# bowtie_index.sh

THREADS="16"
for FASTA in scaffolds/*_scaffolds.fasta; do
 bowtie2-build "$FASTA" \
  "bowtie2_index/$(basename "${FASTA%_scaffolds.fasta}")"
\
  --threads "$THREADS"
done
```

Now, each set of reads that went into metagenome assembly can be mapped to their respective indices with the `bowtie2` command using another shell script `bowtie_map.sh`. The screen output from the Bowtie2 runs will be used below. To save the screen output and run Bowtie2 in the background, the following script will be executed with `nohup sh bowtie_map.sh > bowtie_map.nohup.out &`:

```
#!/bin/sh
# bowtie_map.sh

INDEX_DIR="bowtie2_index"
READS_DIR="corrected_reads"
OUTPUT_DIR="map_files"
THREADS="16"

mkdir -p "$OUTPUT_DIR"

for FWD_READ in ${READS_DIR}/*_F.fastq.gz; do
```

```
# Extract the base name without the forward suffix
 BASE_NAME=$(basename "$FWD_READ" _F.fastq.gz)

# Specify the matching reverse read
 REV_READ="${READS_DIR}/${BASE_NAME}_R.fastq.gz"
# Specify the singleton read
 UNPAIRED_READ="${READS_DIR}/${BASE_NAME}_U.fastq.gz"

# Run bowtie2 with paired-end and unpaired (singleton)
reads
 bowtie2 -x "${INDEX_DIR}/${BASE_NAME}" \
 -1 "$FWD_READ" -2 "$REV_READ" \
 -U "$UNPAIRED_READ" \
 -S "${OUTPUT_DIR}/${BASE_NAME}.sam" \
 --threads "$THREADS"

# Convert SAM to BAM, sort, and index
 samtools view \
 -bS "${OUTPUT_DIR}/${BASE_NAME}.sam" \
 --threads "$THREADS" \
| samtools sort \
 -o "${OUTPUT_DIR}/${BASE_NAME}.bam" \
 --threads "$THREADS"
rm "${OUTPUT_DIR}/${BASE_NAME}.sam"
  samtools  index  "${OUTPUT_DIR}/${BASE_NAME}.bam"  -@
"$THREADS"
done
```

There are four things that the above script does:

1. The command `bowtie2` is executed for each sample among the reads and indices. The flag `-x` specifies the target index, while `-1`, `-2`, and `-U` specify the forward, reverse, and unpaired reads for that same sample. The flag `-S` writes the output alignment file in SAM format. The output SAM file is a large, human-readable file with mapping statistics for every read.
2. Since SAM files are seldom used on their own, they are converted in BAM format to save on space using the `samtools view -bS` command from SAMtools [42]. BAM is a binary version of SAM. It contains the same information but in binary rather than human-readable text.
3. The BAM-formatted output is provided as an input (using the *pipe* operator |) to `samtools sort` which will alphanumerically sort the information before writing the final BAM file using the sample name and the `.bam` extension. Sorting BAM files this way is a common prerequisite for

many tools that handle them. The SAM-formatted file is removed since it is now no longer needed.

4. The sorted BAM file is indexed. Indexing BAM files is another common prerequisite for tools that handle them.

Now that the reads that went into metagenome assembly have been mapped to the filtered scaffolds, read recruitment can be assessed. Read recruitment can be calculated at any point after read mapping using tools such as CoverM (github.com/wwood/CoverM). However, Bowtie2 automatically reports the read recruitment to each index after mapping in its screen output. To obtain these values from the `bowtie_map.nohup.out` file, the following can be executed:

```
grep "overall alignment rate" bowtie_map.nohup.out
89.33% overall alignment rate
91.36% overall alignment rate
88.37% overall alignment rate
88.65% overall alignment rate
90.32% overall alignment rate
86.22% overall alignment rate
89.64% overall alignment rate
92.86% overall alignment rate
87.90% overall alignment rate
88.61% overall alignment rate
83.33% overall alignment rate
78.23% overall alignment rate
73.42% overall alignment rate
73.52% overall alignment rate
89.11% overall alignment rate
50.99% overall alignment rate
85.38% overall alignment rate
73.98% overall alignment rate
73.57% overall alignment rate
77.73% overall alignment rate
```

Every index built from the SPAdes scaffolds recruited at least 50% of the reads, and several much more than that. This is a good sign that the previous sequence read QC and subsequent assembly methods worked as intended. However, read recruitment is not the only way to measure the assembly quality and perform metagenome QC. Another important factor to consider is the contiguity of an assembly. An ideal metagenome assembly would have one scaffold per genetic element (chromosome, plasmid, virus, etc.) that existed in the actual sample. In reality, these elements often appear highly fragmented in metagenomes due to the deliberate fragmentation of nucleic acids during DNA sequencing (an approach known as "shotgun" sequencing) and from the DNA extraction process itself [43]. Greater fragmentation can lead to issues with metagenome binning and sequence analysis later [43]. The magnitude of this problem can typically be reduced with greater sequencing depth [43] or with high-fidelity long-read sequencing [43], but these options are not always financially or technically viable.

To measure the contiguity of the metagenomes, the tool metaQUAST [44] will be used. metaQUAST will calculate several summary statistics to assess metagenome quality and provide tabular as well as interactive, graphical outputs. Here, metaQUAST is used because of its speed when calculating basic statistics. Below is an example of a basic metaQUAST command:

```
metaquast scaffolds/* --min-contig 1000 --max-ref-number
0 --threads 16 --output-dir metaquast
```

Using `scaffolds/*` after the metaquast command will run the MetaQUAST workflow on all scaffolds in the `scaffolds` folder. The flag `--min-contig 1000` sets the minimum contig (in this case, scaffold) length for analysis to 1000 bp. The flag –`max-ref-number 0` will skip downloading reference genomes and comparing to the input assemblies, which will reduce runtime. The maximum number of threads to use is specified with `--threads 16`, and the output folder "metaquast' is specified with `--output-dir metaquast`.

A table of summary statistics for each input metagenome will be provided in the output `report.tsv.` An interactive report `report.html` will also be generated, which can be opened in a web browser. This interactive report is very powerful and will provide many different visualizations of metagenome assembly statistics. There are several tables and graphs generated by metaQUAST that are useful to consider when performing metagenome QC. One of the most useful reports is a table with length and contiguity statistics for every input metagenome (only six assemblies shown; *see* Fig. 1).

The number of contigs (again, scaffolds in this case) is the first reported statistic. The number of scaffolds in an assembly is not a sufficient statistic by itself to assess metagenome quality. For example, sample ERR599044 has more scaffolds than ERR594407, but the former is composed of shorter scaffolds than the latter. How is this known? For one, the largest scaffold in ERR599044 is reported to be 580 kb, while the largest in ERR594407 is 1.46 Mb. However, the most informative statistics here are N50, L50, N90, and L90. These four statistics are direct measurements of contiguity.

N50 is defined as "the length of the shortest contig for which longer and equal length contigs cover at least 50% of the assembly" [45]. In other words, if one were to order all scaffolds in an

QUAST

Quality Assessment Tool for Genome Assemblies by CAB

28 February 2024, Wednesday, 09:31:36

View in Icarus contig browser

All statistics are based on contigs of size >= 1000 bp, unless otherwise noted (e.g., "# contigs (>= 0 bp)" and "Total length (>= 0 bp)" include all contigs).

Worst Median Best ☑ Show heatmap

Statistics without reference	ERR599017_scaffolds	ERR594415_scaffolds	ERR599174_scaffolds	ERR594391_scaffolds	ERR599044_scaffolds	ERR594407_scaffolds
# contigs	57 644	52 382	63 593	62 195	89 318	24 038
# contigs (>= 0 bp)	591 973	241 357	636 346	278 064	2 444 899	142 226
# contigs (>= 1000 bp)	57 644	52 382	63 593	62 195	89 318	24 038
# contigs (>= 5000 bp)	3245	6229	3557	8886	8412	4057
# contigs (>= 10000 bp)	592	2293	699	3372	3297	1995
# contigs (>= 25000 bp)	32	508	49	850	920	642
# contigs (>= 50000 bp)	1	102	3	234	311	265
Largest contig	55 361	179 113	83 971	1 693 895	579 666	1 456 704
Total length	124 160 796	163 041 179	136 780 541	230 111 134	268 616 874	133 149 004
Total length (>= 0 bp)	342 737 291	249 413 992	373 173 543	326 789 782	859 304 933	184 723 123
Total length (>= 1000 bp)	124 160 796	163 041 179	136 780 541	230 111 134	268 616 874	133 149 004
Total length (>= 5000 bp)	26 062 594	74 408 487	29 619 535	125 279 743	123 941 686	95 604 064
Total length (>= 10000 bp)	8 535 172	47 437 714	10 444 199	87 457 523	89 087 341	81 230 736
Total length (>= 25000 bp)	984 448	20 840 835	1 638 237	49 922 511	53 457 785	60 619 367
Total length (>= 50000 bp)	55 361	7 108 834	193 303	29 281 462	32 369 796	47 737 913
N50	2321	4257	2296	5946	4206	19 200
N90	1152	1295	1149	1397	1224	1603
auN	3831.9	11 374	4049	52 178	25 632	167 085
L50	14 439	7774	15 685	7013	10 676	917
L90	46 052	38 007	50 814	42 612	64 924	13 362
GC (%)	43.97	37.12	44.98	44.04	48.67	55.07
Mismatches						
# N's per 100 kbp	0	119.08	0	143.08	62.22	111.23
# N's	0	194 151	0	329 239	167 133	148 102

Fig. 1 A subset of metaquast statistics for assembled metagenomes. Table captured from the metaQUAST html output

assembly by decreasing length, the N50 of that assembly would be the size of the scaffold in the very middle of that list. Here, ERR599017 has a lower N50 than ERR599044, suggesting that the former is less contiguous than the latter. N90 is like N50, except that the value gives the length of the contig that is longer than 90% of an assembly. L50 and L90 are also like N50 and N90, but instead of the values representing sequence lengths, they represent the rank order of a scaffold by length. In other words, the L50 value of 7774 for ERR594415 means that 7774 scaffolds make up 50% of the length of the entire metagenome. Lower L50/L90 values are indicative of greater contiguity. Additionally, metaQUAST has counted the number of Ns in the scaffolds. In some cases, more Ns are met with greater contiguity because of scaffolding. For example, ERR594407 has 79% more Ns per 100 kb than ERR599044, but it has a considerably higher N50 and lower L50. While a smaller fraction of Ns in a metagenome assembly is desired, the fact that Ns are present may be indicative that the genomic structure was preserved during assembly, maintaining contiguity.

So, what do the reported values here mean for the quality control of the assembled *Tara* metagenomes? Generally, more contiguous metagenomes are of a higher quality than discontiguous metagenomes. However, it is not easy to provide a specific N50/N90 or L50/L90 value that should be used as a cutoff to determine whether a metagenome should be analyzed or not. This is especially the case with the dataset being used here because there is a biological reason to only compare viromes to viromes and metagenomes to metagenomes. Half of the assemblies are viromes (shown here are ERR599017, ERR599174, and ERR599044), representing the virus-sized fraction of a sample. Viruses have genome sizes orders of magnitude smaller than that of their cellular hosts, so it makes biological sense for a virome to appear more discontiguous than a cellular metagenome. In this case, it may not necessarily be true that the viral genomes are more fractured than the total metagenomes, but instead they are just smaller genomes. Additionally, without knowing what the true community composition is in each sample, it is improbable to come up with a desired N- or L- statistic for any given sample since there will be real biological variation in genome composition within samples. Having prior knowledge of sampling design such as that in Table 2 is useful here. If it is assumed that biological and or methodological replicates should have similar microbial communities, by extension, they should have similar N- or L-statistics and scaffold lengths. Assemblies within these replicates that deviate from the rest could be candidates for removal or further curation.

It is generally recommended to remove scaffolds <1000 bp from metagenome assemblies before genome binning and further analysis. Retaining scaffolds <1000 bp in metagenomes can lead to increased computation times for downstream analyses with little informational reward [2, 28]. They can also make binning difficult [46, 47]. Therefore, removing scaffolds <1000 bp is often beneficial and a good step to perform in metagenome QC. Doing so with the assembled *Tara* metagenomes here will result in the removal of a significant fraction of all base pairs in each assembly. This is sometimes the unkind reality of metagenomics. One may assume that all of the DNA in an environmental sample may be chromosomal or otherwise circular and organized. However, environmental DNA, protected or not, may come from a wide variety of unknown sources and can end up making up a large fraction of a final DNA sample [48, 49].

Highly fragmented assemblies like these can sometimes be mitigated by increasing the sequencing depth [43]. Unless there is prior information available about the community composition of a sample, it is difficult to estimate the ideal sequencing depth to result in a more contiguous metagenome assembly [50]. Additionally, metagenome co-assembly can sometimes improve contiguity. Co-assembly is the process of combining separate metagenomic read samples (ideally technical replicates) and assembling them together. On the one hand, the added read depth of similar microbial communities can make assembling rare and variable community members more successful [51]. On the other hand, if there is a lot of intersample variation, co-assembly may not end up improving assembly quality due to the added complexity [52]. Here, not all depths and locations are replicated across the 20 *Tara* samples, and viromes and metagenomes should not be co-assembled. Therefore, the single-assembly approach will be upheld to avoid biasing assembly quality toward any sample location or depth.

In any case, unless there is a clear biological reason to retain very short sequences, one can use a single line of code to achieve this using `bioawk`:

```
mkdir -p filtered_scaffolds

for FASTA in scaffolds/*_scaffolds.fasta; do \
 bioawk -c fastx '{ if(length($seq) >= 1000) \
 print ">"$name"\n"$seq }' $FASTA \
> filtered_scaffolds/$(basename $FASTA); done
```

The filtered scaffolds will be in a new folder filtered_scaffolds. Comparing the sizes of the two folders with du -sh scaffolds filtered_scaffolds shows volumes of 13 GB and 4 GB, a near 70% reduction in file size. That is a lot of low-quality data removed! This should not impact the statistics calculated by metaQUAST since it was already set to only analyze sequences > = 1000 bp with `--min-contig 1000`. On the other hand, read recruitment will surely be affected by removing 70% of the bases across all assemblies. Just as read recruitment was assessed for the unfiltered scaffolds, it can be done again using the filtered scaffolds to see how much of the original sequence data ends up in the filtered assemblies. The binning process below requires BAM files for reads mapped to the filtered scaffolds. It will be assumed that the researcher has generated these and put them in a new folder named `map_files_filt`. To be concise, this will not be shown, but one can expect a drop in read recruitment considering the number of scaffolds removed.

3.6 Binning Microbial Genomes

The fundamental difference between metagenome assembly and single-genome assembly is that metagenomes contain DNA fragments of multiple community members. Metagenomics was first used to extract meaningful information (gene clusters) from uncultivated microbes in a community without genome-level resolution [1]. It later became possible to reconstruct whole genomes from metagenomes de novo [2, 3], which allows for a better understanding of microbial populations in their original environments. Generating *metagenome-assembled genomes* (MAGs) is therefore a powerful step in nearly any metagenomic study. Generating MAGs, a process not unlike to putting together pieces of a puzzle, is called *binning*. There have been a variety of tools developed to bin microbial scaffolds into MAGs [53–57]. The most common ones used today are MetaBAT2 [55], MaxBin2 [53], and CONCOCT [57]. These three binning tools are incorporated into the wrapper called metaWRAP [15], which allows users to bin their scaffolds with all three tools and then refine them into consensus bins. While metaWRAP's multi-tool and refinement approach can be useful, metaBAT2 alone often performs very well, and it is much simpler to run. This is especially the case when mapped read `.bam` files are provided to assist in the binning process. Here, metaBAT2 will be used to bin the filtered mixed-community metagenome scaffolds into MAGs. Scaffolds originating from the viromes will be binned with a different approach below (*see* Section 3.9).

To run metaBAT2 on every mixed-community metagenome, the script `run_metabat2.sh` will be used along with a list of the mixed-community metagenome sample names in a file `metagenome_samples.txt`. It is assumed that reads were mapped to the filtered scaffolds that are to be binned, and the corresponding `.bam` files are in the folder `map_files_filt`.

```
cat metagenome_samples.txt
ERR598984
ERR599005
ERR599006
ERR599017
ERR599044
ERR599126
ERR599133
ERR599148
ERR599165
ERR599174
```

The mapping files are not required to run metaBAT2, but they will substantially increase its accuracy and number of bins generated.

```
#!/bin/sh
# run_metabat2.sh

THREADS="16"
SAMPLES="metagenome_samples.txt"

mkdir -p metabat2_bins && cd metabat2_bins

while IFS= read -r SAMPLE || [ -n "$SAMPLE"]
do
# Skip empty lines
if [ -z "$SAMPLE"]; then
continue
fi
```

```
 echo "Binning sample $SAMPLE"

# Command to run for each sample
 runMetaBat.sh \
 "../filtered_scaffolds/${SAMPLE}_scaffolds.fasta" \
 "../map_files_filt/${SAMPLE}.bam"
 --minContig 2500 \
 --maxP 95 \
 --minS 60 \
 --minCV 1 \
 --minClsSize 200000 \
 --numThreads "$THREADS"
 echo "Done ${SAMPLE}"
done <../"$SAMPLES"
```

Here, the flags `--minContig 2500`, `--maxP 95`, `--minS 60`, `--minCV 1`, `--minClsSize 200000` are the default parameters for metaBAT2, but the user should consider each when running metaBAT2. They each can affect the number of bins generated and the overall quality of the bins, with a trade-off between maximizing recovery and increasing contamination. Assessing the quality of bins and understanding contamination will be addressed below.

The script will create a folder named `metabat2_bins` that will contain one subfolder per sample, each containing the MAGs binned for that sample. Across all 10 mixed-community metagenome samples, a total of 436 MAGs were generated, ranging from as few as 23 to as many as 86 in any given sample (Fig. 2a). The number of MAGs binned by metaBAT2 increased with the number of scaffolds, total number of bases, and the N50 of the input assembly (Fig. 2b).

An individual MAG roughly represents a microbial population at the species or genus level [2, 58, 59]. This does not necessarily mean that each MAG represents a complete genome. In fact, the completeness and contamination (sequences originating from another organism) for MAGs can be estimated quite confidently with MAG QC. It is generally accepted that a MAG with at least 90% estimated completeness and at most 10% estimated contamination is a "high-quality" MAG [60, 61]. High-quality means the MAG is an accurate and near-complete representation of a microbial population [60]. MAGs with at least 50% completeness and at most 10% contamination are of medium-quality. They are still useful and adequate for population- and community-level analyses [61], but interpretations based on them should be formed and considered with caution. Any MAG that has both an estimated completeness less than 50% and an estimated contamination over

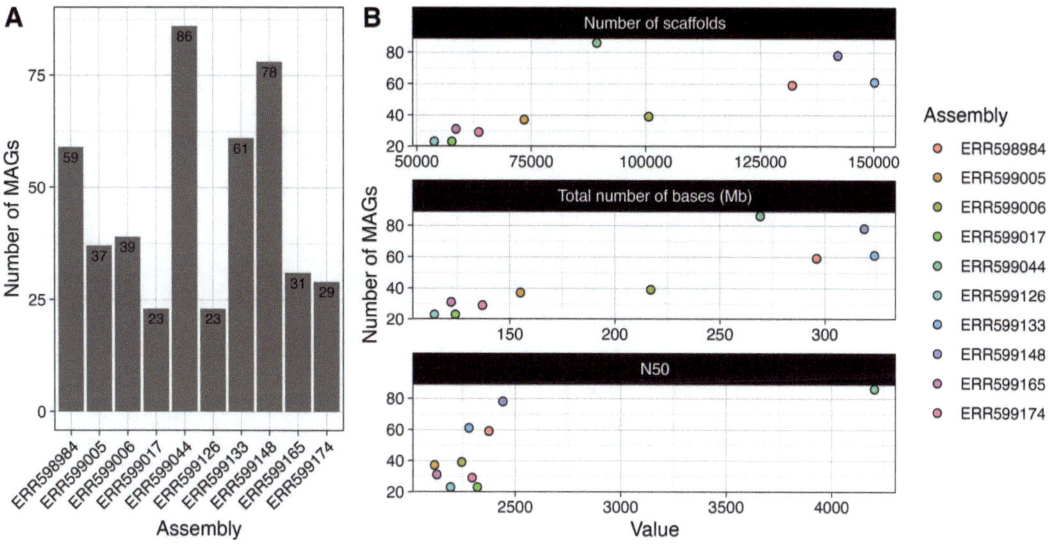

Fig. 2 MAGs binned per assembly. (**a**) Number of MAGs generated for each *Tara* mixed-community metagenome assembly. (**b**) Number of MAG versus total number of scaffolds in the input assembly, total number of bases in the input assembly, and input assembly N50. Statistics calculated for scaffolds > 1000 bp, as demonstrated in Section 3.5

10% is of low-quality [61] and should not be further analyzed with the remaining medium- and high-quality MAGs.

After metaBAT2 has finished binning each mixed-community metagenome sample, the quality of each will be assessed with the tool CheckM [60] using the script `run_checkm.sh`:

```
#!/bin/sh
# run_checkm.sh

THREADS="16"
SAMPLES="metagenome_samples.txt"

mkdir -p checkm

while IFS= read -r SAMPLE || [ -n "$SAMPLE"]
do
# Skip empty lines
if [ -z "$SAMPLE"]; then
continue
fi

# Find the path to the metaBAT2 folder for the current
sample
# Using 'find' with wildcards '*' because folder names can
vary
```

```
# Adjust if your metaBAT2 outputs have a different pattern
 MAG_DIR=$(find metabat2_bins -name ${SAMPLE}_*.metabat-
bins-*)

 echo "Running CheckM on sample $SAMPLE"

# Command to run for each sample
 checkm lineage_wf \
 --extension fa \
 --file "checkm/${SAMPLE}/${SAMPLE}_quality.txt" \
 --tab_table \
 --threads "$THREADS" \
 "$MAG_DIR" \
 "checkm/${SAMPLE}"
 echo "Done ${SAMPLE}"
done < "$SAMPLES"
```

This script will use the sample names in `metagenome_samples.txt` again to run checkM on each output folder created by metaBAT2. The folder names are inferred using the find command since the `runMetaBat.sh` command invoked by the `run_metabat2.sh` script above will create variable folder names based on the dates and times of execution. Here, the lineage workflow of checkM is executed with `lineage_wf`. This workflow obtains completeness and contamination estimates using different sets of marker genes for each "lineage" the query MAGs were placed in by checkM. MetaBAT2 writes the output MAGs to files ending in `.fa`, which is different than the input file extension `.fna` that checkM assumes. This extension is provided to checkM with the `--extension` flag. The flags `--file` and `--tab_table` will write the results to a tab-delimited table that can be read later. As usual, `--threads` sets the maximum number of CPUs to use.

The checkM output files should be in the new folder checkm, with one subfolder per sample. For each set of input MAGs per sample, their quality estimations are found in a table with a name ending in `_quality.txt`. These can each be manually inspected to sort high-, medium-, and low-quality MAGs. Alternatively, one can use a language like Python and write a script to automatically do this. Such a script `sort_mags.py` is provided below:

```
# sort_mags.py

import os
import csv
from pathlib import Path
```

```
def process_sample(sample, checkm_path, metabat2_bin-
s_path, sorted_mags_path, summary_writer):
 quality_file = checkm_path / f"{sample}/{sample}_qual-
ity.txt"

# Check if the quality file exists before proceeding
if not quality_file.exists():
 print(f"Quality file not found for sample {sample},
skipping...")
return

 bins_dir = next(metabat2_bins_path.glob(f"{sample}_*.
metabat-bins-*"), None)

if not bins_dir:
 print(f"Bins directory not found for sample {sample},
skipping...")
return

with quality_file.open() as f:
 reader = csv.DictReader(f, delimiter='\t')
for row in reader:
 bin_id = row['Bin Id']
 new_bin_id = f"{sample}__{bin_id.replace('bin.',
'bin_')}"
 completeness = float(row['Completeness'])
 contamination = float(row['Contamination'])

if completeness >= 90.0 and contamination <= 10.0:
 quality = "high_quality"
elif 50.0 <= completeness < 90.0 and contamination <=
10.0:
 quality = "medium_quality"
else:
 quality = "low_quality"

 fa_file_path = bins_dir / f"{bin_id}.fa"
if fa_file_path.exists():
 quality_dir = sorted_mags_path / quality
 quality_dir.mkdir(parents=True, exist_ok=True)
 symlink_filename = f"{new_bin_id}.fa"
 symlink_path = quality_dir / symlink_filename
if not symlink_path.exists():
 symlink_path.symlink_to(fa_file_path.resolve())

 summary_writer.writerow([bin_id, sample, new_bin_id,
completeness, contamination, quality])
```

```
def main(checkm_path_str, metabat2_bins_path_str, sor-
ted_mags_path_str):
 checkm_path = Path(checkm_path_str)
 metabat2_bins_path = Path(metabat2_bins_path_str)
 sorted_mags_path = Path(sorted_mags_path_str)

# Create sorted_mags directory and its subdirectories
 sorted_mags_path.mkdir(parents=True, exist_ok=True)

 summary_file_path = sorted_mags_path / "checkm_summary.
txt"
with summary_file_path.open('w', newline='') as summar-
y_file:
 writer = csv.writer(summary_file, delimiter='\t')
 writer.writerow(["BinID", "Sample", "NewBinID",
 "Completeness", "Contamination",
 "Quality"])

for sample_dir in checkm_path.iterdir():
if sample_dir.is_dir():
 sample = sample_dir.name
 process_sample(sample, checkm_path,
 metabat2_bins_path,
 sorted_mags_path, writer)

if __name__ == "__main__": checkm_path = "checkm", meta-
bat2_bins_path = "metabat2_bins", sorted_mags_path =
"sorted_mags"
main(checkm_path, metabat2_bins_path, sorted_mags_path)
```

Assuming the user has Python (versions 3 and above) installed, the script can be run with `python3 sort_mags.py`. This script will create links to the original `.fa` files produced by metaBAT2 (so do not remove the originals; they are not copies, just shortcuts!). The links will be created beneath the new folder `sorted_mags` sorted into subfolders `high_quality`, `medium_quality`, and `low_quality` based on the thresholds described above, but they can be adjusted. A table named `checkm_summary.txt` will be written by this script that will display the determined qualities for each bin. The filenames for the links will be prepended with their sample name, so `checkm_summary.txt` also provides a key to convert the old MAG names to the new ones:

```
head sorted_mags/checkm_summary.txt
BinID Sample NewBinID Completeness Contamination Quality
bin.1 ERR599006 ERR599006__bin_1 46.19 1.72 low_quality
```

```
bin.10 ERR599006 ERR599006__bin_10 43.26 3.45 low_quality
bin.11 ERR599006 ERR599006__bin_11 40.52 0.0 low_quality
bin.12 ERR599006 ERR599006__bin_12 13.79 0.0 low_quality
bin.13 ERR599006 ERR599006__bin_13 91.87 1.08 high_qual-
ity
bin.14 ERR599006 ERR599006__bin_14 6.03 0.0 low_quality
bin.15 ERR599006 ERR599006__bin_15 0.0 0.0 low_quality
bin.16 ERR599006 ERR599006__bin_16 53.85 0.43 medium_qu-
ality
bin.17    ERR599006    ERR599006__bin_17    43.97    10.34
low_quality
```

Using this table, the number of low-, medium-, and high-quality MAGs can be counted using `grep`:

```
grep -c "low_quality" sorted_mags/checkm_summary.txt
grep -c "medium_quality" sorted_mags/checkm_summary.txt
grep -c "high_quality" sorted_mags/checkm_summary.txt
324
90
22
```

The total number of medium- and high-quality MAGs is 90 + 22 = 112. For 10 complex, mixed-community metagenome samples, this is a good amount to work with for downstream ecological and functional analyses. These 112 MAGs will be used in the further sections. If one wishes to attempt to get high-quality MAGs out of the low-quality ones, *bin refinement* is worth consideration. Bin refinement describes the manual or automated process of assigning and re-assigning scaffolds to MAGs based on sequence composition, GC skew, and gene annotations [62, 63]. Automated bin refinement can be performed using the tool metaWRAP [15]. Manual bin refinement can be achieved in a visual and interactive interface within the suite of (meta)genomics tools in Anvi'o [18].

The scaffolds from the 10 mixed-community samples that were either unbinned or ended up in low-quality bins are still useful. They contain potentially interesting gene-level information that can be leveraged later; however, they should not be used for genome-based analyses. The links to the medium- and high-quality MAGs can be copied (preserving as links to save on storage) and placed into their own combined folder `sorted_mags/medium_-high_quality` for further analysis.

```
mkdir orted_mags/medium_high_quality
cp -P sorted_mags/medium_quality/*.fa sorted_mags/high_-
quality/*.fa sorted_mags/medium_high_quality
```

3.7 Assigning Taxonomy to Microbial Bins

At this point, the mixed-community metagenomes have been assembled, QC'd, binned into MAGs, and filtered even further to determine which MAGs represent medium- or high-quality genomes. It has not yet been determined just which organisms these MAGs represent. The taxonomy of these MAGs may be informative on their functional potential and possible evolutionary histories. For whole-genome sequences, it is typical to obtain taxonomic assignments using a 16S rRNA marker gene approach [64, 65], a multi-gene tree approach [66, 67], a whole-genome distance-based approach [68, 69], or a combination of these approaches [70]. These approaches are useful for isolate genomes, but they tend to perform poorly when applied to MAGs. Since MAGs tend to be incomplete, there may not be 16S rRNA genes or other marker genes on the genome. Also, since MAGs can vary greatly in length, completeness, and contamination, methods that leverage whole-genome distances will not always work. There is a robust, highly standardized tool that is specifically designed to obtain taxonomic assignments for MAGs. The tool GTDB-Tk [71, 72] can take fragmented, potentially contaminated genomes and assign taxonomy using a comprehensive database of isolate and uncultivated genomes [73] in a high-throughput way.

Here, GTDB-Tk will be run using the `classify_wf` option, which will obtain taxonomic assignments for the input medium- and high-quality MAGs by placing them in the GTDB reference phylogenomic trees. Since all the MAGs to be analyzed are in the folder `sorted_mags/medium_high_quality`, this directory can be given to the `gtdbtk classify_wf` command. The extension of the MAG files will be specified (`.fa`), and the taxonomy results will be written inside the folder `gtdb_taxonomy`:

```
gtdbtk classify_wf \
  --genome_dir sorted_mags/medium_high_quality \
  --out_dir gtdb_taxonomy \
  --extension fa \
  --cpus 16
```

The results will be written to `gtdb_taxonomy/gtdbtk.ar53.summary.tsv` for identified archaea and `gtdb_taxonomy/gtdbtk.bac120.summary.tsv` for identified bacteria.

3.8 Identifying Viruses and Other Mobile Genetic Elements

It is estimated that there are 10^{31} viruses on Earth, primarily viruses of microbes [74]. As predators of microbes, viruses have great ecological impacts on their host populations [75, 76] and, by extension, biogeochemical impacts on their surrounding environments [77–79]. In addition to the 10 mixed-community metagenomes that were used to bin microbial genomes into MAGs, the *Tara* oceans data used here contains 10 viromes that were generated to enrich for viral DNA. However, it should not be assumed that all scaffolds among the viromes are viral sequences. Although viromes generally outperform mixed-community metagenomes in maximizing observable virus abundance and diversity [23], non-viral DNA can still end up in viromes after enrichment [14, 48]. Additionally, when sequencing metagenomes of whole microbial communities, one can expect to sequence viral DNA either from viruses in an extracellular state or from viruses integrated in their host genomes. Because of the imperfect enrichment for viruses in viromes and the presence of viruses among microbes, viral scaffolds need to be identified and sorted from nonviral scaffolds in both viromes and mixed-community metagenomes.

Several tools have been developed in recent years to identify viral sequences in metagenome data, but the most robust, accurate, and user-friendly ones to date are geNomad [80], VIBRANT [81], and VirSorter2 [82]. Each tool performs similarly when benchmarked against each other [80, 81], but each has its own unique qualities that may influence the researcher's decision to use one over the other. For example, VirSorter2 contains different classifiers that are specialized to specific types of viruses [82]. VIBRANT produces additional outputs for the identification of putative *auxiliary metabolic genes* [81]. Additionally, geNomad identifies plasmids as well as viruses, and it attempts to obtain taxonomic assignments to identified viruses [80]. Since plasmids may also be of interest to the researcher in a metagenomic study in addition to viruses, geNomad will be used to identify viral and plasmid scaffolds in all 20 *Tara* assemblies here. When running, geNomad is implemented as a series of modules. They can all be executed sequentially by executing `genomad end-to-end`. This will be done for every assembly of filtered scaffolds using the script `run_genomad.sh`:

```
#!/bin/sh
# run_genomad.sh

SCAFFOLD_DIR="filtered_scaffolds"
OUTPUT_DIR="genomad_out"
DB_DIR="/path/to/genomad/db"
THREADS="16"
```

```
mkdir -p "$OUTPUT_DIR"

for SCAFF in ${SCAFFOLD_DIR}/*fasta; do
# Extract the sample name
 SAMPLE=$(basename "$SCAFF" _scaffolds.fasta)

# Run geNomad
 genomad end-to-end \
 --cleanup \
 --threads "$THREADS" \
 "$SCAFF" \
 "${OUTPUT_DIR}/${SAMPLE}" \
 "$DB_DIR"

 echo "Done ${SAMPLE}"
done
```

The `DB_DIR` variable will need to be set to the full path to the local copy of the geNomad database created during installation. The `--cleanup` flag removes intermediate files after geNomad finishes a run on a sample, to save on storage. The results will be located beneath the folder `genomad_out` with subfolders for each sample. The final virus and plasmid results will be in another subfolder with the suffix `_summary`. For example, the folder `genomad_out/ERR599174/ERR599174_scaffolds_summary` contains virus and plasmid results for sample ERR599174:

```
ls genomad_out/ERR599174/ERR599174_scaffolds_summary
ERR599174_scaffolds_plasmid.fna
ERR599174_scaffolds_plasmid_genes.tsv
ERR599174_scaffolds_plasmid_proteins.faa
ERR599174_scaffolds_plasmid_summary.tsv
ERR599174_scaffolds_virus.fna
ERR599174_scaffolds_virus_genes.tsv
ERR599174_scaffolds_virus_proteins.faa
ERR599174_scaffolds_virus_summary.tsv
```

For this sample, all identified virus nucleotide sequences are in the file `ERR599174_scaffolds_virus.fna`, and all identified plasmid nucleotide sequences are in `ERR599174_scaffolds_plasmid.fna`. Additionally, the files ending with `_summary.tsv` contain scaffold-level information for the identified viruses and plasmids.

To do a quick check of how many viruses and plasmids were identified in this sample, `grep` can be used:

```
grep -c ">" genomad_out/ERR599174/ERR599174_scaffolds_-
summary/ERR599174_scaffolds_virus.fna
grep -c ">" genomad_out/ERR599174/ERR599174_scaffolds_-
summary/ERR599174_scaffolds_plasmid.fna
1170
151
```

A substantial number of viral scaffolds were identified in this sample. While the plasmid sequences identified by geNomad may be of interest to the researcher, they will not be further investigated here.

Combining all `virus_summary.tsv` files can allow for the number of viral scaffolds across all samples to be visualized. Shown in Fig. 3 are the fraction and total number of scaffolds identified as viral for the mixed-community metagenome and virome assemblies of each *Tara* sample. As expected, the viromes in each assembly pair yielded a greater fraction of viral scaffolds and a greater total number of viral scaffolds.

3.9 Binning Viral Genomes

In Section 3.6, scaffolds were binned into MAGs. This was because microbial genomes appear highly fragmented in metagenomic data, and reconstructing genomes into MAGs can improve genomic- and population-level analyses. It is also common for viral genomes to be fragmented in mixed-community metagenomes and viromes. Although viruses typically have genome sizes orders of magnitude smaller than their microbial hosts, genome fragmentation can make metagenomic analyses of viral communities a challenge [83, 84]. Viral genome fragments can be binned into *viral metagenome-assembled genomes* (vMAGs) to overcome some of these challenges.

Binning viral genomes is not (yet) as common as binning microbial genomes into MAGs, primarily due to the small size of viral genomes and the high complexity and mosaicism of viral scaffolds [85]. However, there are several benefits to binning viral scaffolds into vMAGs. First, vMAGs can accurately reconstruct more complete viral genomes than individual scaffolds [86–88]. Second, binning scaffolds of the same viruses into vMAGs can lead to a more accurate representation of viral diversity in a community, since treating single scaffolds as an individual genome may overestimate diversity [89]. Third, there are some viruses with large genomes, such as “jumbo”, “huge”, and “mega” phage [90], and NCLDVs (a.k.a. “giant viruses”) which infect microbial eukaryotes [91]. Binning viral scaffolds may improve one’s ability to detect these unique viruses in metagenomic data [92].

There are some drawbacks to binning viral scaffolds into vMAGs. Most viral metagenomic tools are not designed to handle

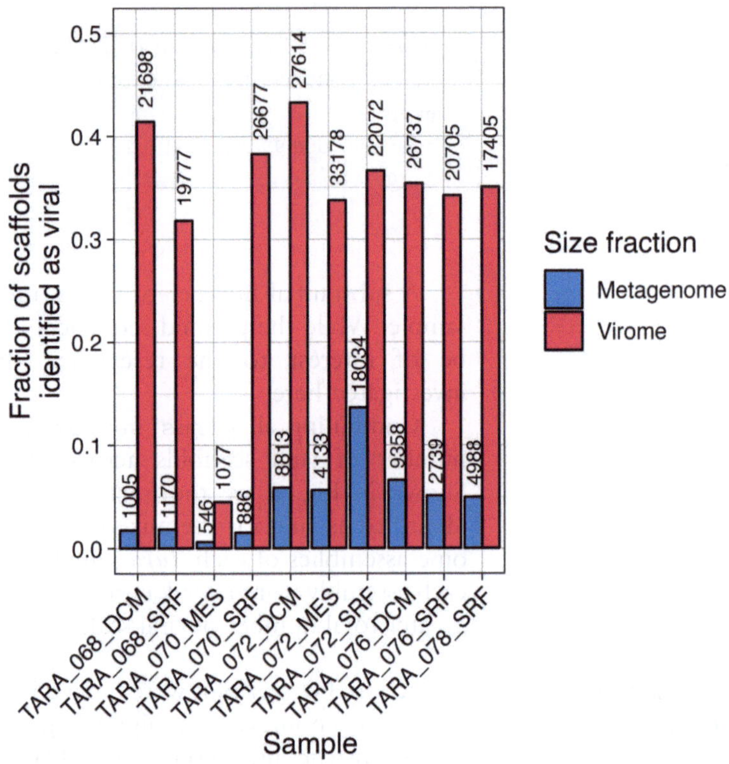

Fig. 3 Viral scaffolds identified per sample. Viral scaffolds identified by geNomad for each *Tara* oceans sample as a fraction of all scaffolds > 1000 bp in the input assembly are given. Bar colors correspond to the virome and mixed-community metagenome size fraction assemblies for each Tara sample. Atop each bar is the total number of viral scaffolds identified in the corresponding assembly

vMAGs, rather, they assume individual scaffolds represent a single viral genome. This requires further processing and manipulation of vMAGs, so they can be compatible with such tools (*see* Sections 3.10, 3.11, and 3.12). Also, just as with binning microbial genomes, binning viral genomes can produce contaminated bins. However, vMAGs can be QC'd like MAGs, both with computational tools and with manual curation.

Although there are many tools available designed to bin microbial scaffolds (*see* Section 3.6), these tools typically do not perform well when binning viral scaffolds [86, 88], except for the kinds of very large viruses mentioned above. This is primarily because viruses do not have universal marker genes like bacteria and archaea, which several microbial binning tools search for in the binning process. Although binning viral scaffolds into vMAGs is mechanistically difficult, there are a handful of tools available specifically designed to generate vMAGs. vRhyme bins viral scaffolds using single- or multi-sample read coverage and sequence features [88]. COBRA bins viruses by resolving metagenome assembly breakpoints in assembly graphs [86]. ViralCC uses metagenomic Hi-C data for genome binning [89]. And, although it is not a tool, PHAMB is a framework for binning viruses using random forest models [87].

Here, vRhyme will be used to bin viral scaffolds into vMAGs since its input requirements (viral scaffolds and read mapping data) were already generated. Although co-assembly was not performed here, it is worth noting that vRhyme is particularly useful for co-assemblies or other metagenomic datasets with biological replicates. This is because of its unique ability to use multisample read coverage metrics in binning [88]. Mapping reads from replicated samples can greatly improve vRhyme's accuracy and recall [88]; the researcher is encouraged to consider this in their own workflow. The script `run_vrhyme.sh`, shown below, will be used to execute vRhyme for every set of viral scaffolds identified by geNomad.

```
#!/bin/sh
# run_vrhyme.sh

THREADS="16"
INDIR="genomad_out"
OUTDIR="vrhyme_bins"
BAMDIR="map_files_filt"

mkdir -p $OUTDIR && cd $OUTDIR

for DIR in../$INDIR/ERR*; do
 SAMPLE="${DIR##*/}"
 echo "Binning sample: $SAMPLE"

# vRhyme command to run for each assembly
 vRhyme \
  -i "../${INDIR}/${SAMPLE}/${SAMPLE}_scaffolds_summary/
${SAMPLE}_scaffolds_virus.fna" \
  -b "../${BAMDIR}/${SAMPLE}.bam" \
  -o "../${OUTDIR}/${SAMPLE}" \
  -l 2000 \
  --prefix "$SAMPLE"__

# Add assembly prefix to vRhyme bin files
for VMAG in../${OUTDIR}/${SAMPLE}/vRhyme_best bins_fas-
ta/*.f*; do
 BASE=$(basename ${VMAG})
mv -- "$VMAG" "../${OUTDIR}/${SAMPLE}/vRhyme_best_bins_-
fasta/${SAMPLE}__${BASE}"
done
```

```
# Extract the unbinned viral sequences with a vRhyme
auxiliary script
 extract_unbinned_sequences.py \
  -i../${OUTDIR}/${SAMPLE}/vRhyme_best_bins.*.membership.
tsv \
  -f "../${INDIR}/${SAMPLE}/${SAMPLE}_scaffolds_summary/
${SAMPLE}_scaffolds_virus.fna" \
  -o "../${OUTDIR}/${SAMPLE}/${SAMPLE}_unbinned_viruses.
tmp.fna"

# Add assembly prefix to headers of unbinned sequences
  UNBINNED_TMP="../${OUTDIR}/${SAMPLE}/${SAMPLE}_unbin-
ned_viruses.tmp.fna"
  UNBINNED="../${OUTDIR}/${SAMPLE}/${SAMPLE}_unbinned_-
viruses.fna"
awk -v PREFIX="$SAMPLE" '/^>/{print ">" PREFIX "__"
substr($0, 2); next} {print}' \
 "$UNBINNED_TMP" \
>> "$UNBINNED"
rm "$UNBINNED_TMP"

# Split the unbinned sequences into individual fasta
files
  UNBINNED_DIR="../${OUTDIR}/${SAMPLE}/vRhyme_unbinned_-
viruses_fasta"
mkdir -p "$UNBINNED_DIR"
if [ -f "$UNBINNED"]; then
 COUNTER=1
awk -v SAMPLE="$SAMPLE" -v OUTDIR="$UNBINNED_DIR" \
 '/^>/ {if(x>0) close(OUTNAME); \
  OUTNAME=sprintf("%s/%s__vRhyme_unbinned_%d.fasta",OUT-
DIR,SAMPLE,++x);} \
 {print > OUTNAME}' "$UNBINNED"
fi

# 'N-link' bins for CheckV with a vRhyme auxiliary script
 link_bin_sequences.py \
 -i "../${OUTDIR}/${SAMPLE}/vRhyme_best_bins_fasta" \
  -o "../${OUTDIR}/${SAMPLE}/vRhyme_best_bins_fasta_-
linked" \

# Add assembly prefix to headers of linked bin sequences
find "../${OUTDIR}/${SAMPLE}" \
 -type f \
 -path "*/vRhyme_best_bins_fasta_linked/*.fasta" | while
read -r FASTA; do
 TMP=$(mktemp) # Create a temporary file
```

```
awk -v SAMPLE="$SAMPLE" '/^>/ {$0=">" SAMPLE "__" substr
($0, 2)} 1' \
 "$FASTA" > "$TMP"
mv "$TMP" "$FASTA"
done

 echo "Done ${SAMPLE}"
done
```

There are multiple components to this script:

1. An output folder `vrhyme_bins` is created with one subfolder per metagenome/virome assembly. vRhyme is executed to bin all viral scaffolds identified by geNomad, for each assembly. The flag `-i` provides the input viral scaffolds, `-b` provides the mapped read `.bam` files previously generated that are required by vRhyme, `-o` sets the output directory for that assembly, `-l 2000` sets the input scaffold length requirement to 2000 bp as recommended by vRhyme, and `--prefix $SAMPLE__` adds the assembly ID as a prefix to the output bin sequence names (with an added double-underscore to easily distinguish the assembly IDs from the scaffold names).
2. The same assembly ID prefix is added to the output bin filenames, since vMAGs from all assemblies will be combined and used below.
3. The viral scaffolds that were not binned by vRhyme are extracted using the vRhyme auxiliary script `extract_unbinned_sequences.py`. They are split up into individual files with assembly ID prefixes in their names and placed into a folder named `vRhyme_unbinned_viruses_fasta`.
4. Scaffolds in each bin are joined by a string of Ns into a single sequence using the vRhyme auxiliary script `link_bin_sequences.py` and are placed into a separate folder `vRhyme_best_bins_fasta_linked`. This is to QC the bins using a tool that expects a single scaffold per virus, below. Assembly ID prefixes are added to their sequence headers, too.

For most samples, around half of all input viral scaffolds at least 2000 bp were binned (Fig. 4a), with an average of around five scaffolds per bin (Fig. 4b). This yields over 156,000 vMAGs across all samples and assemblies (here, both binned and unbinned viral genomes are referred to as vMAGs).

The completeness and contamination of each vMAG can be estimated with the tool CheckV [84]. CheckV estimates completeness and contamination with a combined marker gene, amino-acid identity, and sequence feature approach [84]. CheckV was designed

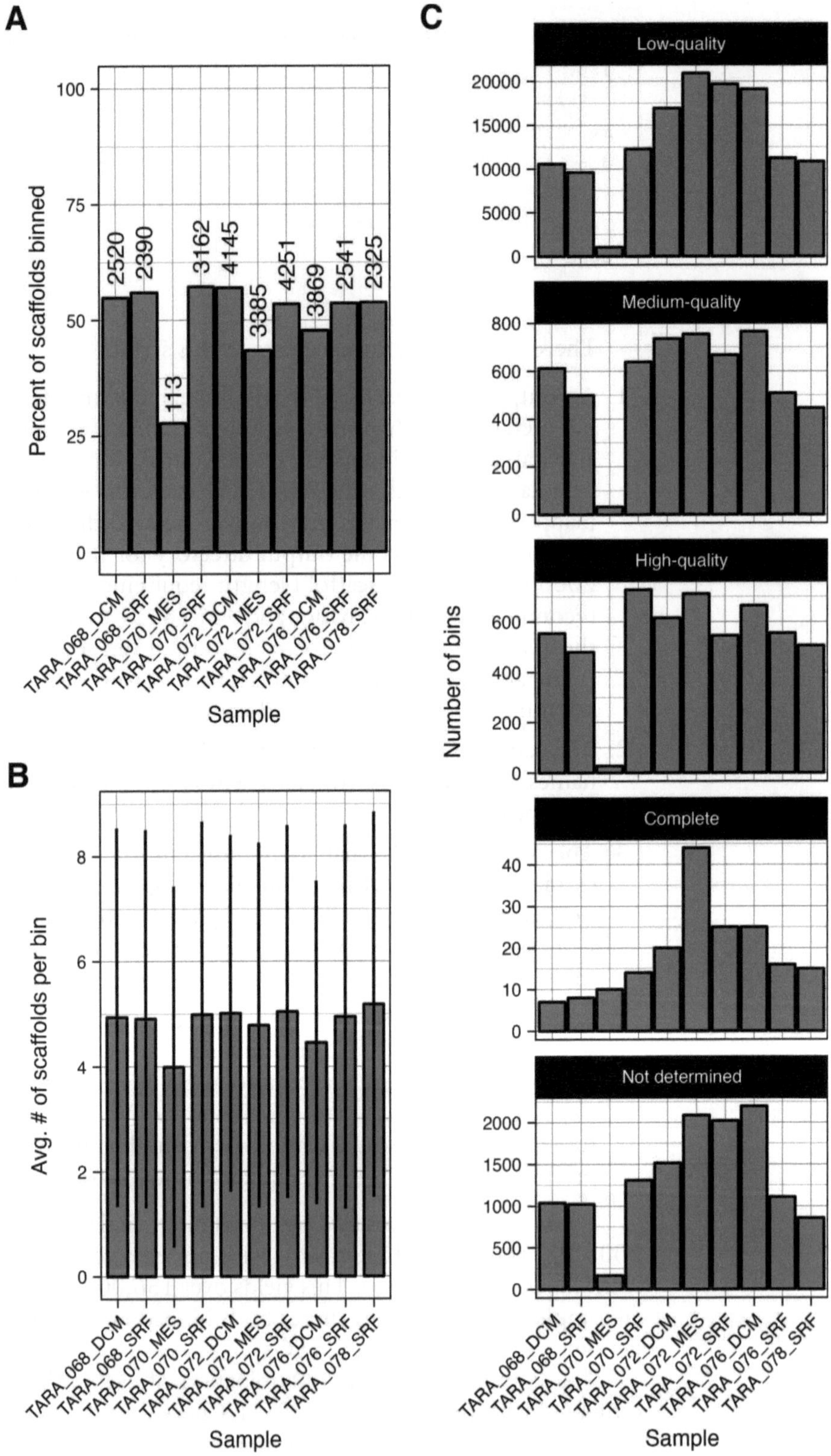

Fig. 4 vRhyme binning and vMAG quality summary. (**a**) Percent of viral scaffolds in each *Tara* sample that were binned. Values atop each bar give the total number of multiscaffold bins in that sample. (**b**) Average number of scaffolds assigned to each bin for every *Tara* sample. Whiskers represent standard deviations. (**c**) Total number of vMAGs (binned and unbinned) assigned to each CheckV quality for every *Tara* sample

to handle single-scaffold viral genomes rather than vMAGs. To run CheckV, the N-linked vMAGs generated by the `run_vrhyme.sh` script above will be combined with the single-scaffold vMAGs with the following command:

```
find vrhyme_bins/ \
 -type f \( \( \
 -path "*/vRhyme_best_bins_fasta_linked/*.fasta" \) \
 -o \( -name "*_unbinned_viruses.fna" \) \) \
 -print0 \
| xargs -0 cat > virus_genomes_linked.fna
```

Now, the quality of every vMAG can be estimated with CheckV:

```
checkv end_to_end \
 virus_genomes_linked.fna \
 checkv_out \
 -t 16
```

The CheckV results will be deposited in a folder `checkv_out`. The final quality estimates for every vMAG are given in the table `quality_summary.tsv` (Fig. 4c). If one wants to take a conservative approach when analyzing viral genomes, just the medium-quality, high-quality, and/or complete vMAGs can be retained for further analysis. It is not always necessary to filter low-quality vMAGs from analyses. For instance, since there is no universal marker gene for viruses, it is not possible to assess the diversity of viral communities with an approach like 16S amplicon sequencing. To obtain accurate estimates of viral diversity or community composition, it is best to retain all identified viral scaffolds (if they are dereplicated and at least 10 kb, *see* Section 3.10). Additionally, CheckV's marker-based completion estimates are biased toward well-studied and characterized viruses. This is no fault of CheckV itself, which is why CheckV assigns confidence levels to each assignment. This means, when analyzing understudied environments, some "low-quality" designations may be false negatives that are the result of unidentifiable marker genes in rare and highly novel viruses [84]. For these reasons, the vMAGs will not be sorted and filtered by CheckV quality like the MAGs were using CheckM quality.

3.10 Microbial and Viral Genome Dereplication

In metagenomics/viromics, genome dereplication is the process of identifying and consolidating redundant genomes that represent the same (or very similar) microbial/viral species within a dataset [52, 93]. This is crucial in studies with multiple metagenomes and/or viromes because it can greatly reduce computational redundancy. Dereplication also improves the clarity of downstream abundance-based microbial and viral diversity assessments since it ensures that each unique genomic entity is represented once. Also, by selecting the best representatives among genome species clusters, dereplication ensures that subsequent analyses are based on the best available data. However, a potential downside is the risk of oversimplification. The removal of closely related genomes could inadvertently eliminate valuable information about fine-scale genetic diversity and evolutionary dynamics within microbial and viral populations [93]. On the other hand, strain-level variation can still be analyzed with dereplicated MAGs/vMAGs with read-mapping-based diversity calculations [59, 94]. Overall, dereplication is highly recommended for analyses of multiple microbial and viral communities.

Tools like dRep [52] have been specifically designed for MAG dereplication and selection of the "best" genomes. Although dRep was primarily designed for microbial genomes, it can be used with viruses/vMAGs by altering some parameters. The vMAGs will be dereplicated first. Before dRep is executed, a list of the vMAG file paths need to be created so that dRep can identify and read the sequences within each individual vMAG. This can be done with the script `make_genome_list_virus.sh`, below:

```
#!/bin/sh
# make_genome_list_virus.sh

BASE_DIR="vrhyme_bins"
OUTPUT_FILE="virus_genome_paths.txt"

# Clear the output file to start fresh
> "$OUTPUT_FILE"

# Find and add paths from vRhyme_best_bins_fasta folders
find "$BASE_DIR" \
 -type f \
 -path "*/vRhyme_best_bins_fasta/*.fasta" \
>> "$OUTPUT_FILE"

# Find and add paths from vRhyme_unbinned_viruses_fasta
folders
find "$BASE_DIR" \
 -type f \
 -path "*/vRhyme_unbinned_viruses_fasta/*.fasta" \
>> "$OUTPUT_FILE"
```

The script will create the file `virus_genome_paths.txt`. This file can be given to the `dRep dereplicate` command:

```
dRep dereplicate \
 --processors 16 \
 --genomes virus_genome_paths.txt \
 --length 10000 \
 --S_algorithm skani \
 --ignoreGenomeQuality \
 -pa 0.8 -sa 0.95 -nc 0.85 \
 -comW 0 -conW 0 -strW 0 \
 -N50W 0 -sizeW 1 -centW 0 \
 --skip_plots \
 dereplicated_viruses
```

By default, dRep forms species-level clusters at an average nucleotide identity of 95%. It considers MAG completeness, contamination, heterogeneity, N50, centrality, and genome size for selecting the best MAG representative of a species-level cluster. Except for genome size, these metrics are not optimal for dereplicating viral genomes (*see* "Comparing and dereplicating non-bacterial genomes" from the dRep manual [52]), so they need not be considered. To achieve this, the flags `--ignoreGenomeQuality -comW 0 -conW 0 -strW 0 -N50W 0 -sizeW 1` and `-centW 0` are added. This will effectively make the chosen representative in each species cluster the largest vMAG. Since over 156,000 vMAGs from all samples and assemblies will be dereplicated at once, the flags `--S_algorithm skani` and `--skip_plots` are added to improve computational efficiency and decrease runtime. The flags `-pa 0.8 -sa 0.95 -nc 0.85` set the primary and secondary cluster ANI cutoffs to 80% and 95%, respectively. A primary cluster ANI cutoff of 80% is recommended because of the high variability of viral genomes, and the secondary cluster ANI cutoff of 95% will result in (roughly) viral species-level clusters [28, 95]. The flag `-nc 0.85` sets a genome overlap of 85% when comparing genomes.

Importantly, the flag `--length 10000` is included. This sets a minimum length requirement of 10 Kkb for input genomes, which means any vMAG less than 10 kb will not end up in the set of dereplicated vMAGs. This minimum can be set to whatever the user desires; however, a minimum genome length of 10 kb has been demonstrated to be an optimal requirement for viral community analyses [28]. Unless a significant fraction of the researcher's vMAGs is below this cutoff, or the study in question is focused on viruses with ultra-small genomes, we recommend using this 10 kb minimum.

The dRep output will be in the folder dereplicated_-viruses. Beneath, the final set of vMAG species representatives are found in the folder dereplicated_genomes. These dereplicated vMAGs can be combined into one single file for later use:

```
cat dereplicated_viruses/dereplicated_genomes/*.fasta \
> virus_genomes_dereplicated.fna
```

For the microbial genomes, a list of the paths to the medium- and high-quality MAGs can be created using a similar script as used for the vMAGs, make_genome_list_mags.sh:

```
#!/bin/sh
# make_genome_list_mags.sh

BASE_DIR="sorted_mags/medium_high_quality"
OUTPUT_FILE="mag_genome_paths.txt"

# Clear the output file to start fresh
> "$OUTPUT_FILE"

# Find and add paths from medium and high quality MAGs
find "$BASE_DIR" \
 -type l \
 -name "*.fa" \
>> "$OUTPUT_FILE"
```

The file mag_genome_paths.txt can be given to the dRep dereplicate command. Since MAGs are being dereplicated this time instead of vMAGs, most of the default dRep options are used:

```
dRep dereplicate \
 --processors 16 \
 --genomes mag_genome_paths.txt \
 --skip_plots \
 dereplicated_mags
```

The microbial dRep output will be in the folder dereplicated_mags. Beneath, the final set of MAG species representatives are found in the folder dereplicated_genomes. Assembly ID prefixes can be added to the scaffold names in these MAGs using the script mag_header_prefix.sh, below, so all MAGs be combined into one file and used later as needed.

```
#!/bin/sh
# mag_header_prefix.sh

EXT=".fa"
INDIR="dereplicated_mags/dereplicated_genomes"
OUTDIR="dereplicated_mags/dereplicated_genomes_prefixed"
OUTFILE="dereplicated_mags.fna"

mkdir -p "$OUTDIR"

> "$OUTFILE"

for MAG in $INDIR/*$EXT; do
 BASE=$(basename "$MAG" "$EXT")
 SAMPLE=$(echo "$BASE" | awk -F '__|_bin' '{print $1}')
 MODIFIED_MAG="$OUTDIR/${BASE}$EXT"
awk -v PREFIX="$SAMPLE" \
 '/^>/{print ">" PREFIX "__" substr($0, 2); next} \
 {print}' "$MAG" > "$MODIFIED_MAG"
cat "$MODIFIED_MAG" >> "$OUTFILE"
done
```

3.11 Assigning Taxonomy to Viruses

Viruses are assigned taxonomy based on genome similarity [96]. To obtain taxonomic assignments for viruses from metagenomes, the tool vConTACT2 [97] can be used. vConTACT2 clusters input viral genomes with reference viral genomes that have taxonomic assignments by building gene-sharing networks [97]. The tool used to identify viral scaffolds earlier, geNomad, provides its own marker-based taxonomic classifications in its output [80]. These may be used instead of vConTACT2, but they were made prior to vMAG binning and may not be complete. Also, a unique feature of vConTACT2 is its generation of viral clusters (VCs). The VCs generated by vConTACT2 roughly represent virus genera or subfamilies [97]. VCs are taxonomically and functionally useful, as the viruses within the same VC are closely related in evolution. Even if a set of viral genomes in a VC was not able to be assigned to a described virus taxon, VC-level analyses can be useful when virus species-level analyses are too high in resolution. vConTACT2 does have a notable drawback; it does not scale so well with large (> 100,000) viral datasets. However, with patience, vConTACT2 can successfully be run on large datasets. vConTACT3 is also in development, which is reported to be more accurate and scalable than vConTACT2.

We will form VCs and attempt to obtain taxonomic assignments for the dereplicated, species-level vMAGs with vConTACT2. vConTACT2 takes `.faa` protein sequences as input. Genes on viral scaffolds were predicted and translated earlier by geNomad, but the geNomad `.faa` outputs do not contain the reformatted headers with assembly IDs and vMAG names. They also include many proteins from viral scaffolds that were filtered during binning and dereplication. For these reasons, genes will be predicted and translated on the file with dereplicated vMAGs, `virus_genomes_dereplicated.fna`, generated above in Section 3.10. The tool prodigal-gv [80] will be used to annotate the vMAGs in this file. Prodigal-gv is based on the robust and accurate gene prediction tool, Prodigal [98], but it is optimized for viruses (particularly "giant viruses" and viruses with stop codon reassignments, but it works well for all kinds of viruses of microbes). The following command is used to run prodigal-gv on the file `virus_genomes_derepliated.fna`:

```
prodigal-gv \
 -i virus_genomes_dereplicated.fna \
 -a virus_genomes_dereplicated.faa \
 -d virus_genomes_dereplicated.ffn \
 -p meta
```

The predicted genes are written to `virus_genomes_dereplicated.ffn` and are translated in the file `virus_genomes_dereplicated.faa`. The flag `-p meta` is necessary to run prodigal-gv in metagenome mode.

Since proteins encoded on multiple scaffolds can belong to a single vMAG, a gene-to-genome table must be made for vConTACT2. This is a three-column table with gene IDs, genome names, and optional keywords in the third. Using the dRep auxiliary script `parse_stb.py` and the Python script `make_g2g.py` shown below, this file can be created:

```
parse_stb.py \
 --reverse \
   --fasta  dereplicated_viruses/dereplicated_genomes/*.
fasta \
 --output virus_genomes_dereplicated
```

```
# make_g2g.py

# Define file paths
faa_file_path = 'virus_genomes_dereplicated.faa'
```

```
stb_file_path = 'virus_genomes_dereplicated.stb'
output_file_path = 'virus_genomes_dereplicated.gene_to_-
genome.csv'

# Load contig to genome mapping from STB file
contig_to_genome = {}
with open(stb_file_path, 'r') as stb_file:
for line in stb_file:
 contig, genome = line.strip().split()
  genome = genome.rsplit('.', 1)[0] # Remove the.fasta
extension
 contig_to_genome[contig] = genome

# Process FAA file and write gene to genome mapping
with open(faa_file_path, 'r') as faa_file, open(output_-
file_path, 'w') as output_file:
 output_file.write('protein_id,contig_id,keywords\n')
for line in faa_file:
if line.startswith('>'):
# Extract gene name
 gene_name = line.split()[0][1:]
# Extract contig name from gene name
 contig_name = '_'.join(gene_name.split('_')[:-1])
# Map contig to genome name
  genome_name = contig_to_genome.get(contig_name, "Un-
known")
 output_file.write(f'{gene_name},{genome_name},\n')

print("Done", output_file_path)
```

Finally, vConTACT2 can be executed using the protein sequences in `virus_genomes_dereplicated.faa` and the newly generated gene-to-genome file `virus_genomes_dereplicated.gene_to_genome.csv`:

```
vcontact2 \
 --raw-proteins virus_genomes_dereplicated.faa \
  --proteins-fp virus_genomes_dereplicated.gene_to_gen-
ome.csv \
 --db ProkaryoticViralRefSeq207-Merged \
 --output-dir vcontact \
 --threads 16
```

When vConTACT2 finished running, there will be two main output files of interest in the folder `vcontact`. The file genome_by_genome_overview.csv is a table with all reference and user-provided viral genomes. The taxonomy for every genome will be given from the Kingdom to the Genus level, if available, as well as information for the VC in which the genomes were clustered. The file c1.ntw is a network table, wherein the first two columns are viral genome names, and the third column is an edge weight that connects them. This weight is associated with the clustering results. Using the genome_by_genome_overview.csv output as a node table and the c1.ntw file as an edge table, a gene-sharing network of viral genome relatedness can be visualized in a desktop program such as Cytoscape [99].

3.12 Predicting the Microbial Hosts of Viruses

Identifying host–virus interactions in microbial communities can offer significant insights on the impact of viruses on microbial community dynamics, the roles of viruses in the biogeochemistry of ecosystems [4, 100], and the flow of genetic information in a community through horizontal gene transfer [10]. In studies with only metagenomic sequence data available, it is difficult to predict the host of viruses, especially highly novel ones. Approaches like single-cell sequencing and Hi-C metagenomics can be used to supplement metagenomic data for predicting the hosts of uncultivated viruses [101, 102], but predicting the microbial hosts of viruses from metagenomic data alone represents a fundamental challenge in microbial and viral ecology. There have been a variety of strategies used to predict the hosts of viruses in metagenomes. Host-based tools detect relationships through sequence alignments between phage and host genomes, which can indicate prophages in host genomes [103] or match phage genomes to host CRISPR spacers [104]. Other, alignment-free host-based tools compare k-mer frequencies between viruses and microbial genomes, which may indicate an adaptation in viral genomes to their host genomes [105, 106]. Phage-based tools, on the other hand, leverage databases of reference viruses with known hosts, offering predictions based on the similarity to the known viruses [107, 108].

The tool iPHoP (integrated Phage-Host Prediction) [109] integrates both phage-based and host-based predictive approaches from multiple tools to gain predictions down to the host genus level for both known and novel phages. It comes with an expansive database of prokaryotic genomes from GTDB [73] to use as potential hosts. If the user has their own set of MAGs, perhaps from the same samples as the query viruses, the MAGs can be combined with iPHoP's database to potentially obtain precise host predictions to the MAGs of interest. This is a powerful feature of iPHoP. To build this custom database, iPHoP requires the MAGs to be placed into a phylogenomic tree with the GTDB genomes using the GTDB-tk de novo workflow. This is different from the GTDB-tk classify workflow used to assign taxonomy to the MAGs, above.

Two trees will need to be built with the medium- and high-quality, dereplicated, *Tara* MAGs, one for archaea and one for bacteria. The command `gtdbtk de_novo_wf` will be used to create these in a new folder, `host_predictions`:

```
mkdir host_predictions
gtdbtk de_novo_wf \
 --archaea \
 --cpus 16 \
 --genome_dir dereplicated_mags/dereplicated_genomes \
 --extension fa \
 --out_dir host_predictions/GTDB-tk \
 --outgroup_taxon p__Nanoarchaeota
```

Here, the flag `--archaea` is used to build a tree with GTDB archaeal sequences, `--genome_dir` provides the path to the dereplicated MAGs, `--extension fa` indicates that the MAG filenames end with the extension `.fa`, `--out_dir host_predictions/GTDB-tk` will put the GTDB-tk results in the folder GTDB-tk beneath the parent folder `host_predictions`, and importantly, `--outgroup_taxon p__Nanoarchaeota` sets the outgroup taxon of the tree to genomes in the phylum Nanoarchaeota. When running `gtdbtk de_novo_wf`, it is important to pick an outgroup phylum that is not represented in the dataset of input MAGs. In several cases, Nanoarchaeota may be a safe choice, but this should not be blindly trusted. If one is studying an environment where it is likely to recover Nanoarchaeota MAGs, a different phylum must be chosen. To be certain of which phyla are present, the previous taxonomy results generated in Section 3.7 can be used, or a simple BLAST of 16S sequences and/or other marker genes may be helpful given their high conservation at the phylum level.

For the bacterial tree, the same command can be executed with some modifications:

```
gtdbtk de_novo_wf \
 --bacteria \
 --cpus 16 \
 --genome_dir dereplicated_mags/dereplicated_genomes \
 --extension fa \
 --out_dir host_predictions/GTDB-tk \
 --outgroup_taxon p__Campylobacterota
```

The results will be deposited in the same folder as the archaeal tree. Here, the outgroup phylum Campylobacterota was chosen, since it was not found in the GTDB taxonomy results from Section 3.7.

Once the two GTDB de novo trees have been built with the dereplicated MAGs, they can be added to the existing iPHoP database with the `iphop add_to_db` command:

```
iphop add_to_db \
 --fna_dir dereplicated_mags/dereplicated_genomes \
 --gtdb_dir host_predictions/GTDB-tk \
 --out_dir host_predictions/iphop_db_with_custom_mags \
 --db_dir /path/to/default/iphop/database
```

The dereplicated MAGs are given to iPHoP with the flag `--fna_dir dereplicated_mags/dereplicated_genomes`, the GTDB-tk results are given with the flag `--gtdb_dir host_predictions/GTDB-tk`, the output custom database location is specified with `--out_dir host_predictions/iphop_db_with_custom_mags`, and the existing default iPHoP database is given with the `--db_dir` flag, which will have to be set to the path of the user's local copy of the iPHoP database.

Since iPHoP expects one virus per scaffold when predicting hosts, the dereplicated vMAGs will have to be linked together using the vRhyme auxiliary script `link_bin_sequences.py`, just as the non-dereplicated vMAGs were when running CheckV in Section 3.9:

```
link_bin_sequences.py \
 -i dereplicated_viruses/dereplicated_genomes \
 -o dereplicated_viruses/dereplicated_genomes_linked

cat dereplicated_viruses/dereplicated_genomes_linked/*.
fasta \
> virus_genomes_dereplicated_linked.fna
```

Here, the `link_bin_sequences.py` script will join together the scaffolds in each dereplicated vMAG and then write each joined vMAG file to a new folder `dereplicated_genomes_linked` within the `dereplicated_viruses` folder. The `cat` command will combine all of the N-linked dereplicated vMAGs into a single file `virus_genomes_dereplicated_linked.fna`, which contains one "scaffold" per vMAG. Now, to predict the microbial hosts of the dereplicated vMAGs, this file can be given to the `iphop predict` command:

```
iphop predict \
  --fa_file virus_genomes_dereplicated_linked.fna \
  --out_dir host_predictions/iphop_output \
  --db_dir host_predictions/iphop_db_with_custom_mags \
  --no_qc \
  --num_threads 16
```

The input vMAG file is specified with the `--fa_file` flag, the custom database with GTDB genomes, and the dereplicated *Tara* MAGs are given with the `--db_dir` flag, and the output folder with predictions `iphop_output` will be created beneath the `host_predictions` folder using the `--out_dir flag`. Importantly, the flag `--no_qc` is included here. By default, iPHoP excludes input viral genomes with many Ns. Since many of the input vMAGs will have long stretches of Ns from linking scaffolds together, this QC step can be skipped with the `--no_qc` option. This flag is not recommended when running `iphop predict` on single-scaffold viral genomes.

iPHoP will produce two main output files, `Host_prediction_to_genus_m90.csv` and `Host_prediction_to_genome_m90.csv`. The former has the host predictions to the host genus level, integrated from the host-based and phage-based prediction tools. The latter has host predictions at the host genome level, integrated from the host-based tools only. There will (likely) be at most one host genus prediction for a given input vMAG in `Host_prediction_to_genus_m90.csv`, but there can be multiple reference or user-provided genomes that are assigned to that genus in `Host_prediction_to_genome_m90.csv`.

To get a simple count of how many vMAG to host genus pairs were predicted, the number of lines in `Host_prediction_to_genus_m90.csv` can be counted with `wc -l`:

```
wc -l host_predictions/iphop_output/Host_prediction_to_genus_m90.csv
3350
```

To get a count of how many vMAG to host genome pairs were predicted from the *Tara* MAGs, the number of occurrences of a regex pattern that follows the MAG naming scheme can be counted using `grep -cP`:

```
grep -cP "ERR\d\d\d\d\d\d__bin_" host_predictions/iphop_output/Host_prediction_to_genomes_m90.csv
56
```

Out of the 156,671 dereplicated vMAGs, host predictions were only obtained for 3350 of them. A fraction this small is quite typical for environmental MAGs and vMAGs, yet the identification of 3350 host predictions indicates a substantial number of interactions. Although the host genera for most of the vMAGs here are unknown, the identified host–virus pairs are very valuable when leveraged in downstream analyses. The output from iPHoP `Host_prediction_to_genus_m90.csv` can be converted into a network diagram and visualized with a program like Cytoscape [99] to qualitatively assess host–virus relationships across the *Tara* samples (Fig. 5).

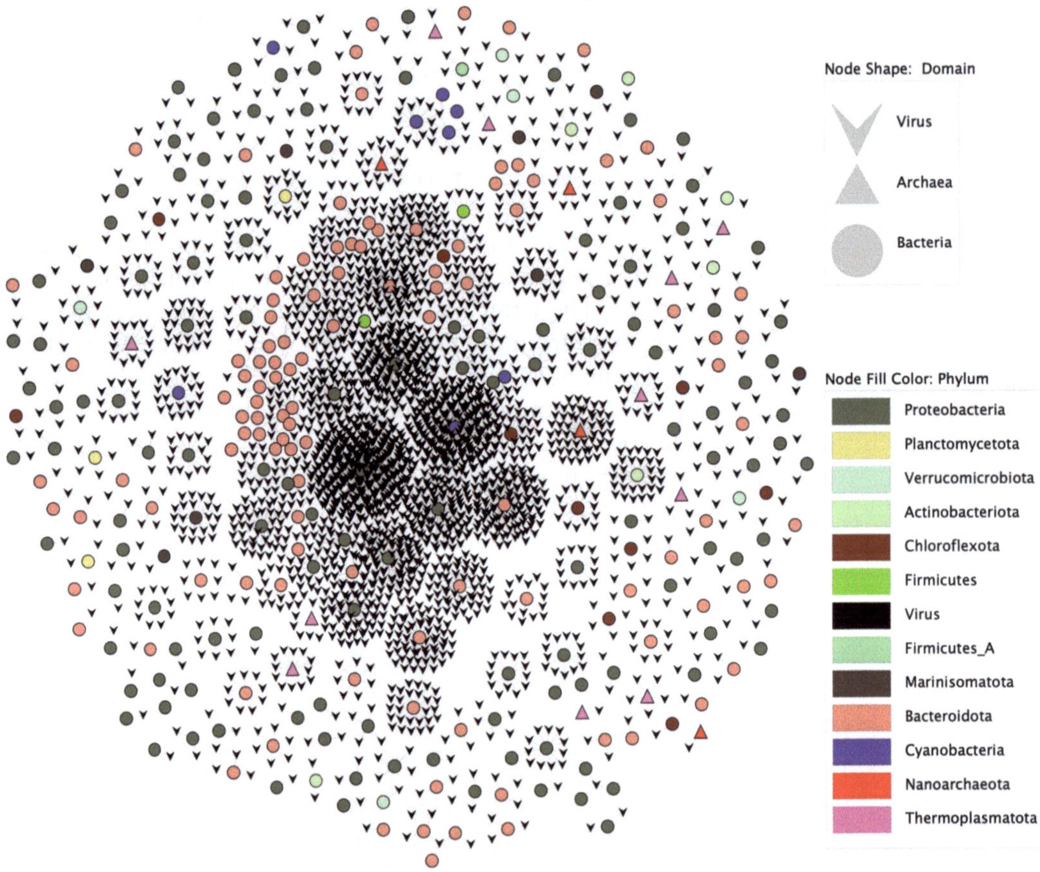

Fig. 5 iPHoP host predictions for *Tara* vMAGs. Host predictions to individual genera are shown, as they appear in the iPHoP output `Host_prediciton_to_genus_m90.csv`. Host phyla that made up < 1% of all predictions were removed. Network built and visualized with Cytoscape [99]

4 Future Directions for Analyses of Microbial and Viral Diversity, Ecology, and Evolution

Section 3 began with a massive volume of bulk, unfiltered, unsorted, and unassembled raw DNA sequence data that, on its own, held very little biological meaning. Its value came from the fact that, as the sections progressed, these sequences were systematically refined and distilled into biologically interpretable information. Now that microbial and viral genomes were reconstructed, filtered, clustered, classified, and associated with each other, we hope that the researcher is equipped with enough data to test their hypotheses. Below, we list a few additional approaches that may be worth pursuing depending on what the researcher is interested in. However, we note that there are virtually endless opportunities that large, multisample metagenomic datasets present, so these approaches are only a small set of the directions one may take with their data.

4.1 Microbial and Viral Species Abundance

For microbial communities, alpha-diversity and beta-diversity are typically estimated using 16S amplicon sequencing and inference workflows [110]. While marker-based sequencing is the best way to estimate the macro-diversity of microbes, it is not possible with viruses since they lack universal marker genes. In metagenomics and viromics, species abundance can be estimated by mapping sequence reads to dereplicated genomes. This can be done for both microbes and viruses [4, 111, 112]. Although filtered DNA sequence reads were already mapped to their respective assemblies in Section 3.5, these mapping data can only be used to analyze the species within each sample, separately. By creating a mapping index from the MAGs/vMAGs dereplicated across all assemblies, all sequence reads can be mapped to the same index. This will allow the assessment of the abundance and distribution of all microbial/viral species in every sample. Mapping reads to the same, multisample genome index can also reveal the presence of species that were not initially detected in a sample due to poor assembly or filtering. Multisample read mapping is therefore a critical step to measure microbial and viral species abundance patterns both within and across communities.

4.2 Comparative Genomics and Microdiversity

Species abundance patterns for microbes and viruses can reveal a lot about the diversity of microbiomes at the community level. To bioinformatically analyze the genetic diversity of microbes and viruses within populations or at the genome level (microdiversity), additional approaches need to be employed. Just as they were useful for estimating species abundance patterns, dereplicated genomes and their read-mapping data can be applied here as well. Single-nucleotide variants, insertions, and deletions can be detected via read alignments in BAM files. Some useful tools that employ a pipeline to analyze microdiversity and calculate evolutionary statistics are inStrain [59] and MetaPop [94]. They are specifically designed to handle large, multisample metagenomic datasets and report many useful statistics on the genetic variation of individual genes and genomes. With these tools, or similar tools, the researcher may see how the genomes of specific microbial or viral taxa in their data vary across their samples. If the researcher has metagenomic data across multiple time points, these tools can be very powerful to test hypotheses on how their microbes and viruses of interest are evolving.

4.3 Functional Analyses of Microbial and Viral Biogeochemistry

Both microbes and viruses have significant impacts on their surrounding environments [79, 113, 114]. Although noncomputational approaches are well suited to test hypotheses on the biogeochemistry of microbiomes, there are several bioinformatic approaches that can assess the contributions of microbes and viruses to the chemistry of their environment. These bioinformatic approaches are particularly useful when studying the biogeochemistry of microbiomes with uncultivated members. METABOLIC [115] is a powerful tool that can take (meta)genomes and predict the metabolic pathways of microbes at the community- and population-level. It can also predict the contribution of individual microbial taxa to specific metabolic pathways in a community. Also, DRAM and DRAM-v [116] are pipelines that identify metabolic genes encoded by microbial and viral genomes from metagenomes. Like METABOLIC, it can assess the presence of key metabolic genes and infer the completeness of specific metabolic pathways in the taxa of interest. Leveraging tools like METABOLIC or DRAM/DRAM-v can be very useful in microbiome studies with biogeochemically or metabolically related hypotheses.

Conflict of Interest Statement

The authors have no conflicts of interest to declare that are relevant to the content of this chapter.

References

1. Handelsman J, Rondon MR, Brady SF, Clardy J, Goodman RM (1998) Molecular biological access to the chemistry of unknown soil microbes: a new frontier for natural products. Chem Biol 5:R245–R249
2. Dick GJ, Andersson AF, Baker BJ, Simmons SL, Thomas BC, Yelton AP, Banfield JF (2009) Community-wide analysis of microbial genome sequence signatures. Genome Biol 10:R85
3. Sharon I, Banfield JF (2013) Genomes from metagenomics. Science 342:1057–1058
4. ter Horst AM, Santos-Medellín C, Sorensen JW, Zinke LA, Wilson RM, Johnston ER, Trubl G, Pett-Ridge J, Blazewicz SJ, Hanson PJ et al (2021) Minnesota peat viromes reveal terrestrial and aquatic niche partitioning for local and global viral populations. Microbiome 9:233

5. Ji M, Fan X, Cornell CR, Zhang Y, Yuan MM, Tian Z, Sun K, Gao R, Liu Y, Zhou J (2023) Tundra soil viruses mediate responses of microbial communities to climate warming. *mBio* 14:e00316-23
6. Jacquet S, Miki T, Noble R, Peduzzi P, Wilhelm S (2010) Viruses in aquatic ecosystems: important advancements of the last 20 years and prospects for the future in the field of microbial oceanography and limnology. Adv Oceanogr Limnol 1:97–141
7. Wilhelm SW, Suttle CA (1999) Viruses and nutrient cycles in the sea. Bioscience 49:781–788
8. Rowan-Nash AD, Korry BJ, Mylonakis E, Belenky P (2019) Cross-domain and viral interactions in the microbiome. Microbiol Mol Biol Rev 83:e00044-18
9. Borodovich T, Shkoporov AN, Ross RP, Hill C (2022) Phage-mediated horizontal gene transfer and its implications for the human gut microbiome. *Gastroenterol Rep (Oxf)* 10: goac012
10. Kosmopoulos JC, Campbell DE, Whitaker RJ, Wilbanks EG (2023) Horizontal gene transfer and CRISPR targeting drive phage-bacterial host interactions and coevolution in “pink berry” marine microbial aggregates. Appl Environ Microbiol 89:7
11. Koskella B, Brockhurst MA (2014) Bacteria–phage coevolution as a driver of ecological and evolutionary processes in microbial communities. FEMS Microbiol Rev 38:916–931
12. Touchon M, Moura de Sousa JA, Rocha EP (2017) Embracing the enemy: the diversification of microbial gene repertoires by phage-mediated horizontal gene transfer. Curr Opin Microbiol 38:66–73
13. Kieft K, Anantharaman K (2022) Virus genomics: what is being overlooked? Curr Opin Virol 53:101200
14. Santos-Medellin C, Zinke LA, ter Horst AM, Gelardi DL, Parikh SJ, Emerson JB (2021) Viromes outperform total metagenomes in revealing the spatiotemporal patterns of agricultural soil viral communities. ISME J 15: 1956–1970
15. Uritskiy GV, DiRuggiero J, Taylor J (2018) MetaWRAP—a flexible pipeline for genome-resolved metagenomic data analysis. Microbiome 6:158
16. Zhou Z, Martin C, Kosmopoulos JC, Anantharaman K (2023) ViWrap: a modular pipeline to identify, bin, classify, and predict viral–host relationships for viruses from metagenomes. *iMeta*. https://doi.org/10.1002/imt2.118
17. Clum A, Huntemann M, Bushnell B, Foster B, Foster B, Roux S, Hajek PP, Varghese N, Mukherjee S, Reddy TBK et al (2021) DOE JGI metagenome workflow. *mSystems* 6:D723–D733
18. Eren AM, Kiefl E, Shaiber A, Veseli I, Miller SE, Schechter MS, Fink I, Pan JN, Yousef M, Fogarty EC et al (2020) Community-led, integrated, reproducible multi-omics with anvi’o. Nat Microbiol 6:3–6
19. Afgan E, Nekrutenko A, Grüning BA, Blankenberg D, Goecks J, Schatz MC, Ostrovsky AE, Mahmoud A, Lonie AJ, Syme A et al (2022) The galaxy platform for accessible, reproducible and collaborative biomedical analyses: 2022 update. Nucleic Acids Res 50:W345–W351
20. Mölder F, Jablonski KP, Letcher B, Hall MB, Tomkins-Tinch CH, Sochat V, Forster J, Lee S, Twardziok SO, Kanitz A et al (2021) Sustainable data analysis with Snakemake. F1000Res 10:33
21. Pesant S, Not F, Picheral M, Kandels-Lewis S, Le Bescot N, Gorsky G, Iudicone D, Karsenti E, Speich S, Troublé R et al (2015) Open science resources for the discovery and analysis of Tara Oceans data. Sci Data 2: 150023
22. Sunagawa S, Acinas SG, Bork P, Bowler C, Acinas SG, Babin M, Bork P, Boss E, Bowler C, Cochrane G et al (2020) Tara Oceans: towards global ocean ecosystems biology. Nat Rev Microbiol 18:428–445
23. Kosmopoulos JC, Klier KM, Langwig MV, Tran PQ, Anantharaman K (2023) Viromes vs. mixed community metagenomes: choice of method dictates interpretation of viral community ecology. *bioRxiv*
24. Knight R, Vrbanac A, Taylor BC, Aksenov A, Callewaert C, Debelius J, Gonzalez A, Kosciolek T, McCall L-I, McDonald D et al (2018) Best practices for analysing microbiomes. Nat Rev Microbiol 16:410–422
25. Liu Y, Zhou J, White KP (2014) RNA-seq differential expression studies: more sequence or more replication? Bioinformatics 30:301–304
26. Gweon HS, Shaw LP, Swann J, De Maio N, AbuOun M, Niehus R, Hubbard ATM, Bowes MJ, Bailey MJ, Peto TEA et al (2019) The impact of sequencing depth on the inferred taxonomic composition and AMR gene content of metagenomic samples. Environ Microbiome 14:7
27. Lynch MDJ, Neufeld JD (2015) Ecology and exploration of the rare biosphere. Nat Rev Microbiol 13:217–229

28. Roux S, Emerson JB, Eloe-Fadrosh EA, Sullivan MB (2017) Benchmarking viromics: an *in silico* evaluation of metagenome-enabled estimates of viral community composition and diversity. PeerJ 5:e3817
29. Zaheer R, Noyes N, Ortega Polo R, Cook SR, Marinier E, Van Domselaar G, Belk KE, Morley PS, McAllister TA (2018) Impact of sequencing depth on the characterization of the microbiome and resistome. Sci Rep 8: 5890
30. Andrews S (2010) FastQC: a quality control tool for high throughput sequence data
31. Bolger AM, Lohse M, Usadel B (2014) Trimmomatic: a flexible trimmer for Illumina sequence data. Bioinformatics 30:2114–2120
32. Bushnell B, Rood J, Singer E (2017) BBMerge—accurate paired shotgun read merging via overlap. PLoS One 12:e0185056
33. Mitchell K, Brito JJ, Mandric I, Wu Q, Knyazev S, Chang S, Martin LS, Karlsberg A, Gerasimov E, Littman R et al (2020) Benchmarking of computational error-correction methods for next-generation sequencing data. Genome Biol 21:71
34. Heydari M, Miclotte G, Demeester P, Van de Peer Y, Fostier J (2017) Evaluation of the impact of Illumina error correction tools on de novo genome assembly. BMC Bioinformatics 18:374
35. Bankevich A, Nurk S, Antipov D, Gurevich AA, Dvorkin M, Kulikov AS, Lesin VM, Nikolenko SI, Pham S, Prjibelski AD et al (2012) SPAdes: a new genome assembly algorithm and its applications to single-cell sequencing. J Comput Biol 19:455–477
36. Koren S, Walenz BP, Berlin K, Miller JR, Bergman NH, Phillippy AM (2017) Canu: scalable and accurate long-read assembly via adaptive *k* -mer weighting and repeat separation. Genome Res 27:722–736
37. Kolmogorov M, Yuan J, Lin Y, Pevzner PA (2019) Assembly of long, error-prone reads using repeat graphs. Nat Biotechnol 37:540–546
38. Li D, Liu C-M, Luo R, Sadakane K, Lam T-W (2015) MEGAHIT: an ultra-fast single-node solution for large and complex metagenomics assembly via succinct *de Bruijn* graph. Bioinformatics 31:1674–1676
39. Vollmers J, Wiegand S, Kaster A-K (2017) Comparing and evaluating metagenome assembly tools from a microbiologist's perspective – Not only size matters! PLoS One 12:e0169662
40. Langmead B, Salzberg SL (2012) Fast gapped-read alignment with Bowtie 2. Nat Methods 9:357–359
41. Li H (2018) Minimap2: pairwise alignment for nucleotide sequences. Bioinformatics 34: 3094–3100
42. Danecek P, Bonfield JK, Liddle J, Marshall J, Ohan V, Pollard MO, Whitwham A, Keane T, McCarthy SA, Davies RM et al (2021) Twelve years of SAMtools and BCFtools. Gigascience 10:giab008
43. Quince C, Walker AW, Simpson JT, Loman NJ, Segata N (2017) Shotgun metagenomics, from sampling to analysis. Nat Biotechnol 35: 833–844
44. Mikheenko A, Saveliev V, Gurevich A (2016) MetaQUAST: evaluation of metagenome assemblies. Bioinformatics 32:1088–1090
45. Alhakami H, Mirebrahim H, Lonardi S (2017) A comparative evaluation of genome assembly reconciliation tools. Genome Biol 18:93
46. Chen L-X, Anantharaman K, Shaiber A, Eren AM, Banfield JF (2020) Accurate and complete genomes from metagenomes. Genome Res 30:315–333
47. Meziti A, Rodriguez-R LM, Hatt JK, Peña-Gonzalez A, Levy K, Konstantinidis KT (2021) The reliability of metagenome-assembled genomes (MAGs) in representing natural populations: insights from comparing MAGs against isolate genomes derived from the same fecal sample. Appl Environ Microbiol 87:e02593-20
48. Lücking D, Mercier C, Alarcón-Schumacher T, Erdmann S (2023) Extracellular vesicles are the main contributor to the non-viral protected extracellular sequence space. ISME Commun 3:112
49. Deatherage BL, Cookson BT (2012) Membrane vesicle release in bacteria, eukaryotes, and archaea: a conserved yet underappreciated aspect of microbial life. Infect Immun 80: 1948–1957
50. Tremblay J, Schreiber L, Greer CW (2022) High-resolution shotgun metagenomics: the more data, the better? Brief Bioinform 23: bbac443
51. Delgado LF, Andersson AF (2022) Evaluating metagenomic assembly approaches for biome-specific gene catalogues. Microbiome 10:72
52. Olm MR, Brown CT, Brooks B, Banfield JF (2017) dRep: a tool for fast and accurate genomic comparisons that enables improved

genome recovery from metagenomes through de-replication. ISME J 11:2864–2868
53. Wu Y-W, Simmons BA, Singer SW (2016) MaxBin 2.0: an automated binning algorithm to recover genomes from multiple metagenomic datasets. Bioinformatics 32:605–607
54. Wu Y-W, Tang Y-H, Tringe SG, Simmons BA, Singer SW (2014) MaxBin: an automated binning method to recover individual genomes from metagenomes using an expectation-maximization algorithm. Microbiome 2:26
55. Kang DD, Li F, Kirton E, Thomas A, Egan R, An H, Wang Z (2019) MetaBAT 2: an adaptive binning algorithm for robust and efficient genome reconstruction from metagenome assemblies. PeerJ 7:e7359
56. Kang DD, Froula J, Egan R, Wang Z (2015) MetaBAT, an efficient tool for accurately reconstructing single genomes from complex microbial communities. PeerJ 3:e1165
57. Alneberg J, Bjarnason BS, de Bruijn I, Schirmer M, Quick J, Ijaz UZ, Lahti L, Loman NJ, Andersson AF, Quince C (2014) Binning metagenomic contigs by coverage and composition. Nat Methods 11:1144–1146
58. Sangwan N, Xia F, Gilbert JA (2016) Recovering complete and draft population genomes from metagenome datasets. Microbiome 4:8
59. Olm MR, Crits-Christoph A, Bouma-Gregson K, Firek BA, Morowitz MJ, Banfield JF (2021) inStrain profiles population microdiversity from metagenomic data and sensitively detects shared microbial strains. Nat Biotechnol 39:727–736
60. Parks DH, Imelfort M, Skennerton CT, Hugenholtz P, Tyson GW (2015) CheckM: assessing the quality of microbial genomes recovered from isolates, single cells, and metagenomes. Genome Res 25:1043–1055
61. Bowers RM, Kyrpides NC, Stepanauskas R, Harmon-Smith M, Doud D, Reddy TBK, Schulz F, Jarett J, Rivers AR, Eloe-Fadrosh EA et al (2017) Minimum information about a single amplified genome (MISAG) and a metagenome-assembled genome (MIMAG) of bacteria and archaea. Nat Biotechnol 35:725–731
62. Tyson GW, Chapman J, Hugenholtz P, Allen EE, Ram RJ, Richardson PM, Solovyev VV, Rubin EM, Rokhsar DS, Banfield JF (2004) Community structure and metabolism through reconstruction of microbial genomes from the environment. Nature 428:37–43
63. Eren AM, Esen ÖC, Quince C, Vineis JH, Morrison HG, Sogin ML, Delmont TO (2015) Anvi'o: an advanced analysis and visualization platform for 'omics data. PeerJ 3: e1319
64. Woese CR, Stackebrandt E, Macke TJ, Fox GE (1985) A phylogenetic definition of the major eubacterial taxa. Syst Appl Microbiol 6: 143–151
65. Schloss PD, Handelsman J (2005) Introducing DOTUR, a computer program for defining operational taxonomic units and estimating species richness. Appl Environ Microbiol 71:1501–1506
66. Guo Y, Zheng W, Rong X, Huang Y (2008) A multilocus phylogeny of the Streptomyces griseus 16S rRNA gene clade: use of multilocus sequence analysis for streptomycete systematics. Int J Syst Evol Microbiol 58:149–159
67. Alanjary M, Steinke K, Ziemert N (2019) AutoMLST: an automated web server for generating multi-locus species trees highlighting natural product potential. Nucleic Acids Res 47:W276–W282
68. Henz SR, Huson DH, Auch AF, Nieselt-Struwe K, Schuster SC (2005) Whole-genome prokaryotic phylogeny. Bioinformatics 21:2329–2335
69. Meier-Kolthoff JP, Auch AF, Klenk H-P, Göker M (2013) Genome sequence-based species delimitation with confidence intervals and improved distance functions. BMC Bioinformatics 14:60
70. Meier-Kolthoff JP, Göker M (2019) TYGS is an automated high-throughput platform for state-of-the-art genome-based taxonomy. Nat Commun 10:2182
71. Chaumeil P-A, Mussig AJ, Hugenholtz P, Parks DH (2022) GTDB-Tk v2: memory friendly classification with the genome taxonomy database. Bioinformatics 38:5315–5316
72. Chaumeil P-A, Mussig AJ, Hugenholtz P, Parks DH (2020) GTDB-Tk: a toolkit to classify genomes with the genome taxonomy database. Bioinformatics 36:1925–1927
73. Parks DH, Chuvochina M, Rinke C, Mussig AJ, Chaumeil P-A, Hugenholtz P (2022) GTDB: an ongoing census of bacterial and archaeal diversity through a phylogenetically consistent, rank normalized and complete genome-based taxonomy. Nucleic Acids Res 50:D785–D794
74. Hendrix RW, Smith MCM, Burns RN, Ford ME, Hatfull GF (1999) Evolutionary relationships among diverse bacteriophages and prophages: all the world's a phage. Proc Natl Acad Sci 96:2192–2197

75. Breitbart M, Bonnain C, Malki K, Sawaya NA (2018) Phage puppet masters of the marine microbial realm. Nat Microbiol 3:754–766
76. Wommack KE, Colwell RR (2000) Virioplankton: viruses in aquatic ecosystems. Microbiol Mol Biol Rev 64:69–114
77. Fuhrman JA (1999) Marine viruses and their biogeochemical and ecological effects. Nature 399:541–548
78. Weitz JS, Wilhelm SW (2012) Ocean viruses and their effects on microbial communities and biogeochemical cycles. *F1000 Biol Rep* 4:17
79. Tran PQ, Anantharaman K (2021) Biogeochemistry goes viral: towards a multifaceted approach to study viruses and biogeochemical cycling. *mSystems* 6:e00046-21
80. Camargo AP, Roux S, Schulz F, Babinski M, Xu Y, Hu B, Chain PSG, Nayfach S, Kyrpides NC (2023) Identification of mobile genetic elements with geNomad. Nat Biotechnol 42: 1303–1312. https://doi.org/10.1038/s41587-023-01953-y
81. Kieft K, Zhou Z, Anantharaman K (2020) VIBRANT: automated recovery, annotation and curation of microbial viruses, and evaluation of viral community function from genomic sequences. Microbiome 8:90
82. Guo J, Bolduc B, Zayed AA, Varsani A, Dominguez-Huerta G, Delmont TO, Pratama AA, Gazitúa MC, Vik D, Sullivan MB et al (2021) VirSorter2: a multi-classifier, expert-guided approach to detect diverse DNA and RNA viruses. Microbiome 9:37
83. Mallawaarachchi V, Roach MJ, Decewicz P, Papudeshi B, Giles SK, Grigson SR, Bouras G, Hesse RD, Inglis LK, Hutton ALK et al (2023) Phables: from fragmented assemblies to high-quality bacteriophage genomes. Bioinformatics 39:btad586
84. Nayfach S, Camargo AP, Schulz F, Eloe-Fadrosh E, Roux S, Kyrpides NC (2021) CheckV assesses the quality and completeness of metagenome-assembled viral genomes. Nat Biotechnol 39:578–585
85. Belcaid M, Bergeron A, Poisson G (2010) Mosaic graphs and comparative genomics in phage communities. J Comput Biol 17:1315–1326
86. Chen L, Banfield JF (2024) COBRA improves the completeness and contiguity of viral genomes assembled from metagenomes. Nat Microbiol 9:737–750
87. Johansen J, Plichta DR, Nissen JN, Jespersen ML, Shah SA, Deng L, Stokholm J, Bisgaard H, Nielsen DS, Sørensen SJ et al (2022) Genome binning of viral entities from bulk metagenomics data. Nat Commun 13:965
88. Kieft K, Adams A, Salamzade R, Kalan L, Anantharaman K (2022) vRhyme enables binning of viral genomes from metagenomes. Nucleic Acids Res 50:e83–e83
89. Du Y, Fuhrman JA, Sun F (2023) ViralCC retrieves complete viral genomes and virus-host pairs from metagenomic Hi-C data. Nat Commun 14:502
90. Al-Shayeb B, Sachdeva R, Chen L-X, Ward F, Munk P, Devoto A, Castelle CJ, Olm MR, Bouma-Gregson K, Amano Y et al (2020) Clades of huge phages from across earth's ecosystems. Nature 578:425–431
91. Ku C (2021) Giant virus-eukaryote interactions as ecological and evolutionary driving forces. *mSystems* 6:e00318-21
92. Schulz F, Roux S, Paez-Espino D, Jungbluth S, Walsh DA, Denef VJ, McMahon KD, Konstantinidis KT, Eloe-Fadrosh EA, Kyrpides NC et al (2020) Giant virus diversity and host interactions through global metagenomics. Nature 578:432–436
93. Evans JT, Denef VJ (2020) To dereplicate or not to dereplicate? *mSphere* 5:e00971-19
94. Gregory AC, Gerhardt K, Zhong Z-P, Bolduc B, Temperton B, Konstantinidis KT, Sullivan MB (2022) MetaPop: a pipeline for macro- and microdiversity analyses and visualization of microbial and viral metagenome-derived populations. Microbiome 10:49
95. Deng L, Ignacio-Espinoza JC, Gregory AC, Poulos BT, Weitz JS, Hugenholtz P, Sullivan MB (2014) Viral tagging reveals discrete populations in Synechococcus viral genome sequence space. Nature 513:242–245
96. Turner D, Kropinski AM, Adriaenssens EM (2021) A roadmap for genome-based phage taxonomy. Viruses 13:506
97. Bin Jang H, Bolduc B, Zablocki O, Kuhn JH, Roux S, Adriaenssens EM, Brister JR, Kropinski AM, Krupovic M, Lavigne R et al (2019) Taxonomic assignment of uncultivated prokaryotic virus genomes is enabled by gene-sharing networks. Nat Biotechnol 37:632–639
98. Hyatt D, Chen G-L, LoCascio PF, Land ML, Larimer FW, Hauser LJ (2010) Prodigal: prokaryotic gene recognition and translation initiation site identification. BMC Bioinformatics 11:119
99. Su G, Morris JH, Demchak B, Bader GD (2014) Biological network exploration with Cytoscape 3. Curr Protoc Bioinformatics 47: 8.13.1–8.13.24

100. Guidi L, Chaffron S, Bittner L, Eveillard D, Larhlimi A, Roux S, Darzi Y, Audic S, Berline L, Brum JR et al (2016) Plankton networks driving carbon export in the oligotrophic ocean. Nature 532:465–470
101. Wu R, Davison MR, Nelson WC, Smith ML, Lipton MS, Jansson JK, McClure RS, McDermott JE, Hofmockel KS (2023) Hi-C metagenome sequencing reveals soil phage–host interactions. Nat Commun 14:7666
102. Labonté JM, Swan BK, Poulos B, Luo H, Koren S, Hallam SJ, Sullivan MB, Woyke T, Eric Wommack K, Stepanauskas R (2015) Single-cell genomics-based analysis of virus–host interactions in marine surface bacterioplankton. ISME J 9:2386–2399
103. Edwards RA, McNair K, Faust K, Raes J, Dutilh BE (2016) Computational approaches to predict bacteriophage–host relationships. FEMS Microbiol Rev 40:258–272
104. Zhang R, Mirdita M, Levy Karin E, Norroy C, Galiez C, Söding J (2021) SpacePHARER: sensitive identification of phages from CRISPR spacers in prokaryotic hosts. Bioinformatics 37:3364–3366
105. Galiez C, Siebert M, Enault F, Vincent J, Söding J (2017) WIsH: who is the host? Predicting prokaryotic hosts from metagenomic phage contigs. Bioinformatics 33:3113–3114
106. Ahlgren NA, Ren J, Lu YY, Fuhrman JA, Sun F (2017) Alignment-free d_2^* oligonucleotide frequency dissimilarity measure improves prediction of hosts from metagenomically-derived viral sequences. Nucleic Acids Res 45:39–53
107. Coutinho FH, Zaragoza-Solas A, López-Pérez M, Barylski J, Zielezinski A, Dutilh BE, Edwards R, Rodriguez-Valera F (2021) RaFAH: host prediction for viruses of bacteria and archaea based on protein content. Patterns 2:100274
108. Villarroel J, Kleinheinz K, Jurtz V, Zschach H, Lund O, Nielsen M, Larsen M (2016) HostPhinder: a phage host prediction tool. Viruses 8:116
109. Roux S, Camargo AP, Coutinho FH, Dabdoub SM, Dutilh BE, Nayfach S, Tritt A (2023) iPHoP: An integrated machine learning framework to maximize host prediction for metagenome-derived viruses of archaea and bacteria. PLoS Biol 21:e3002083
110. Callahan BJ, McMurdie PJ, Rosen MJ, Han AW, Johnson AJA, Holmes SP (2016) DADA2: high-resolution sample inference from Illumina amplicon data. Nat Methods 13:581–583
111. Pérez-Cobas AE, Gomez-Valero L, Buchrieser C (2020) Metagenomic approaches in microbial ecology: an update on whole-genome and marker gene sequencing analyses. Microb Genom 6:mgen000409
112. Santos-Medellín C, Blazewicz SJ, Pett-Ridge-J, Firestone MK, Emerson JB (2023) Viral but not bacterial community successional patterns reflect extreme turnover shortly after rewetting dry soils. Nat Ecol Evol 7:1809–1822. https://doi.org/10.1038/s41559-023-02207-5
113. Zimmerman AE, Howard-Varona C, Needham DM, John SG, Worden AZ, Sullivan MB, Waldbauer JR, Coleman ML (2020) Metabolic and biogeochemical consequences of viral infection in aquatic ecosystems. Nat Rev Microbiol 18:21–34
114. Sokol NW, Slessarev E, Marschmann GL, Nicolas A, Blazewicz SJ, Brodie EL, Firestone MK, Foley MM, Hestrin R, Hungate BA et al (2022) Life and death in the soil microbiome: how ecological processes influence biogeochemistry. Nat Rev Microbiol 20:415–430
115. Zhou Z, Tran PQ, Breister AM, Liu Y, Kieft K, Cowley ES, Karaoz U, Anantharaman K (2022) METABOLIC: high-throughput profiling of microbial genomes for functional traits, metabolism, biogeochemistry, and community-scale functional networks. Microbiome 10:33
116. Shaffer M, Borton MA, McGivern BB, Zayed AA, La Rosa SL, Solden LM, Liu P, Narrowe AB, Rodríguez-Ramos J, Bolduc B et al (2020) DRAM for distilling microbial metabolism to automate the curation of microbiome function. Nucleic Acids Res 48:8883–8900

Chapter 7

Identifying Differential Network Properties and Driver Microbes in Microbial Association Networks Using CompNet and NetShift

Kuntal Kumar Bhusan, Tungadri Bose, and Anirban Dutta

Abstract

This chapter describes CompNet and NetShift, two network analysis tools, which can help in comparison and analysis of microbial association networks. Biological interaction networks are basic representations of entities present in a biological system and their inter-relationships, which lies at the foundation of any modeling approaches. Inter-microbial associations define the community structure of a given microbiome and understanding these interaction networks are pivotal to modeling the microbiome. CompNet helps in the comparison of various network properties across multiple different microbial association networks that may represent communities inhabiting different/contrasting environments. NetShift helps in the identification of key microbes (driver organisms) in the network that can drive changes in community composition and interactions (network topology) characteristic of communities residing in different environments.

Key words Microbiome, Association networks, Microbial co-occurrence, Keystone species, Driver microbes, Community shuffling

1 Introduction

Microorganisms inhabit every niche of the living world. On the one hand, they need to adapt themselves to the host environment and any changes therein. On the other hand, they can influence their neighbors in the ecosystem as well as impact their habitat. It needs to be recognized that individual microorganisms alone may not cause measurable changes. Rather, it is the collective behavior of microbial communities that shapes ecological outcomes. For instance, a burgeoning population of pathogenic cells can disrupt a healthy microbial community and lead to infection, while beneficial interactions can enhance the host's ability to extract energy from food through co-metabolism. By probing these emergent behaviors, researchers can gain insights into the ecosystems, and

Karthik Raman and Gayathri Sambamoorthy (eds.), *Modelling the Microbiome*, Methods in Molecular Biology, vol. 3006,
https://doi.org/10.1007/978-1-0716-5080-6_7,

microbial association networks are essential tools in these studies [1, 2].

Microbial association networks provide invaluable insights into the complex rules governing ecosystems, serving as a primary abstraction for understanding the relationships and interactions among microorganisms [1, 3]. These networks can help researchers measure the stability of these ecosystems and monitor anomalies that may indicate ecological disturbances. These networks can provide preliminary clues toward tracing the paths from association to causation. Network-driven approach also facilitates the inference of missing links and pathways, helping to fill in gaps in our understanding of microbial dynamics and symbiotic interactions. Furthermore, the knowledge gained from these studies has significant applications in bioengineering, where it can be leveraged to create stable and feasible bacterial communities tailored for specific bioprocesses. Ultimately, exploring microbial association networks not only deepens our understanding of ecological systems but also opens new avenues for innovation in health, industry, and environmental sustainability.

While different network properties of a microbial interaction network can reflect aspects like stability/robustness of the community and possible paths for flow of biochemical entities or information, it would be understood that the contribution or importance of different microbes in the network may not be equally significant. It is therefore important to pinpoint organisms that participate in critical interactions to maintain ecological balance [4, 5]. Some of these may as well play a pivotal role in shaping community composition and behavior, acting as catalysts for change within the network [4, 6]. Furthermore, the roles of these microbes can be contextual, i.e., in an alternate environment, or in vicinity of a different set of neighbors, the central role played by a species can vary significantly [7].

Given the importance of understanding the above aspects of microbial association networks, this chapter will focus on techniques that enable the following:

1. Comparison of microbial association networks representing different/ contrasting environments in terms of global and local network properties.
2. Identification of key microbes (driver organisms) in the network that can drive the change in community composition and interactions (network topology) characteristic of different environmental microbiomes.

The abovementioned aspects can help elucidate how shifts in microbial populations can impact ecosystem health, resilience, and response to environmental changes.

The following sections are arranged in order to first exemplify data sources, how to build networks from source data, and then demonstrate how CompNet and NetShift can be used to analyze and compare multiple microbial interaction networks.

2 Materials

2.1 Dataset

2.1.1 Microbiome Sample Source

Study of microbial communities is transforming our understanding of human health, environment, and industrial bioprocesses. When it comes to studying human microbiome, noninvasive sampling of fecal matter is one of the most common ways to probe the microbial community residing in the human gut. Other commonly analyzed human specimens for microbiome analysis include swabs from oral cavity, nasopharynx, skin, vagina, as well as body fluids like blood and urine. Samples obtained through biopsy and endoscopic procedures are also often subjected to microbiome studies. Environmental samples, such as those from soil, and water, are other sources for microbiome studies aimed at understanding biodiversity of soil and water bodies or monitoring impacts of environmental pollution and climate change. One of the most common objectives of a microbiome study is to understand the variation in microbial community in samples derived from case (patients, treatment group, polluted environment, etc.) versus control group (healthy individuals, treatment naïve group, pristine environment, etc.). This section describes the methods/protocols that are available for building microbial association networks from such case versus control microbiome experimental designs/studies.

2.1.2 Microbiome Sequencing Experiments

Experiments aimed at deciphering the microbial content in a collected sample either follow an amplicon sequencing approach or a whole genome shotgun (WGS), i.e., a metagenomic sequencing approach [8–10]. The primary advantage of amplicon sequencing is its ability to identify and quantify microbial taxa present in a sample specimen at a relatively lower cost. Taxonomic marker genes such as 16S rRNA gene (for bacteria) and internal transcribed spacer (ITS) region of the ribosomal cistron (for fungi) are commonly used for PCR amplification, followed by high-throughput DNA sequencing, for this purpose. In contrast, WGS protocols involve deeper sequencing of randomly sheared DNA fragments from the entirety of genetic material present in a sample. Although expensive, this approach can provide a view of the microbial community at a deeper resolution including information on the genes and pathways encoded by different microbial species belonging to the community. For creating microbial association networks from data pertaining to co-occurrence of microbes across different samples, information related to the taxonomic distribution of microbes in each sample is usually adequate [11, 12]. Such data can be generated from both amplicon and metagenomic (WGS) sequencing methods. Some studies may additionally choose to use bipartite graphs to concomitantly discern inter- and intra-relationships between taxonomic groups and their functional potential. However, this chapter will only focus on the aspect of microbial association networks inferred through studying microbial co-occurrence.

In case versus control observational studies, microbiota samples are usually collected from two groups of individuals. For example, a study designed to observe microbiota variations associated with breast cancer would collect sample specimens from a cohort of breast cancer patients and an age-matched healthy group of participants. Alternatively, a study designed to assess the impact of an intervention on the gut microbiota would obtain samples from a group treated with the intervention and another treatment naïve or a placebo-treated group. During experimentation, the DNA is first extracted from the sample specimens using suitable extraction kits or protocols and further processed as per chosen sequencing protocol (metagenomic or amplicon sequencing). For amplicon sequencing, the specific regions of interest (such as 16S rRNA or ITS) from extracted DNA is amplified using PCR. In the WGS approach, sequencing libraries are prepared with random short DNA fragments generated through ultrasonic shearing or enzymatic digestion. Subsequently, high-throughput sequencing is performed using one of the available NGS platforms (e.g., Illumina, Ion-torrent, and PacBio) to obtain raw sequence data in FASTQ format [13, 14].

2.1.3 Sequenced Microbiome Data

The sequenced microbiome data (also called reads) in a FASTQ file is then analyzed to decipher the presence of the microorganisms as well as the function encoded by their genome. The bioinformatic steps involved in this process comprise filtering raw read data to remove low-quality sequences or sequences generated due to experimental artifacts as well as trimming of reads to remove primers and low-quality sequence fragments. The filtered (and trimmed) reads are subsequently used for generating the taxonomy and/or functional counts data. In case of amplicon sequencing experiments, the abundance (count) of each of the taxonomic units (e.g., species, sub-species, or strains) the microbes might belong to is ascertained through Operational Taxonomic Unit (OTU) clustering protocol or Amplicon Sequence Variant (ASV) calling protocol. DADA2, QIIME/QIIME2, and Mothur are popular tools/protocols used for such analysis [15–17]. In case of metagenomic approach, the reads are assembled into longer contiguous sequences (contigs) and binned into genomic fragments/scaffolds or metagenome-assembled genomes (MAGs) or assigned to reference genomes [18, 19]. The taxonomic composition of the microbiota as well as the functional potential of the metagenomes can subsequently be determined by tools/pipelines like MetaPhlAn, HUMAnN, and Kraken [20–22] by comparing against existing taxonomic and functional databases. Taxonomy databases like SILVA and Greengenes and protein functional databases such as KEGG and Pfam are commonly employed for inferring microbial taxonomy composition and its functional potential from microbiome data [23–25]. Count data obtained from each of the multiple samples analyzed in a study/experiment are subsequently combined to generate "microbiome abundance tables" ("taxonomic abundance tables" and/or "functional abundance tables"). These tables can be further analyzed to investigate co-occurrence patterns of microbial taxa/functions and used for building microbial interaction networks. Overlaying the metadata associated with the samples, such as whether the sample was obtained from a patient (case) or a healthy individual (control), can then help in comparing the microbial association networks of the case and control subjects.

2.1.4 Microbiome Data from Curated Repositories

Alternatively, for exploring microbial interactions with data available from existing literature, case versus control microbiome abundance tables can be obtained from repositories like MGnify and curatedMetagenomicData [26, 27]. These repositories make available microbiome abundance tables from previously published studies which can directly be used for microbiome association network studies, minimizing efforts required for generation and curation of metagenomic data from new experiments.

In this chapter, we demonstrate network construction and analysis steps using a case versus control microbiome dataset corresponding to microbiome data from stroke patients (case) and heathy subjects (control) from MGnify (ID MGYS0000050). The data pertaining to the study can be accessed from either by searching the ID in Mgnify or by directly using the following link: https://www.ebi.ac.uk/metagenomics/studies/MGYS0000050. The "taxonomic assignments file" corresponding to each sample is available under the "Analysis summary" section of the above webpage. The examples in this chapter are generated using the outputs corresponding to Pipeline version: 2.0. The taxonomic assignments files of multiple samples can be combined/grouped into "taxonomic abundance tables." Metadata corresponding to the sample IDs can be obtained by opening the corresponding project pages at https://www.ebi.ac.uk/ena/browser/view/PRJEB9365. The exact metadata can be obtained by selecting the "run_accession" and "sample_alias" checkboxes under "show column selection" and downloading the corresponding "TSV" generated by the webservice. Authors of the original study reported changes in the gut microbiome of stroke and healthy patients using differential abundance analysis [28]. This chapter intends to demonstrate a network-based perspective of the changes happening at the microbial community level using specialized techniques on the same case versus control dataset.

2.2 Dependencies and Software Tools

In this chapter, we focus our analysis using two tools, namely, CompNet and NetShift for performing all the network analysis [6, 29]. CompNet is a GUI-based standalone tool for visualization and comparison of multiple interaction networks mainly designed for Linux operating system but also available for Windows as a binary file. On the other hand, NetShift is a web server implementation for "driver analysis" which is platform independent. CompNet has two important dependencies, namely, R and "igraph" [30, 31]. Following the installation of R, the "igraph" package can be easily installed using the following command inside the R terminal:

```
install.packages("igraph")
```

The CompNet tool can be downloaded from the link https://web.rniapps.net/compnet/ which will be available as a zipped folder. On the other hand, NetShift is available as a web server at https://web.rniapps.net/netshift/. In a linux-based system, the CompNet executable file needs to be given appropriate read/write permissions (using the chmod command) to make it functional. Both CompNet and NetShift require edge-lists as input which can be generated from the "taxonomic abundance tables" referred to in the "Dataset" section. Each of the edges in the edge-list indicates an association/interaction between a pair of microbial taxa. As an example, the association or interaction represented by the edges can be as simple as co-occurrence of taxa, inferred by measuring correlation of relative abundance of two taxa across a given number of samples. Numerous tools are available for generation of microbial association network from "taxonomic abundance tables" [32, 33]. For simplicity and ease of installation, we will use a popular R package named "ccrepe" [34] that can be installed as a R Bioconductor package using the following command:

```
if (!require("BiocManager", quietly = TRUE))
 install.packages("BiocManager")
BiocManager::install("ccrepe")
```

Node
In graph theory, nodes (also called vertices) are the building blocks of a network and represent the connected points where information can be

received, created, stored, and transmitted. In the case of a microbial network, nodes represent the resident microbes constituting a microbial community. Often the size of the node is used to represent the abundance of the corresponding microbe in the community or its centrality measure within the network.

Edge *In a graph, an edge is a connection between any two nodes representing the relationship between the nodes. In the case of a microbial network, edges usually represent the association/ correlation/ relationship between two microbes in the microbial community. While undirected edges in a microbial network represent association between a pair of nodes, directed edges might represent causations, such as a symbiotic effect. Additionally edges in some microbial networks may possess edge-weights representing the strength of association/ correlation/ relationship between two microbes in the community.*

3 Methods

3.1 Building Association Networks Using CCREPE

Taxonomic abundance tables, also referred to as microbial abundance matrix, often contain information/count of reads mapped at different levels of taxonomic hierarchy. Before proceeding to association network construction and analysis, it is a usual practice to cumulate abundance/counts at a specific taxonomic level, such as a "family," "genus," and "species". The dataset used for demonstration in this chapter was originally obtained using amplicon sequencing, which allows decent taxonomic classification accuracies till "genus" level. Hence, a "genus" level analysis is suggested for data from similar sources and has been used here for demonstration. Deep WGS-based metagenomic experiments may allow accurate classifications at deeper taxonomic levels like "species" and "strains" and the choice of taxonomic level at which the microbial association network is to be created can be chosen accordingly. It may also be noted that microbiome count matrices can often be "sparse", i.e., may contain a high proportion of zero counts, and hence it is suggested that a prevalence-based filtering be also performed on the data to avoid spurious correlations. For the current example, the data was partitioned into case- and control-specific taxonomic abundance tables using the metadata information, and a filtration step was performed to retain only those genera that were present in at least 30% of samples corresponding to the control- and case-specific abundance tables. This step can be easily performed using an R-script or an MS-Excel-based operation. Once the data is segregated in two tab-delimited tables, say "Stroke_Control.tsv" and "Stroke_Patient.tsv," where rows represent taxonomic groups and columns represent samples, the following R script can be used as a reference to filter the data:

```
filter_cutoff = 30
all_data = c("Stroke_Control.tsv","Stroke_Patient.tsv")
for(this_file in all_data)
{
data=read.table(file = this_file,sep="\t",header=T, row.
names = 1, check.names = F)
data_t = t(data)
TaxaFil <- as.integer((filter_cutoff*dim(data_t)[1])/100)
filtered_data <- data_t[,colSums(data_t != 0) >= TaxaFil]
write.table(t(filtered_data),paste("filtered",this_file,
sep='_'),sep='\t',quote=F)
}
```

The obtained filtered matrices can now be used as an input to the "ccrepe" program to generate microbial association networks. The CCREPE method analyzes relative read counts by permuting and renormalizing them to uncover correlations driven exclusively by compositionality. In other words, it determines statistically significant correlations between pairs of taxa by repeatedly sampling a subset of their abundance values through bootstrapping. This process is applied to all possible pairs within a dataset, and the resulting connections form the microbial association network. For our analysis, we used the default values (in "ccrepe") to assess the statistical significance of a correlation (Spearman's rank correlation in this case) between a given pair of genera using a re-sampling and null distributions of correlation values (using a default of 1000 iterations). An edge is assigned between a pair of genera for every predicted significant positive correlation ($p < 0.005$). A sample R-script to generate the case and control network from the filtered control and case tables ("filtered_Stroke_Control.tsv" and "filtered_Stroke_Patient.tsv") is provided below:

```
all_data = c(
"filtered_Stroke_Control.tsv",
"filtered_Stroke_Patient.tsv"
)

library("ccrepe")
library("igraph")

for(this_file in all_data)
{
data=read.table(file = this_file,sep="\t",header=T, row.
```

```
names = 1, check.names = F)
data = t(data)
data.rowsum <- apply(data, 1, sum)
data.norm <- data/data.rowsum
ccrepe.output <- ccrepe(x = data.norm)
ccrepe.output.sig<-ifelse(ccrepe.output$sim.score> 0 &
ccrepe.output$p.values<0.005,1,0)

ccrepe.output.sig.upper <- graph_from_adjacency_matrix
(ccrepe.output.sig,mode="upper")
ccrepe.output.sig.upper <- simplify(ccrepe.output.sig.
upper)
ccrepe.output.sig.upper.edges<-as_edgelist(ccrepe.out-
put.sig.upper)
write.table(ccrepe.output.sig.upper.edges,paste("edge-
list",this_file,sep='_'),sep='\t',quote=F, row.names = F,
col.names = F)
}
```

The expected output from this step are two edge-lists, one list corresponding to the case (microbial association network for stroke patients, namely, edgelist_filtered_Stroke_Patient.tsv) and another list corresponding to control (microbial association network for healthy controls, namely, edgelist_filtered_Stroke_Control.tsv). An example output of few lines of a tab-delimited edge-list is shown below where each lines represent a pair of microbes, wherein their names are separated by tabs. It may be noted that in certain cases edge-lists can be used to depict directed networks (not applicable in this case study), and for such edge-lists, the tool CompNet considers the microbes listed under the first column as the "source" microbes and those under the second column as the "target" microbes.

```
g__Acidaminococcus	g__Megasphaera
g__Acinetobacter	g__Actinomyces
g__Acinetobacter	g__Rothia
g__Acinetobacter	g__Streptococcus
g__Actinomyces	g__Rothia
g__Actinomyces	g__Streptococcus
g__Akkermansia	g__Oxalobacter
g__Alistipes	g__Bilophila
g__Anaerostipes	g__Bacteroides
```

3.2 Visualization and Comparison Using CompNet

CompNet provides a user-friendly graphical interface for visualization and comparison of multiple interaction networks [29]. The generated case and control networks can be loaded in CompNet using the "File- > Open- > Edge-lists" option, which upon selection of the two networks (edge-lists) from the file location, presents the user with a preview of the edge-list files. As mentioned above, by default, the software assumes the first column as "source" node and the second column as "target" node which are delimited by a tab. Given the current analysis workflow generates a tab-delimited edge-list corresponding to an undirected network, users can proceed without any changes by simply pressing the "ok" button. In the current example workflow, to start with, individual networks are loaded in CompNet and three important graph properties, namely, degree, betweenness, and coreness are computed. This can be achieved by using the option "Find current network statistics" (available at the right bottom corner of the application) followed by clicking the "ok" button. The tool also automatically identifies communities in the loaded network and colors them distinctly (the node names/labels are colored to indicate the community). Figures 1 and 2 show the visual output of the control and case networks, respectively, along with the three key network properties.

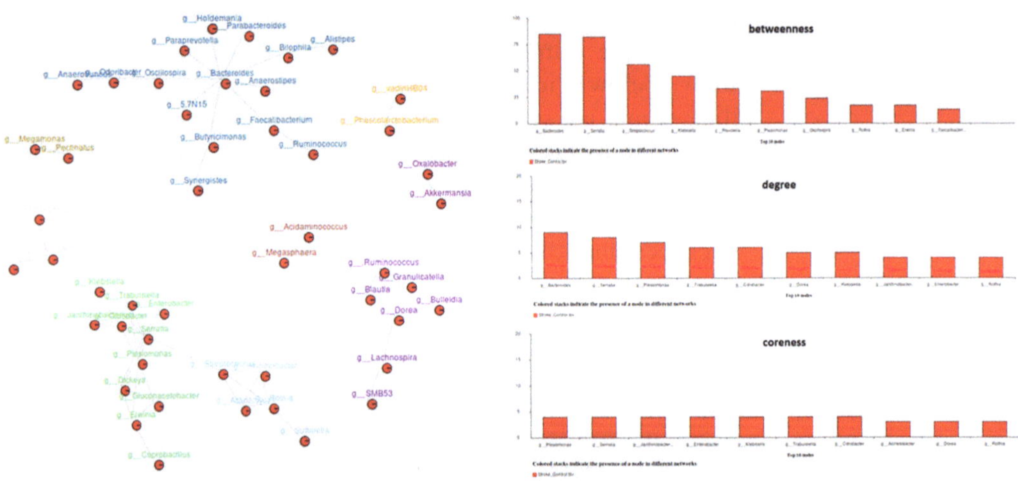

Fig. 1 Microbial association network (constructed from control dataset) visualized on CompNet canvas. Nodes represent different microbial genera and edges represent their mutual associations inferred from co-occurrence of the genera across multiple samples belonging to the control dataset. Densely connected communities are indicated with their constituent node names presented in the same color. To the right of the figure, bar graphs depict betweenness, degree, and coreness measures of select nodes (top 10 nodes for each measure) from the graph

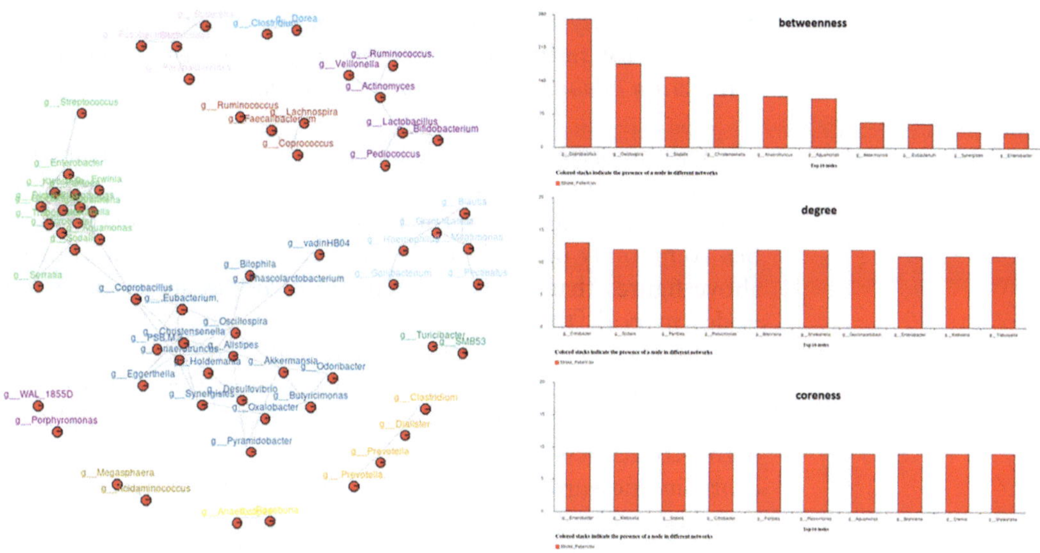

Fig. 2 Microbial association network (constructed from case dataset) visualized on CompNet canvas. Nodes represent different microbial genera and edges represent their mutual associations inferred from co-occurrence of the genera across multiple samples belonging to the case dataset. Densely connected communities are indicated with their constituent node names presented in the same color. To the right of the figure, bar graphs depict betweenness, degree, and coreness measures of select nodes (top 10 nodes for each measure) from the graph

Degree *The number of edges connected to a node in a network is the measure of its degree centrality. A higher degree is characteristic of an important/ central node in a network. In a directed network, it has two sub-types: in-degree (number of edges pointing toward the node) and out-degree (number of edges pointing away from the node). In microbial networks, it is a measure of the number of microbes with which a given microbe co-occurs/ interacts among samples derived from a particular environment.*

Betweenness *For a particular node it is measured by counting the number of shortest paths between all pairs of nodes passing through it. For a microbial network, betweenness centrality is used to identify 'gatekeepers' whose loss disproportionately affects the clustering coefficient of the network (how well the nodes in a network tend to cluster together). Microbes with high betweenness centralities often aid in joining distinct sub-communities or modules. In microbial networks nodes depicting both high degree and betweenness centralities are sometimes considered as 'hub' or 'keystone' microbes.*

Coreness *This centrality measure helps identify tightly interlinked groups within a network. The coreness of a node is defined as the maximum k-core the node participates in a given network, where k-core is a sub-network within the network in which every node has a*

degree centrality of at least 'k'. In microbial community networks coreness is an indicator of the ability of a microbe to form close interacting colonies with other microbes within the community.

The importance of a node in any network is usually defined by its centrality measures. Degree, betweenness, and coreness are most widely used for describing the importance of microbes in a microbial community network [6, 35, 36]. It can be observed from the above figures that a very different set of nodes appear as central nodes in both the networks. While *Bacteroidetes*, *Serratia*, and *Streptococcus* emerge as the top three high betweenness nodes in the control network, the case network harbors the genera *Coprobacillus*, *Oscillospira*, and *Sodalis* as the top three nodes exhibiting high betweenness. A similar pattern is observed for the two other centrality measures, namely, degree and coreness. It may be appreciated here that although all the microbiome samples used to build the compared case and control networks are derived from the same site/environment (human stool samples) and constituted of a similar set of organisms, the way these organisms are interconnected to each other, as evident from their centrality properties, are in a way characteristic to the case and control conditions. CompNet also provides a unique option to load multiple interaction networks at the same time and compare each of them with the union network (union of all the compared networks) as a background. This feature can be utilized while loading the networks using the "File- > Open- > Edge-lists" option. Unlike the earlier case where a single network was chosen, users can select both the case and control edge-lists from the file explorer and CompNet will create and load the union network of the selected files. From a microbiome perspective, such union networks can be viewed to be representing the entire spectrum of associations possible between the microbes constituting case and control networks combined. Biologically, such a union network can provide insights into overall changes in the microbial community structure. Figure 3 provides a screenshot of the union network generated by CompNet, where nodes are represented as "pie-charts," indicating their presence or absence in the compared (case and control) networks.

On comparing betweenness, degree, and coreness measures for the nodes in the union network (using the same "Find current network statistics" option as described earlier), a stacked bar plot of the top 10 nodes for each measure is generated, colored based on their presence in both or distinct networks. A closer look into the union network reveals that a lot of nodes show distinct association with either the case or the control network, which can be identified easily from the nodes showing up in a single color. The "global graph property" comparison feature in CompNet further allows to compare the overall statistics related to the nodes/edges in the two networks as shown in Fig. 4.

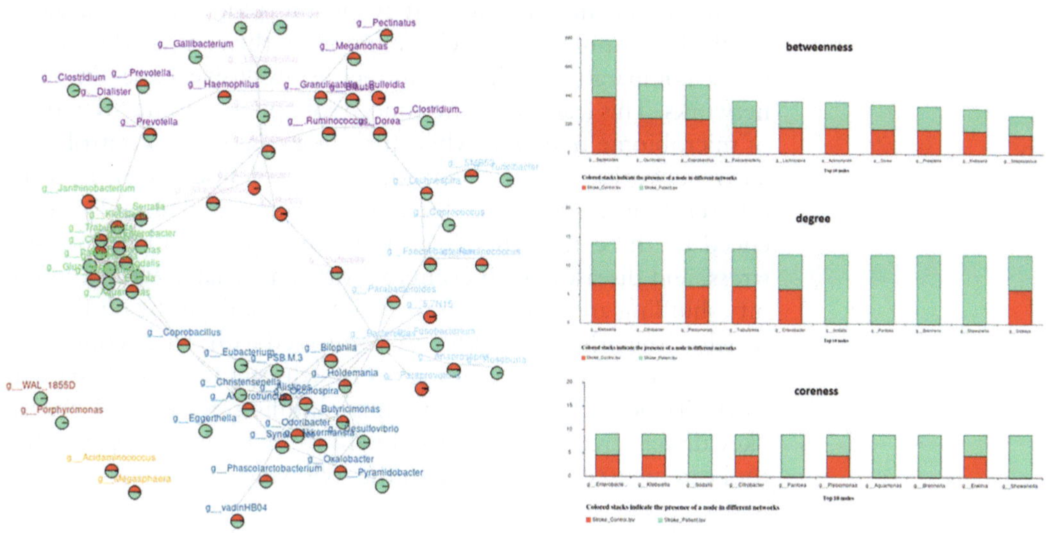

Fig. 3 Union of the case and control microbial association networks visualized on CompNet canvas. Nodes are represented as a pie chart wherein the pie-proportion and colors depict the affiliation of the node to the case network (green), the control network (red), or both. Densely connected communities are indicated with their constituent node names presented in the same color. To the right of the figure, stacked bar graphs depict betweenness, degree, and coreness measures of select nodes (top 10 nodes for each measure) from the union network, while the colors of the stacks indicate the affiliation of the nodes to the case (green) or control (red) networks

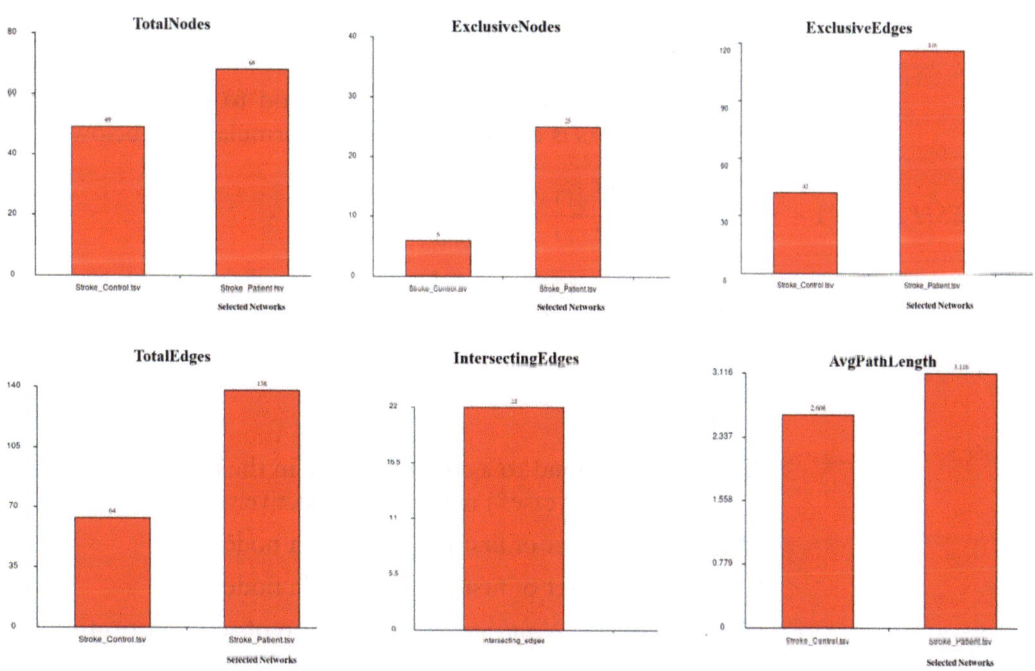

Fig. 4 Comparison of different global network properties between case and control networks by CompNet

It can be noted here that while the plot (Fig. 4) shows a relatively low number of exclusive nodes in the compared networks but a comparatively higher number of exclusive edges in those networks. This leads us to the concept of understanding "rewiring" of microbiome networks while analyzing case versus control data. Two networks having a comparable set of nodes can associate differently in response to different environmental stimuli experienced by the community (e.g., nutrient depletion or abundance, stress, and disease). Such events are often associated with community shuffling and changes in the neighborhoods of the nodes (i.e., which of the other nodes they are connected with) when comparing between a control (healthy) and case (diseased/ perturbed) network. To identify such community shuffling events and predict "key driver" microbes associated with those changes, the NetShift tool can be used as described in the following section.

3.3 Identification of "Driver Nodes" Using NetShift

The underlying principle of NetShift assumes that a microbial node that experiences a high change in its network neighborhood (the way it is connected to other nodes in vicinity) when present in a case associated community with respect to a control associated community, but is also observed to remain as a central node in both communities, can be regarded as a "key driver" node [6]. Although graph properties like degree can quantify the associations of a node, it cannot differentiate the actual change in neighbors. NetShift introduces a novel scoring scheme called Neighbor Shift score or "NESH" to quantify the change in a node's neighborhood from a "control" to a "case" state.

The NESH score for a node common to the "control" and "case" networks is calculated using the formula given below:

$$NESH_{A\rightarrow B} = 1 - \left\{ \left(\frac{[Neighbors]^A \cap [Neighbors]^B}{[Neighbors]^A \cup [Neighbors]^B} \right) - \left(\frac{[Neighbors]^B - [Neighbors]^A}{Max\ degree\ in\ B} + \frac{[Neighbors]^B - [Neighbors]^A}{[Neighbors]^B \cup [Neighbors]^A} \right) \right\}$$

where:

A and B correspond to a common node in the healthy ("control") and disease ("case") networks, respectively.

$[Neighbors]^A$: A set of first neighbors of a node A.

$[Neighbors]^B$: A set of first neighbors of a node B.

$[Neighbors]^A \cap [Neighbors]^B$: Intersection of sets constituting first neighbors of A and B.

$[Neighbors]^A \cup [Neighbors]^B$: Union of sets constituting first neighbors of A and B.

Max degree in B: Maximum degree of any node in the network containing B (i.e., "case" network).

[*Neighbors*]^{B}– [*Neighbors*]^{A}: Relative complement of sets constituting first neighbors of B and A, i.e., objects belonging to set [*Neighbors*]^{B} and not to set [*Neighbors*]^{A}.

The NESH score aims to provide a "directional" score to the association networks from a reference state ("control" or state "A") to a disease state ("case" or state "B"). There are three components of the scoring equation. The first component provides a measure of the extent of neighborhood similarity irrespective of the direction of change (A- > B or B- > A), while the second and third components penalize the score over exclusive enrichment in the set of first neighbors corresponding to the disease/case ("B") set over the healthy/control ("A") set. The second component quantifies the exclusive enrichments over the maximum interacting partners a node can have in the disease (quantified by the maximum degree observed for a node in the case network or "Max degree in B"). The third component quantifies the exclusive set enrichments over the union of the interacting partners of the node in the two compared networks. Overall, the second and third components, taken together, act as a holistic representation of the changes in the disease/case state with respect to the reference/control. It may be noted here that NESH score does not consider the directionality of the edges in a network and is meant to capture the change in associations of key nodes.

The NESH score is designed to be lowest when a node's associated partners are identical in both the "case" and "control" groups. It is higher when only a subset of these partners is present in the case group, and it reaches its highest value when an entirely new set of associated partners appears in the "case" network. Another aspect pertains to quantifying whether there is an increase in the importance of a node in the disease (case) state, which is addressed by NetShift by comparing the weighted change in betweenness as "DelBet" (or delta betweenness). An increase in a node's importance within the disease (case) context serves as an indicator of it playing an important role in disease progression. Hence, when a microbial node in the disease (case) displays a modified set of associations (as identified with a high NESH score) and simultaneously exhibits growing importance (indicated by an increase in weighted betweenness) for the overall network, it is seen as having a vital role in microbial interactions and is classified as a "driver." However, such a comparison requires a comparable set of nodes between the two states (networks) for minimizing bias in scoring. Subsequently, NetShift performs the comparison between the case and control using a subset network formed by the edges connecting the common nodes between the two compared networks. The methodology for obtaining the "subset network" is described in the original NetShift publication [6].

In continuation to the example workflow, the case and control networks (i.e., edgelist_filtered_Stroke_Patient.tsv and edgelist_filtered_Stroke_Control.tsv) can be uploaded to the NetShift web application using the appropriate "choose file" option. Upon clicking the "Run NetShift" button, the tool performs the analysis in the backend and presents the results in a single dashboard output. In addition to displaying some important graph properties of the control and case counterparts of the subset network, the first section of the result presents the Jaccard Edge Index between the two networks.

Jaccard Edge Index *The Jaccard Edge Index (or Jaccard Similarity for edges) between two networks is a measure used to compare the similarity of the edge sets between the two networks. Specifically, it quantifies the proportion of common edges shared by the networks (intersection) relative to the total number of distinct edges (union) across both networks.*

where, A_E *and* B_E *represent the edge set in network A and B respectively.*

In this example, the Jaccard Edge Index was found to be 0.262 indicating a reasonably high amount of network rewiring with each network having a significantly different set of associations when compared against a similar set of common nodes. It is noteworthy to mention that NetShift analysis should only be performed if the compared networks show a low (ideally less than 0.7) Jaccard Edge Index between the compared pair of networks. The second section of the result dashboard displays the NetShift plot with identified drivers. The nodes in the identified common subset network are positioned along the periphery of the circle. "Nodes-labels" for nodes belonging to the same community are assigned similar colors (randomly) and their sizes are proportional to the NESH scores. The nodes themselves are colored red if their betweenness has increased from control to case. Therefore, it may be understood that larger red nodes are particularly significant "drivers." In our dataset, it can be observed that the genera *Oscillospira*, *Anaerotruncus*, and *Alistipes* are identified as the top three drivers by NetShift as shown in Fig. 5.

An accompanying table is generated along with the plot containing the exact details of each node including their count of neighbors in each state, NESH scores, and the DelBet values. In the final section, details of the overall community changes are depicted using a "community shuffling" plot as shown in Fig. 6. This plot is generated by first identifying the communities for the

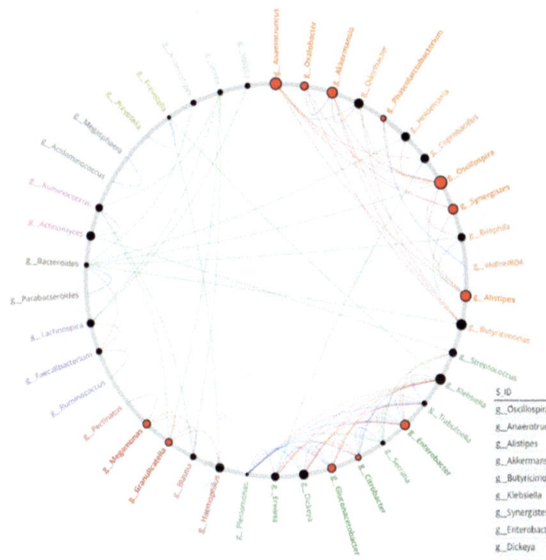

S_ID	n(control)	n(case)	core(case)	Union	Intersect	Exclusive	Jaccard-score	NESH-score	DelBet	COM
g_Oscillospira	2	6	2	8	0	6	0.000	2.500	0.667	1
g_Anaerotruncus	1	4	2	5	0	4	0.000	2.300	0.429	1
g_Alistipes	1	3	2	4	0	3	0.000	2.125	0.067	1
g_Akkermansia	1	5	2	5	1	4	0.200	2.100	0.563	1
g_Butyricimonas	2	3	2	5	0	3	0.000	1.975	-0.158	1
g_Klebsiella	4	7	6	9	2	5	0.222	1.958	-0.540	2
g_Synergistes	1	2	2	3	0	2	0.000	1.917	0.067	1
g_Enterobacter	4	8	6	9	3	5	0.333	1.847	0.212	2
g_Dickeya	4	6	6	8	2	4	0.250	1.750	-0.125	2
g_Odoribacter	2	2	2	4	0	2	0.000	1.750	-0.183	1

Showing 1 to 10 of 38 entries   Previous 1 2 3 4 Next

Fig. 5 Depiction of the NetShift plot with "driver microbes" which were identified to drive the changes in the case-associated community with respect to the control-associated community. Nodes in the identified common subset network are positioned along the periphery of the circle and the node labels for nodes belonging to the same community are assigned similar colors. Node sizes are proportional to the NESH scores and nodes for which the betweenness centrality in the subset network has increased from control to case are colored red. The associated table lists the key characteristics of the node featuring in the subset network and is sorted in descending order of their NESH scores

control and case networks independently, and subsequently assigning community memberships to each node as identified by a distinct color for the node-label and community identifier (e.g., C1 and C2). The nodes corresponding to the control network are placed along the periphery of the left half of the circle. Similarly, the right half of the circle contains the case nodes with an edge connecting the case node to its counterpart control node present in the left half of the circle. Additionally, key node-properties, namely, degree, betweenness, and coreness, for the "subset networks" considered for NetShift analysis, are calculated independently for each network, that can be selected using a button for proportionally depicting the node sizes over the circular plot. This allows one to easily identify the extent of community shuffling (or resilience) between the control and case state. The mapped property (degree in this example) also enables us to visualize important central nodes for each community. In the analyzed dataset, it can be observed that majority of nodes in the "case community C2" show a high degree

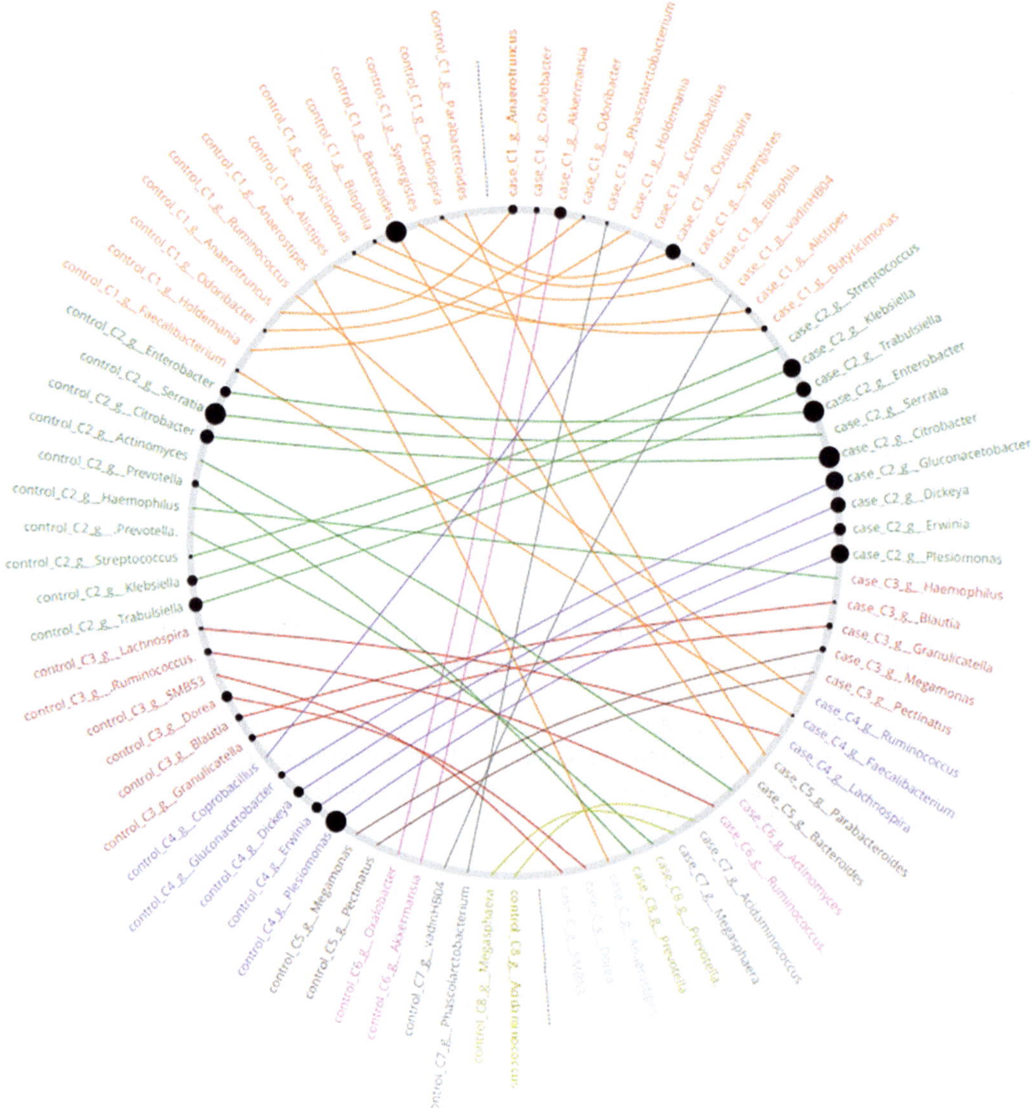

Fig. 6 Community shuffling plot depicting the change in community memberships in the case-associated community with respect to the control-associated community. Nodes corresponding to the control- and case-associated networks are placed along the periphery of the left and right half of the circle, respectively, with the node-label colors indicating community memberships. Node sizes are proportional to the degree centrality of the nodes in the control- and case-associated networks

indicating an important role in the disease state. Additionally, it can be observed that a few nodes which were earlier showing a high degree in the network representing the microbial interactions in the control group, have lost their importance in the disease state.

3.4 Interpretation and Applicability of NetShift

The human body's resident microbiome creates a complex environment that supports the development of the host's immune system and guards against the invasion of pathogens, a concept known as colonization resistance [37]. Despite this, the specific ways in which pathogens manage to bypass these protective mechanisms remain largely unclear [38]. The "keystone pathogen" hypothesis suggests that certain low abundant microbial pathogens can trigger inflammatory diseases by altering or driving a healthy microbiome into a dysbiotic state [39]. When a microbe interacts with a distinct group of partners in a disrupted environment, it plays a significant role in reshaping the entire microbial community. These microbes are often seen as crucial components or potential "drivers" of infection. Previous research has labeled these organisms as "alpha-bugs," highlighting their involvement in the process of carcinogenesis [40]. The identified driver microbes can hence be considered to be associated with disease progression even in the absence of significant differential abundance counts. Alternatively, if the differential abundance profiles also concur, the identified drivers can be further probed for causal association with disease. This use is not limited to understanding human diseases, as it is equally applicable to understanding "community change" events in different ecosystems. For example, one of the studies has used NetShift to identify changes in the interaction patterns of microbes, although they showed similar diversity in the two soil states at different pH levels [41]. "Driver genera" were identified that could produce secondary metabolites to help bacterial community increase its resistance to hostile environments. Similarly, in another study [42], it was seen that bacterial communities modify their ecological interactions to help the plant in moderating the abiotic stress and mutual survival. Taxonomic groups and specific OTUs known to include species relevant for plant protection were identified as potential drivers. These driver taxa appear to be good candidates as indicators not only of environmental perturbations but also of candidate bacteria with plant beneficial traits. Table 1 summarizes some recent studies where NetShift has been a primary component in the analysis workflow, and thus indicates typical use case scenarios.

Conflict of Interest Statement

The authors are salaried employees of TCS Research, the research and development division of Tata Consultancy Services Ltd.

Table 1
Summary of some recent studies which used NetShift as a primary component in the analysis workflow

Area of research	Why NetShift was used	Related references (DOI)
Human health	Identification of key drivers from the microbiome in diseases: Colorectal cancer, thyroid carcinoma, chronic gastritis, diabetes, hypertensive disorders of pregnancy (HDP), dyslipidemia in pregnant women, bone fragility, influenza infection, schizophrenia, SARS-CoV-2, periodontitis, obesity, *Clostridium difficile* infection, autism, sleep disorders, and diarrhea Study the relationship between gut microbiome stability and BMI Uncover predictive factors for restoring gut microbiota affected by antibiotic treatment Analyze drivers in intestinal microbiota linked to allograft stability after lung transplantation Understand the association between oral and gut microbiota in patients with alcohol dependence	10.1002/ijc.33596 10.3389/fmicb.2023.1238829 10.3389/fmicb.2023.1112709 10.3389/fcimb.2023.1123544 10.1128/msphere.00162-21 10.12938/bmfh.2022-047 10.3390/nu15183925 10.1038/s41392-023-01515-3 10.3389/fmicb.2023.1203678
Plant health	Identify drivers in plant disease-associated microbiomes Evaluate the impact of various fertilizers and fungicides on root bacterial communities Study shifts in the soil microbiome for environmentally sustainable potato production Analyze soil bacterial communities in desert crops Investigate the role of key microbial species in pathogen suppression and cropping styles in the rhizosphere Conduct soil microbiome analysis in apple orchards Examine medicinal and commercially important plants affected by diseases (e.g., dwarf bunt of wheat, root rot, wilt disease in tomato, Dollar spot) Evaluate the effect of initial soil microbiome composition and functioning on future plant health Assess the effectiveness of phage combination therapies for improving soil health using microbial drivers Explore manipulation of soil microbial drivers with prebiotic-like compounds	10.1007/s42832-022-0142-6 10.1038/s41467-022-35452-6 10.1038/s43705-022-00094-8 10.1126/sciadv.aaw0759 10.1038/s41587-019-0328-3 10.1016/j.molp.2023.08.004 10.1007/s00248-020-01535-4

(continued)

Table 1 (continued)

Area of research	Why NetShift was used	Related references (DOI)
Environmental health	Study the impact of a dam on ecological communities Study relationship between environmental stress and microbiomes in extreme environments Analyze the effect of oil and gas extraction on marine microbiomes Evaluate the influence of mixed plantations and cropping techniques on soil microbiomes Study methane emissions in rice paddies under different watering regimes Understand spatial characteristics of dissolved organic matter in river microbiomes Study fungal microbiome associated with agricultural waste	10.3390/su15054465 10.1016/j.ecolind.2021.108064 10.1002/ldr.5099 10.1016/j.apsoil.2023.105229 10.1016/j.jenvman.2023.118881 10.1016/j.biortech.2023.128843 10.1016/j.apsoil.2021.104107 10.3390/ani14010097 10.1038/s41598-023-43345-x 10.1016/j.aquaculture.2022.738643
Animal health	Evaluate the importance of microbiome associated with postpartum immunosuppression in periparturient dairy cows having excessive lipolysis Identify driver microbes in phage-linked microbiome of housefly larvae Identify aging-related microbiome drivers in monkeys Study diet-associated microbiome community changes in sea urchins Analyze polymicrobial pathogens in shrimps by identifying driver microbes Analysis of aquaculture-related microbiome communities in fish and shrimp during various infections	10.1186/s40168-023-01492-3 10.3389/fimmu.2022.1101130 10.3389/fimmu.2022.885722 10.1016/j.gpb.2021.09.009 10.1186/s42523-021-00140-1 10.1128/spectrum.00599-21 10.1007/s00248-020-01511-y

References

1. Gallardo-Navarro ́O, Aguilar-Salinas B, Rocha J, Olmedo-Álvarez G (2024) Higher-order interactions and emergent properties of microbial communities: the power of synthetic ecology. Heliyon 10:e33896. https://doi.org/10.1016/j.heliyon.2024.e33896
2. Geesink P, ter Horst J, Ettema TJG (2024) More than the sum of its parts: uncovering emerging effects of microbial interactions in complex communities. FEMS Microbiol Ecol 100:fiae029. https://doi.org/10.1093/femsec/fiae029
3. Röttjers L, Faust K (2018) From hairballs to hypotheses–biological insights from microbial networks. FEMS Microbiol Rev 42:761. https://doi.org/10.1093/femsre/fuy030
4. Berry D, Widder S (2014) Deciphering microbial interactions and detecting keystone species with co-occurrence networks. Front Microbiol 5:219. https://doi.org/10.3389/fmicb.2014.00219
5. Wang Q, Nute M, Treangen TJ (2023) Bakdrive: identifying a minimum set of bacterial

species driving interactions across multiple microbial communities. Bioinformatics 39: i47–i56. https://doi.org/10.1093/bioinformatics/btad236
6. Kuntal BK, Chandrakar P, Sadhu S, Mande SS (2019) "NetShift": a methodology for understanding "driver microbes" from healthy and disease microbiome datasets. ISME J 13:442–454. https://doi.org/10.1038/s41396-018-0291-x
7. Singh R, Dutta A, Bose T, Mande SS (2022) A compendium of predicted growths and derived symbiotic relationships between 803 gut microbes in 13 different diets. Curr Res Microb Sci 3:100127. https://doi.org/10.1016/j.crmicr.2022.100127
8. Bharti R, Grimm DG (2021) Current challenges and best-practice protocols for microbiome analysis. Brief Bioinform 22:178–193. https://doi.org/10.1093/bib/bbz155
9. Cao Y, Fanning S, Proos S et al (2017) A review on the applications of next generation sequencing technologies as applied to food-related microbiome studies. Front Microbiol 8:1829. https://doi.org/10.3389/fmicb.2017.01829
10. Wensel CR, Pluznick JL, Salzberg SL, Sears CL (2022) Next-generation sequencing: insights to advance clinical investigations of the microbiome. J Clin Invest 132:e154944. https://doi.org/10.1172/JCI154944
11. Guseva K, Darcy S, Simon E et al (2022) From diversity to complexity: microbial networks in soils. Soil Biol Biochem 169:108604. https://doi.org/10.1016/j.soilbio.2022.108604
12. Junker R, Valence F, Mistou M-Y et al (2024) Integration of metataxonomic data sets into microbial association networks highlights shared bacterial community dynamics in fermented vegetables. Microbiol Spectrum 12: e00312. https://doi.org/10.1128/spectrum.00312-24
13. Atkinson L, Lee JC, Lennon A et al (2023) Untargeted metagenomics protocol for the diagnosis of infection from CSF and tissue from sterile sites. Heliyon 9:e19854. https://doi.org/10.1016/j.heliyon.2023.e19854
14. Bose T, Wasimuddin, Acharya V et al (2023) A cross-sectional study on the nasopharyngeal microbiota of individuals with SARS-CoV-2 infection across three COVID-19 waves in India. Front Microbiol 14:1238829. https://doi.org/10.3389/fmicb.2023.1238829
15. Bolyen E, Rideout JR, Dillon MR et al (2019) Reproducible, interactive, scalable and extensible microbiome data science using QIIME 2. Nat Biotechnol 37:852–857. https://doi.org/10.1038/s41587-019-0209-9
16. Callahan BJ, McMurdie PJ, Rosen MJ et al (2016) DADA2: high-resolution sample inference from Illumina amplicon data. Nat Methods 13:581–583. https://doi.org/10.1038/nmeth.3869
17. Schloss PD, Westcott SL, Ryabin T et al (2009) Introducing mothur: open-source, platform-independent, community-supported software for describing and comparing microbial communities. Appl Environ Microbiol 75:7537–7541. https://doi.org/10.1128/AEM.01541-09
18. Chen L-X, Anantharaman K, Shaiber A et al (2020) Accurate and complete genomes from metagenomes. Genome Res 30:315–333. https://doi.org/10.1101/gr.258640.119
19. Yang C, Chowdhury D, Zhang Z et al (2021) A review of computational tools for generating metagenome-assembled genomes from metagenomic sequencing data. Comput Struct Biotechnol J 19:6301–6314. https://doi.org/10.1016/j.csbj.2021.11.028
20. Beghini F, McIver LJ, Blanco-Míguez A et al (2021) Integrating taxonomic, functional, and strain-level profiling of diverse microbial communities with bioBakery 3. eLife 10:e65088. https://doi.org/10.7554/eLife.65088
21. Blanco-Míguez A, Beghini F, Cumbo F et al (2023) Extending and improving metagenomic taxonomic profiling with uncharacterized species using MetaPhlAn 4. Nat Biotechnol 41:1633–1644. https://doi.org/10.1038/s41587-023-01688-w
22. Wood DE, Lu J, Langmead B (2019) Improved metagenomic analysis with Kraken 2. Genome Biol 20:257. https://doi.org/10.1186/s13059-019-1891-0
23. Balvočiūtė M, Huson DH (2017) SILVA, RDP, Greengenes, NCBI and OTT — how do these taxonomies compare? BMC Genomics 18:114. https://doi.org/10.1186/s12864-017-3501-4
24. Kanehisa M, Furumichi M, Sato Y et al (2023) KEGG for taxonomy-based analysis of pathways and genomes. Nucleic Acids Res 51: D587–D592. https://doi.org/10.1093/nar/gkac963
25. Mistry J, Chuguransky S, Williams L et al (2021) Pfam: the protein families database in 2021. Nucleic Acids Res 49:D412–D419. https://doi.org/10.1093/nar/gkaa913
26. Pasolli E, Schiffer L, Manghi P et al (2017) Accessible, curated metagenomic data through ExperimentHub. Nat Methods 14:1023–1024. https://doi.org/10.1038/nmeth.4468
27. Richardson L, Allen B, Baldi G et al (2023) MGnify: the microbiome sequence data

analysis resource in 2023. Nucleic Acids Res 51:D753–D759. https://doi.org/10.1093/nar/gkac1080

28. Yin J, Liao S-X, He Y et al (2015) Dysbiosis of gut microbiota with reduced trimethylamine-N-oxide level in patients with large-artery atherosclerotic stroke or transient ischemic attack. J Am Heart Assoc 4:e002699. https://doi.org/10.1161/JAHA.115.002699
29. Kuntal BK, Dutta A, Mande SS (2016) CompNet: a GUI based tool for comparison of multiple biological interaction networks. BMC Bioinformatics 17:185. https://doi.org/10.1186/s12859-016-1013-x
30. R: The R Project for Statistical Computing. https://www.r-project.org/. Accessed 8 Nov 2024
31. The igraph software package for complex network research | BibSonomy. https://www.bibsonomy.org/bibtex/bb49a4a77b42229a427fec316e9fe515. Accessed 8 Nov 2024
32. Cappellato M, Baruzzo G, Patuzzi I, Camillo BD (2021) Modeling microbial community networks: methods and tools. Curr Genomics 22:267. https://doi.org/10.2174/1389202921999200905133146
33. Matchado MS, Lauber M, Reitmeier S et al (2021) Network analysis methods for studying microbial communities: a mini review. Comput Struct Biotechnol J 19:2687–2698. https://doi.org/10.1016/j.csbj.2021.05.001
34. Schwager E, Weingart G, Bielski C (2014) CCREPE: compositionality corrected by pErmutation and rEnormalization. Bioconductor
35. Kajihara KT, Hynson NA (2024) Networks as tools for defining emergent properties of microbiomes and their stability. Microbiome 12:184. https://doi.org/10.1186/s40168-024-01868-z
36. Zamkovaya T, Foster JS, de Crécy-Lagard V, Conesa A (2021) A network approach to elucidate and prioritize microbial dark matter in microbial communities. ISME J 15:228–244. https://doi.org/10.1038/s41396-020-00777-x
37. Buffie CG, Pamer EG (2013) Microbiota-mediated colonization resistance against intestinal pathogens. Nat Rev Immunol 13:790. https://doi.org/10.1038/nri3535
38. Stecher B, Hardt W-D (2008) The role of microbiota in infectious disease. Trends Microbiol 16:107–114. https://doi.org/10.1016/j.tim.2007.12.008
39. Hajishengallis G, Darveau RP, Curtis MA (2012) The keystone-pathogen hypothesis. Nat Rev Microbiol 10:717–725. https://doi.org/10.1038/nrmicro2873
40. Sears CL, Pardoll DM (2011) Perspective: alpha-bugs, their microbial partners, and the link to colon cancer. J Infect Dis 203:306–311. https://doi.org/10.1093/jinfdis/jiq061
41. Galvez G, Ortega J, Fredericksen F et al (2022) Co-occurrence interaction networks of extremophile species living in a copper mining tailing. Front Microbiol 12:791127. https://doi.org/10.3389/fmicb.2021.791127
42. Mandakovic D, Aguado-Norese C, García-Jiménez B et al (2023) Testing the stress gradient hypothesis in soil bacterial communities associated with vegetation belts in the Andean Atacama Desert. Environ Microbiome 18:24. https://doi.org/10.1186/s40793-023-00486-w

Chapter 8

Estimating Effective Pairwise Interactions to Predict the Structures of Microbial Communities (EPICS)

Gayathri Sambamoorthy, Aamir Faisal Ansari, and Narendra M. Dixit

Abstract

The engineering of multispecies microbial communities is important to applications in healthcare, biotechnology, and environmental sustainability. Predicting the structures of such communities requires knowledge of the interactions between the species involved. When high-order interactions are present, bottom-up approaches, which rely on the assembly of all possible subcommunities, become prohibitive because the number of such subcommunities scales exponentially with the number of species. Here, we present an alternative, top-down approach, EPICS, which requires the assembly of subcommunities whose number scales linearly with the number of species, hugely reducing experimental effort. EPICS estimates effective pairwise interactions between species, which subsume high-order interactions, using data from monocultures and leave-one-out subcommunities and predicts community structures. The method is efficient and scalable to large communities.

Key words Microbiome, Community structure, High-order interactions, Population dynamics

1 Introduction

Multispecies microbial communities are key constituents of many natural ecosystems, significantly influencing their functioning [1–3]. They also pervade niches in the human body and affect our health [4–7]. Maintaining stable microbial communities is thus important to environmental sustainability and human health. This requires knowledge of the structures of the communities, defined by the identities and the abundances of the species they contain, and their relationship with function [8]. Modern high-throughput sequencing and data analysis techniques have greatly advanced our abilities to deduce the structures of microbial communities and

Gayathri Sambamoorthy and Aamir Faisal Ansari contributed equally with all other contributors.

Karthik Raman and Gayathri Sambamoorthy (eds.), *Modelling the Microbiome*, Methods in Molecular Biology, vol. 3006, https://doi.org/10.1007/978-1-0716-5080-6_8,

elucidate their structure–function relationships [9–11]. This has enabled, together with the coming-of-age of gene editing and other synthetic biology tools, the design and assembly of synthetic microbial communities, with a variety of applications in biotechnology and healthcare [12–14]. For instance, efforts are on the rise to design communities for targeted therapeutics [15, 16]. The success of these efforts relies on the ability to predict the structures of the communities formed by the assembled species.

The structure of a microbial community is dictated by the interactions between the species it contains [14]. Unravelling the interactions can be challenging, especially when high-order interactions are involved [17–20]. High-order interactions are those that occur when more than two species interact simultaneously. For instance, a ternary interaction occurs when two species secrete different metabolites that react in the extracellular medium to form a product that affects the growth of a third species. Similarly, a quaternary interaction would involve four species, and so on. In an n species community, interactions up to the nth order can exist. The classical approach to estimating these interactions involves assembling subcommunities comprising different species combinations and comparing the species abundances realized. For instance, comparing the abundances of species 1, in its monoculture and in co-culture with species 2 would provide an estimate of the effect of species 2 on species 1. Thus, to estimate pairwise interactions in an n species community, all possible monocultures and two-species cultures must be assembled and compared. For ternary interactions, all possible three-species communities must additionally be assembled. In this way, to estimate the entire interaction map of the community, all possible species combinations must be assembled, thus requiring a total of $(2^n - 1)$ experiments. Knowledge of the interactions can then be used to predict the structure of the community. This approach, classified as a "bottom-up" approach, can be prohibitive with large communities because the number of experiments scales exponentially with the number of species. It has thus only been applied to synthetic communities containing a small number (<10) of species [20, 21].

In this chapter, we present an alternative, "top-down" approach that enables prediction of microbial community structures, with the number of experiments scaling linearly with the number of species [22]. The method estimates effective pairwise interactions between species using data from monocultures and leave-one-out subcommunity cultures and a computational algorithm. The effective pairwise interactions subsume higher-order interactions and enable prediction of the community structure. We call the method EPICS: *E*ffective *P*airwise *I*nteractions for predicting *C*ommunity *S*tructures. The method not only reduces experimental effort, because of its linear scaling with the number of

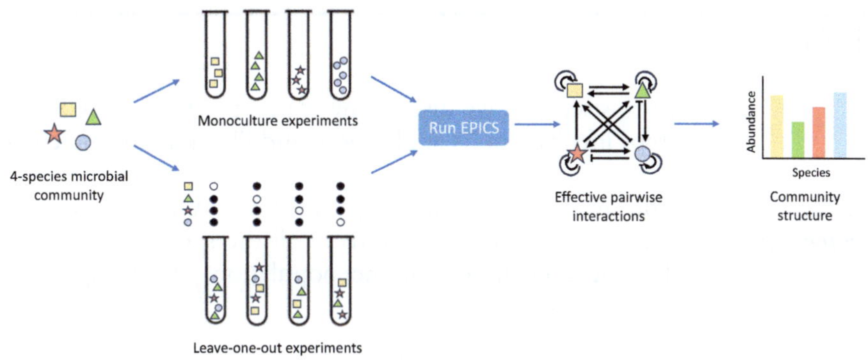

Fig. 1 Schematic of EPICS. Using data from monoculture and leave-one-out cultures, EPICS estimates effective pairwise interactions and predicts community structures

species, but is also expected to become increasingly accurate as the number of species grows. The method is thus efficient and scalable to large communities. A schematic of the method is provided in Fig. 1.

The chapter is organized as follows. In the next Subheading, we briefly describe the monoculture and leave-one-out datasets, which serve as experimental inputs to the algorithm. In Subheading 3, we present the algorithm and illustrate its implementation in MATLAB. We end with a listing of the limitations of the method in Subheading 4. Further details of the method and its application to different communities can be found in the original paper [22].

2 Experimental System and Input Datasets

We consider a system of n different microbial species culturable in a given nutrient environment. We first require that monoculture experiments be performed, where the individual species are cultured separately in the medium for a sufficiently long period (usually >72 h) that the species populations reach steady states. The nutrient may have to be replenished as necessary. The abundances (or populations) of each of the species must then be measured using qPCR or any other technique of choice. This set of experiments yields the n monoculture abundances.

Next, we require that leave-one-out experiments be performed. In each leave-one-out experiment, one of the n species is left out and the remaining species are co-cultured. There are thus n such experiments. The species are to be cultured in the same medium as above and the abundances of each species in each experiment measured. We thus have $n - 1$ species abundances for each of the n experiments, amounting to a total of $n(n - 1)$ measurements. Together with the monoculture data, we thus obtain n^2 measurements from $2n$ experiments. This data forms the input to EPICS (*see* **Note 1**).

3 Methods

In this section, we discuss the concept behind EPICS, outline the steps involved in its application, and illustrate its implementation.

3.1 Assume a Population Dynamics Model

The dynamics of a microbial community is defined by the time evolution of the populations of all the species in the community. The time evolution of the population, N_i, of species i in the community is written as:

$$\frac{dN_i}{dt} = N_i \times f_i(\mathbf{N}) \quad (1)$$

where $f_i(\mathbf{N})$ is the per-capita growth rate of species i. $f_i(\mathbf{N})$ depends on the abundances of all the species in the community, contained in the column vector $\mathbf{N} = (N_1 N_2 \ldots N_n)^T$.

The initial step is to assume a suitable population dynamics model, which defines $f_i(\mathbf{N})$. One typically assumes that the community follows the Generalised Lotka–Volterra (GLV) model with high-order interactions, widely used to describe microbial communities [23]. (EPICS works with other models too [22]; *see* **Note 2**). According to the GLV model,

$$f_i(\mathbf{N}) = r_i + \sum_{j=1}^{n} A_{i\leftarrow j} N_j + \sum_{j=1}^{n} \sum_{k=1}^{n} B_{i\leftarrow(j,k)} N_j N_k + \ldots \quad (2)$$

Here, r_i is the growth rate of species i in the absence of any resource constraints. $A_{i \leftarrow i}$ represents self-interactions between the members of species i, which limit their growth and drive the population in monoculture to its steady state. $A_{i \leftarrow i}$ is thus related to the monoculture abundance measured. $A_{i \leftarrow j}$ determines the effect of species j on species i and is thus a binary, or pairwise, interaction coefficient. $B_{i \leftarrow (j,k)}$ represents the effect of species j and k together on species i, beyond their individual effects, and is thus a ternary interaction coefficient. In the same way, quaternary and other higher-order interaction terms can be written.

3.2 Calculate the Effective Pairwise Interactions

If all the above interaction coefficients are known, Eqs. (1) and (2) can be used to predict the community structure. In general, however, the interaction coefficients are not known. In a bottom-up approach, one must perform the full factorial set of experiments, involving all possible subcommunities, to estimate the coefficients. EPICS progresses by calculating effective pairwise interactions that subsume the higher-order terms. Accordingly, in EPICS,

$$f_i(\mathbf{N}) = r_i + \sum_{j=1}^{n} A_{i\leftarrow j}^{\text{eff}} N_j \quad (3)$$

Here, $A^{\text{eff}}_{i \leftarrow j}$ is the effective pairwise interaction coefficient determining the effect of species j on species i in the presence of all the other species in the culture at their respective abundances. $A^{\text{eff}}_{i \leftarrow j}$ thus subsumes the effect of all the higher-order interactions involved. EPICS thus replaces the problem of estimating the full interaction map of the community with that of estimating the effective pairwise interaction coefficients, $A^{\text{eff}}_{i \leftarrow j}$.

EPICS applies to community structures at steady state. Ignoring the trivial, vanishing states, at steady state, $f_i(\boldsymbol{N}) = r_i + \sum_{j=1}^{n} A^{\text{eff}}_{i \leftarrow j} N_j = 0$. Subsuming r_i into the interaction coefficients (or redefining $A^{\text{eff}}_{i \leftarrow j}/r_i$ as $A^{\text{eff}}_{i \leftarrow j}$), the steady-state condition implies $1 + \sum_{j=1}^{n} A^{\text{eff}}_{i \leftarrow j} N_j = 0$. EPICS assumes that this condition must hold for each species in the monoculture and the leave-one-out experiments. Thus,

$$1 + A^{\text{eff}}_{i \leftarrow i} N_i^m = 0 \tag{4}$$

and

$$1 + \sum_{j \neq h}^{n} A^{\text{eff}-\overline{h}}_{i \leftarrow j} N_j^{\overline{h}} = 0 \tag{5}$$

where $i = 1, 2, \ldots, n$ in Eq. (4) and N_i^m are the individual monoculture abundances. In Eq. (5), $N_i^{\overline{h}}$ denotes the steady-state abundance of species i in the subcommunity without species h (i.e., where species h is left out). Thus, $i \in (1, 2, \ldots, h-1, h+1, \ldots, n)$ and $h = 1, 2, \ldots, n$. Equations (4) and (5) contain n^2 equations in as many unknowns, the unknowns being the coefficients $A^{\text{eff}}_{i \leftarrow \text{i}}$ and $A^{\text{eff}-\overline{h}}_{i \leftarrow j}$. The equations are linear algebraic and are therefore readily inverted and yield unique solutions (*see* **Note 3**).

EPICS then makes the approximation that the interaction coefficients $A^{\text{eff}-\overline{h}}_{i \leftarrow j}$, obtained from the data of the leave-one-out experiments, hold in the all-species community. In other words, $A^{\text{eff}}_{i \leftarrow j} \approx A^{\text{eff}-\overline{h}}_{i \leftarrow j}$. This approximation is reasonable because leaving one species out is unlikely to affect the interactions between the other species substantially, especially when the number of species is large [22]. With this approximation, all the effective pairwise interactions are estimated (*see* **Note 4**).

3.3 Estimate the Species Abundances in the Community

Using the effective pairwise interactions, one can estimate the species abundances N_j in the all-species community at steady-state using

$$1 + \sum_{j=1}^{n} A^{\text{eff}}_{i \leftarrow j} N_j = 0 \tag{6}$$

where $i = 1, 2, \ldots, n$. Equation (6) is a set of n equations in the n unknown abundances, N_j. The equations are again linear algebraic and thus readily inverted to yield N_j and, hence, the community structure.

3.4 Implementation

We have implemented EPICS in MATLAB. The code can be downloaded from the GitHub repository https://github.com/narendradixit/EPICS/tree/v1. The code requires the number of species in the community, monoculture abundances, and leave-one-out abundances as inputs. The effective pairwise interactions and the structure of the community are outputs.

The detailed steps to run the EPICS code are as follows:

1. Download EPICS.m from https://github.com/narendradixit/EPICS/tree/v1.
2. In MATLAB, navigate to the folder containing the EPICS.m file.
3. In the MATLAB command prompt, type EPICS.

```
>> EPICS
```

4. The code starts execution and displays a message to input the number of species in the community.

```
Welcome to EPICS...
Input number of species in the community=
```

5. Once the number of species is input, the code displays a message to input the monoculture abundances.

```
Input monoculture abundances
(Put your entries wrapped in square brackets. Put a
space or a comma after each entry)=
```

6. Next, input the abundances of the leave-one-out species as a matrix, with rows pertaining to each leave-one-out experiment and columns containing the abundance of each species in the corresponding experiments.

```
Input abundances in leave-one-out subcommunities
(Put your entries wrapped in square brackets.
(Rows are subcommunities, columns are species)=
```

7. The code calculates the effective pairwise interactions and estimates the abundances of the species in the community following the algorithm above.

4 Notes

1. The leave-one-out datasets must contain non-zero values of the abundances. The algorithm fails otherwise. In a recent study, we have generalized the algorithm (termed gEPICS) to handle missing species in leave-one-out datasets [24].
2. The implementation assumes that the community follows the GLV model with high-order interactions. If other models are better suited, the implementation must be adapted accordingly [22].
3. EPICS relies on a system of linear equations and thus does not capture multiple steady states that the system may exhibit.
4. The effective pairwise interactions estimated do not subsume the *n*th order interaction. This would be a limitation in the rare cases where an *n*th order interaction exists in a community of *n* species. The highest order of interaction is typically far smaller than *n*.

Acknowledgments

We thank Yugandhar B.S. Reddy and Janhavi Raut for discussions and comments. GS acknowledges support from the IISc-IoE program.

References

1. Baldrian P (2017) Forest microbiome: diversity, complexity and dynamics. FEMS Microbiol Rev 41:109–130
2. Pankhurst C, Ophel-Keller K, Doube B, Gupta V (1996) Biodiversity of soil microbial communities in agricultural systems. Biodivers Conserv 5:197–209
3. Moran MA (2015) The global ocean microbiome. Science 350:aac8455
4. Ursell LK, Metcalf JL, Parfrey LW, Knight R (2012) Defining the human microbiome. Nutr Rev 70:S38–S44
5. Lloyd-Price J, Abu-Ali G, Huttenhower C (2016) The healthy human microbiome. Genome Med 8:1–11
6. Young VB (2017) The role of the microbiome in human health and disease: an introduction for clinicians. BMJ 356:j831
7. Janssens Y, Nielandt J, Bronselaer A, Debunne N, Verbeke F, Wynendaele E, Van Immerseel F, Vandewynckel Y-P, De Tré G, De Spiegeleer B (2018) Disbiome database: linking the microbiome to disease. BMC Microbiol 18:1–6
8. Fuhrman JA (2009) Microbial community structure and its functional implications. Nature 459:193–199
9. Bier RL, Bernhardt ES, Boot CM, Graham EB, Hall EK, Lennon JT, Nemergut DR, Osborne BB, Ruiz-Gonzalez C, Schimel JP (2015) Linking microbial community structure and

microbial processes: an empirical and conceptual overview. FEMS Microbiol Ecol 91:fiv113

10. Di Bella JM, Bao Y, Gloor GB, Burton JP, Reid G (2013) High throughput sequencing methods and analysis for microbiome research. J Microbiol Methods 95:401–414
11. Wei LQ, Cheong IH, Yang GH, Li XG, Kozlakidis Z, Ding L, Liu NN, Wang H (2021) The application of high-throughput technologies for the study of microbiome and cancer. Front Genet 12:699793
12. Tsoi R, Dai Z, You L (2019) Emerging strategies for engineering microbial communities. Biotechnol Adv 37:107372
13. Großkopf T, Soyer OS (2014) Synthetic microbial communities. Curr Opin Microbiol 18:72–77
14. Lawson CE, Harcombe WR, Hatzenpichler R, Lindemann SR, Löffler FE, O'Malley MA, García Martín H, Pfleger BF, Raskin L, Venturelli OS (2019) Common principles and best practices for engineering microbiomes. Nat Rev Microbiol 17:725–741
15. Sharma A, Das P, Buschmann M, Gilbert JA (2020) The future of microbiome-based therapeutics in clinical applications. Clin Pharmacol Ther 107:123–128
16. Sorbara MT, Pamer EG (2022) Microbiome-based therapeutics. Nat Rev Microbiol 20: 365–380
17. Ansari AF, Acharya NI, Kumaran S, Ravindra K, Reddy YB, Dixit NM, Raut J (2019) High-order interactions can eclipse pairwise interactions in shaping the structure of microbial communities. Ind Eng Chem Res 58:23508–23518
18. Grilli J, Barabás G, Michalska-Smith MJ, Allesina S (2017) Higher-order interactions stabilize dynamics in competitive network models. Nature 548:210–213
19. Bairey E, Kelsic ED, Kishony R (2016) High-order species interactions shape ecosystem diversity. Nat Commun 7:12285
20. Gould AL, Zhang V, Lamberti L, Jones EW, Obadia B, Korasidis N, Gavryushkin A, Carlson JM, Beerenwinkel N, Ludington WB (2018) Microbiome interactions shape host fitness. Proc Natl Acad Sci USA 115:E11951–E11960
21. Guo X, Boedicker JQ (2016) The contribution of high-order metabolic interactions to the global activity of a four-species microbial community. PLoS Comput Biol 12:e1005079
22. Ansari AF, Reddy YBS, Raut J, Dixit NM (2021) An efficient and scalable top-down method for predicting structures of microbial communities. Nat Comput Sci 1:619–628
23. Kuntal BK, Gadgil C, Mande SS (2019) Web-gLV: a web based platform for Lotka-Volterra based modeling and simulation of microbial populations. Front Microbiol 10:288
24. Basith AA, Sambamoorthy G, Aditya K, Ansari AF, Cheviti B, Reddy YBS, Raut J, Dixit NM (2026) Generalized EPICS (gEPICS): Handling species extinctions in predicting microbial community structures using effective pairwise interactions Ecol Model 514:111505

Chapter 9

Evaluating Metabolic Support in Pairwise Microbial Communities Using MetQuest

Pratyay Sengupta, Sandhya Vasudevan, and Karthik Raman

Abstract

This chapter delves into MetQuest, a computational tool that enumerates all possible reaction pathways that are feasible in genome-scale metabolic networks given a set of available metabolites. MetQuest accomplishes this by leveraging a guided breadth-first search framework, followed by dynamic programming. It enables us to perform interesting microbial community analyses, such as identifying metabolic exchanges between community members. Here, we specifically demonstrate how MetQuest can be employed to determine the Metabolic Support Index (MSI) for a pairwise microbial community. MSI is a metric that reflects an organism's benefit from the other member in co-culture, shedding light on the dynamics governing community interactions. Here, we present a pipeline to compute the MSI for two-membered communities from their genome-scale metabolic models. We illustrate this approach using the case of *Acinetobacter baumannii* and *Klebsiella pneumoniae*, two bacteria known to cross-feed metabolites. The MetQuest Python protocol used here is available from https://github.com/RamanLab/metquest/tree/master/MSI_Protocol.

Key words Microbial communities, Genome-scale metabolic models, Metabolic cross-feeding, Metabolic Support Index, MetQuest

1 Introduction

Microorganisms are the key drivers of essential functions that underpin ecosystem maintenance, from governing the health of humans to the colossal biogeochemical cycles of our planet [1]. The complex interactions within microbial communities, as well as with their hosts, serve as the cornerstone for the community's functionality, structure, and colonization strategies. Elucidating the dynamics of these interactions will be an invaluable step in characterizing each member's role, unraveling the community's response to different perturbations, and ultimately harnessing their potential for our benefit [2, 3]. In this context, systems and network biology can help decipher microbial associations and provide significant insights into community dynamics [4]. Recent

Karthik Raman and Gayathri Sambamoorthy (eds.), *Modelling the Microbiome*, Methods in Molecular Biology, vol. 3006, https://doi.org/10.1007/978-1-0716-5080-6_9,

developments in genome-scale metabolic reconstructions facilitate the modeling of microbes using their genomes and enable the uncovering of plausible metabolites pivotal for satisfying the metabolic goals of each individual and the community [5]. A systematic approach revealing the metabolic preferences of the community is pivotal in the post-genomic era for re-engineering communities to serve designated functional deliverables.

To address this, MetQuest emerges as a robust and scalable graph-theoretic algorithm that determines a community's metabolic exchanges and interactions [6]. For a given metabolic network, it identifies all possible pathways of a specified size by connecting a given set of seed (input) metabolites to target metabolites. The algorithm employs an exhaustive two-phase framework in which, first, a guided breadth-first search extracts the set of triggerable reactions (whose inputs are available) in the network along with the scope of metabolites they can produce. A novel dynamic programming approach follows this, where pathways of specified sizes (or sub-networks) are assembled, connecting the precursors to the target metabolites. The detailed algorithm can be found in the original publication [6]. Despite the robustness of the MetQuest algorithm, it has inherent limitations, as is the case with any modeling approach (*see* **Note 1**). MetQuest is provided as a Python package and can be accessed at https://github.com/RamanLab/metquest.

Identifying feasible pathways and accessible reactions within genome-scale metabolic networks paves the way for calculating the Metabolic Support Index (MSI), which provides meaningful perspectives into community interaction dynamics. For a pair of microorganisms, MSI quantifies the additional support or benefit one organism derives from its fellow member in co-culture, relative to its monoculture [7]. The metric accurately ascertains which organism gains the maximum advantage from the interaction. It accomplishes this by assessing the number of "stuck" reactions (i.e., reactions that cannot occur due to the unavailability of certain metabolites) that get enabled in the presence of the other organism.

In this chapter, we present a detailed pipeline to compute MSI for two microorganisms utilizing MetQuest. As a case study, we consider *Acinetobacter baumannii* and *Klebsiella pneumoniae*, which are known to share a mutualistic relationship, as reported by Semenec et al. [8]. When co-cultured with *K. pneumoniae*, *A. baumannii* benefits by cross-feeding on the by-products of sugar fermentation that *K. pneumoniae* releases. In return, it confers protection to *K. pneumoniae*, for instance, from antibiotics. Here, we demonstrate the utility of MetQuest in predicting microbial interactions by comparing its results with the findings of Semenec et al.

2 Materials

2.1 Obtaining the Genome-Scale Metabolic Models of the Organisms

We can obtain genome-scale metabolic models from multiple resources with varied curation approaches. One such centralized repository is the BiGG Models Knowledgebase (http://bigg.ucsd.edu), which offers 118 high-quality, manually curated genome-scale metabolic models (GEMs) [9]. BioModels complements this by hosting 6753 patient-specific GEMs derived from The Cancer Genome Atlas (TCGA) datasets alongside 641 manually curated small-scale models from the literature [10]. The Virtual Metabolic Human (VMH) database further extends the scope with 7302 gut microbe models (AGORA2) and a human model (Recon3D) that are made with semi-automated data-driven approaches [11]. Additionally, 41 curated plant models are present in PlantSEED, built through automated annotation and transcriptome data [12].

The manual reconstruction of GEMs is a complex and time-consuming task that necessitates integrating a large number of gene–protein–reaction (GPR) association data and other biological data sources. Over the last decade, multiple advances have been made in an attempt to automate model reconstruction; tools such as AuReMe, CarveMe, *merlin*, ModelSEED, and the RAVEN Toolbox are but a few of the many available tools [13–17]. For example, CarveMe is an automated computational pipeline that transforms manually curated universal reaction datasets into organism-specific GEMs. On the other hand, ModelSEED v.2 optimizes model templates using machine learning, resulting in significantly larger GEMs with more reactions and genes and fewer gap-filled reactions. Furthermore, it assesses core metabolic modules that ensure ATP production and metabolic consistency during expansion or gap-filling and is available through KBase (https://www.kbase.us/). Despite the utility of these tools, draft GEMs often contain inaccuracies, such as improperly constrained reactions or non tailored biomass reactions, requiring gap-filling and improved context specificity [18]. Thus, an amalgamation of gap-filling algorithms and experimental data (e.g., cultivation experiments, transcriptomics, proteomics) is critical for improving draft GEMs, making them biologically accurate and relevant to the application.

In this chapter, we retrieve GEMs for *A. baumannii* and *K. pneumoniae* from the BiGG database. Since the models of the two strains used in the study by Semenec et al. (AB6870155 and KP6870155) are unavailable, we use the closely related strains, AYE and MGH 78578, as substitutes, respectively [19, 20]. We can download the corresponding models in the SBML (Systems Biology Markup Language) format using the command line with the `wget` command.

```
# Collecting the Acinetobacter baumannii AYE (iCN718) and
# Klebsiella pneumoniae subsp. pneumoniae MGH 78578
(iYL1228) models

!wget http://bigg.ucsd.edu/static/models/iCN718.xml -P
"/Path/to/the/directory/1_Models/"
!wget http://bigg.ucsd.edu/static/models/iYL1228.xml -P
"/Path/to/the/directory/1_Models/"
```

2.2 Required Software and Tools

To execute this protocol and calculate the Metabolic Support Index (MSI) between two organisms using MetQuest, Python is required, along with the necessary dependencies listed in the provided YAML file. The YAML file simplifies the setup process by allowing users to create a conda environment preconfigured with all required packages. Users can create the environment by running the following command in the conda shell:

```
conda env create -f metquest-protocol.yml
```

The YAML file, Jupyter Notebook, and additional supporting files are available on the GitHub repository: https://github.com/RamanLab/metquest/tree/master/MSI_Protocol.

3 Methods

This section explains how one can use MetQuest to compute two organisms' Metabolic Support Index (MSI) when they co-occur in a community. For obtaining MSI, the individual metabolic graphs of both microorganisms and their pair graphs, along with seed metabolites in a particular format, are required. Below, we discuss the entire pipeline from acquiring the models of the two organisms to computing their MSI values.

3.1 Evaluating the Quality of the Genome-Scale Metabolic Models

3.1.1 Assessment of Genomic Similarity

It is necessary to ensure that the downloaded GEMs accurately represent the organisms used in the reference study. To accomplish this, we compare the genetic similarity of the strains used in the reference study (AB6870155 and KP6870155) to the strains employed to construct the GEMs (AYE and MGH 78578). The Average Nucleotide Identity (ANI), a metric that evaluates the mean nucleotide identity of orthologous gene pairs shared by two microbial genomes, is estimated with OrthoANI v.0.6.0 [21].

Table 1
Metadata of the GEMs used in this study and genomic comparison with the strains from the study by Semenec et al.

Organism	BiGG model ID	Strain used for GEM reconstruction	# Metabolites	# Reactions	Strain used in the Semenec et al. study	ANI (%)
Acinetobacter baumannii	iCN718	AYE	888	1015	AB6870155	99.8
Klebsiella pneumoniae	iYL1228	MGH 78578	1658	2262	KP6870155	99.0

```
# Calculating pairwise average nucleotide identity (ANI)
# Between Acinetobacter baumannii AYE (GCF_000069245.1)
# and AB6870155 (GCF_000794125.2)
!orthoani -q /Path/to/the/directory/2_Genomes/
GCF_000069245.1.fna -r
/Path/to/the/directory/2_Genomes/GCF_000794125.2.fna -j 12

# Calculating pairwise average nucleotide identity (ANI)
# Between Klebsiella pneumoniae subsp. pneumoniae MGH
78578
# (GCF_000016305.1) and KP6870155 (GCF_027126715.1)
!orthoani -q /Path/to/the/directory/2_Genomes/
GCF_000016305.1.fna -r
/Path/to/the/directory/2_Genomes/GCF_027126715.1.fna -j 12

99.80144101731602
99.0398276003276
```

The ANI results, along with other genome and model-related metadata, are provided in Table 1. ANI scores of over 99% suggest a high level of genetic similarity, indicating that the strains are nearly identical and that the models can serve as reliable alternatives. Such high ANI values suggest that the metabolic capabilities captured in the downloaded GEMs are quite similar to those of the reference strains, allowing for useful downstream investigations.

3.1.2 Quality Evaluation of GEMs Using MEMOTE

Next, we assess the quality of the models employing MEMOTE, an open-source software tool for GEM quality control and benchmarking [22]. It ensures the accuracy of the metabolic models through a structured approach that includes standardized tests and the adoption of community-driven practices. We can obtain

MEMOTE reports for the models via the command line by using MEMOTE v.0.17.0. The overall scores of models iCN718 and iYL1228 are 70% and 84%, respectively.

```
# Running MEMOTE on the GEMs
# Model: iCN718
!memote report snapshot --filename
"/Path/to/the/directory/3_MEMOTE/iCN718.html"
/Path/to/the/directory/1_Models/iCN718.xml

# Model: iYL1228
!memote report snapshot --filename
"/Path/to/the/directory/3_MEMOTE/iYL1228.html"
/Path/to/the/directory/1_Models/iYL1228.xml
```

3.1.3 Validation of SBML Files

To validate the format of the GEMs, we use the `validate_SBML` function. This function accepts the path to the directory containing the models and evaluates all the GEM files in that location. The function uses the `SBMLReader` module from the `libsbml` package (https://github.com/sbmlteam/libsbml) to identify structural inconsistencies in the SBML files. Below is an example of the usage of the function `validate_SBML`:

```
# Validating the SBML files
validate_SBML("/Path/to/the/directory/1_Models/")
```

The output of this validation step should confirm whether each file is in a valid SBML format, as shown below:

```
[SUCCESS] 'iCN718.xml' is a valid SBML file.
[SUCCESS] 'iYL1228.xml' is a valid SBML file.
```

3.2 Preparing the Medium File to Simulate the GEMs

MetQuest determines the Metabolic Support Index (MSI) by categorizing all model reactions into defined pathways. This simulation requires a biologically relevant set of initial compounds known as seed metabolites. These seed metabolites can be obtained either from the constituents of a well-defined medium, such as Luria-Bertani (also known as Lysogeny Broth or LB medium), or from the composition of a community-specific minimal growth medium.

3.2.1 Estimating the Minimal Medium

We can use the `min_medium` function to compute the minimum medium necessary for a list of GEMs. This function accepts two parameters:

(i) `path_to_models`, which specifies the directory containing the model files

(ii) `path_to_medium_file`, which indicates the file path of the output medium file

First, it reads the models using `read_sbml_model` from `COBRApy` v.0.29.1, which loads the models from the given directory (`path_to_models`). Subsequently, it employs the `minimal_medium` function from `COBRApy` to extract the minimal medium for each model. By default, this function constrains the growth rate to 0.1 h^{-1} and minimizes the number of metabolite exchanges to achieve this growth. The function can be called as follows:

```
# Calculating the minimal medium
path_to_models = "/Path/to/the/directory/1_Models/"
path_to_medium_file =
"/Path/to/the/directory/5_Int_dir/curated_minimal_medium.
txt"
min_medium(path_to_models, path_to_medium_file)
```

The function returns the minimal set of metabolites required for the growth of all organisms in the community as a text file (`path_to_medium_file`).

```
[SUCCESS] Minimal medium calculated for 'iCN718.xml'.
[SUCCESS] Minimal medium calculated for 'iYL1228.xml'.
[SUCCESS] Minimal medium metabolites saved to
'/Path/to/the/directory/5_Int_dir/curated_minimal_medium.
txt'.
```

3.2.2 Utilizing Defined Medium

Several defined media are available for simulating community models. In this protocol, we utilize Lysogeny Broth (LB) under aerobic and anaerobic conditions and the M9 medium (with cofactors). We can source the composition of these media from the media database provided in the CarveMe repository (https://github.com/cdanielmachado/carveme/blob/master/carveme/data/input/media_db.tsv). The medium files (in the required format) are also available in the GitHub repository: https://github.com/RamanLab/metquest/tree/master/MSI_Protocol/4_Medium.

3.2.3 Formatting the Medium File

To generate the medium file in the required format, we first use the `get_model_info` function to obtain relevant information about the models. This is later utilized to create the seed metabolites file in the desired format. The input to this function is

(i) `model_dir`, which specifies the path to the folder containing the models

The function returns several outputs:

(i) `model_id_dict`, a dictionary structure mapping the model files with the corresponding model IDs
(ii) `model_compartments_dict`, a dictionary comprising the compartment details of the models (e.g., "c" for "cytosol", "e" for "extracellular space", "p" for "periplasm", "r" for "endoplasmic reticulum")
(iii) `model_e0_dict`, a dictionary mapping the model IDs to exchange reactions
(iv) `model_id_list`, a list of the model IDs
(v) `model_list`, a list of the file names of the models

```
# Getting the details of the models
model_id_dict,  model_compartments_dict,  model_e0_dict,
model_id_list,
model_list = get_model_info(path_to_models)
print(model_id_list)
print(model_compartments_dict)
```

The below output shows the `model_id_list` and `model_compartments_dict` variables:

```
['iCN718', 'iYL1228']
{'iCN718': ['c', 'e', 'p', 'r'], 'iYL1228': ['p', 'c', 'e']}
```

Next, we can use the format_seed function to create the formatted seed metabolite file. The function accepts several arguments:

(i) `seed_dir`, the directory for the minimal medium file
(ii) `site_name`, a base name for the formatted text file
(iii) `model_compartments_dict` and `model_id_list`, obtained from the `get_model_info` function
(iv) `path_to_seed_metabolites_file`, the destination to write the formatted seed metabolites

```
# Formatting the seed metabolites
seed_dir  =  "/Path/to/the/directory/4_Medium/1_Minimal_-
medium/"
site_name = "curated_minimal_medium"
```

```
path_to_seed_metabolites_file=
"/Path/to/the/directory/5_Int_dir/curated_minimal_me-
dium_formatted.txt"
```

The output of this function is a text file with the formatted seed metabolites.

```
[SUCCESS] Formatted seed metabolites saved to
'/Path/to/the/directory/5_Int_dir/curated_minimal_me-
dium_formatted.txt'.
```

If additional cofactors need to be included in the formatted seed metabolites file, we can add them by creating a `cofactors.txt` file containing the necessary cofactors, saved in the `seed_dir`. These cofactors will then be included with the other metabolites from the minimal medium to generate the final formatted seed metabolites file. In this study, we incorporate a set of essential cofactors to calculate the tailored minimal medium. The list of cofactors was previously curated and published by Kumar et al. [23].

3.3 Construction of Single and Pair Graphs

In this step, we create the single and pair graphs of the organisms. A GEM can be represented as a bipartite graph, *G*(*M*, *R*, *E*), comprising two types of nodes: metabolites (*M*) and reactions (*R*), connected by a set of directed edges (*E*), representing the relationship between the nodes and depicting the flow of metabolites through reactions. After obtaining individual graphs for each organism, these graphs can be combined using a common extracellular space (*see* Subheading 3.3.2) and overlapping exchange reactions in both models to build a pair graph [6].

3.3.1 Generating Single Graphs

We use the `get_individual_graph function` to create individual graphs for each organism. It requires the following parameters:

(i) `model_dir`, the path to the model folder

(ii) `intermediate_dir`, the directory where the graph and namemap pickle files will be saved

(iii) `model_id_dict`, `model_compartments_dict`, and `model_list` inputs derived from the `get_model_info` function

```
# Getting the individual graphs
model_dir = "/Path/to/the/directory/1_Models/"
intermediate_dir = "/Path/to/the/directory/5_Int_dir/"
```

```
individual_graph_dict, individual_namemap_dict =
get_individual_graph(model_dir, model_id_dict, model_-
compartments_dict, model_list, intermediate_dir)
```

The outputs of this function include pickle files representing the GEM graph and their corresponding `namemap` variables, which are stored in the specified intermediate directory.

```
[INFO] Creating individual graphs...
Filenames ['iCN718.xml', 'iYL1228.xml']
Number of edges in graph 6194
Number of nodes in graph 2467
Graph and namemap saved for file(s) in /Path/to/the/direc-
tory/5_Int_dir/
Number of edges in graph 11214
Number of nodes in graph 5039
Graph and namemap saved for file(s) in /Path/to/the/direc-
tory/5_Int_dir/
[INFO] Saved graph and namemap for all the models in /
Path/to/the/directory/5_Int_dir/
```

The `get_individual_graph` function intrinsically uses the Python package, `MetQuest`, to create bipartite graphs represented as dictionaries with nodes and directed edges. While nodes are stored with their labels as keys and attributes as values, the directed edges are stored using the source nodes as keys and dictionaries of target nodes as values with attributes such as edge weights. These graphs are then serialized (pickled) into files for storage and later access.

The `namemap` variable for each model maps node labels to biologically relevant identifiers. For example, in *Klebsiella pneumoniae*, the model ID: iYL1228 may have a node label such as Org_iYL1228_IR1, which corresponds to the enzyme 2-acyl-glycerophospho-ethanolamine acyltransferase, associated with the biological identifier, 2AGPEAT120.

3.3.2 Generating the Pair Graph

Next, we employ the `get_pair_graph` function to create the pair graph of the metabolic networks of the two organisms. Pair graphs are formed by joining the individual graphs of two organisms through their shared extracellular space and overlapping exchange reactions. This function requires the same set of arguments as `get_individual_graph`.

```
# Getting the pair graphs
model_dir = "/Path/to/the/directory/1_Models/"
intermediate_dir = "/Path/to/the/directory/5_Int_dir/"

pair_graph_dict, pair_namemap_dict = get_pair_graph(mod-
el_dir,
model_id_dict, model_compartments_dict, model_list, in-
termediate_dir)
```

This function returns pickle files representing the integrated GEM graph of both organisms and the `namemap` variable stored in the specified intermediate directory.

```
[INFO] Creating pairwise graphs...
Filenames ['iCN718.xml', 'iYL1228.xml']
Number of edges in graph 17408
Number of nodes in graph 7400
Graph and namemap saved for file(s) in /Path/to/the/direc-
tory/5_Int_dir/
[INFO] Saved graph and namemap for pair: iCN718_iYL1228
```

3.4 Calculating the Metabolic Support Index (MSI)

This pipeline concludes by calculating the MSI for the given pair of organisms. The index quantifies the metabolic support offered by one microorganism to another within a shared environment. We perform this calculation using the `find_metabolic_suppor-t_index` function that consecutively analyzes the individual metabolic graphs of each microorganism and their combined pair graph to calculate MSI. It determines how many stuck reactions exist in each individual graph and how many of those reactions are relieved when the two species merge into a single shared metabolic network. The MSI for each organism is calculated using the following formula:

$$\mathrm{MSI}\ (A|AB) = 1 - \frac{n_{\mathrm{stuck}\ A|AB}}{n_{\mathrm{stuck}\ A}}$$

The MSI computation takes into account the number of stuck reactions in the individual network of organism A ($n_{\mathrm{stuck}\ A}$) and the number of stuck reactions in the paired community structure where A interacts with B ($n_{\mathrm{stuck}\ A\ |\ AB}$). The MSI value reflects to what extent the metabolic constraints in an individual organism are relieved by the existence of the same organism in a community, and eventually quantifies the metabolic benefit gained from the interaction. Further details about the implementation and specifics of the MSI calculation are available in the original publication [7]. An example calculation of MSI is illustrated in **Note 2**.

The find_metabolic_support_index function requires several arguments:

(i) `intermediate_dir`, the path to the folder containing all the necessary data files, including the single and pair graphs created in the previous steps, as well as the formatted seed metabolites file

(ii) names of the saved graph pickle files, which include the combined (`graph_name_combined`) and individual (`graph_name_firstorg`, `graph_name_secondorg`) graph files

(iii) `seed_metabolite_fname`, the name of the formatted seed metabolites file

```
# Calculating the Metabolic Support Index (MSI)
graph_name_combined = "iCN718_iYL1228_.gpickle"
graph_name_firstorg = "iCN718_.gpickle"
graph_name_secondorg = "iYL1228_.gpickle"
seed_metabolite_fname    =    "curated_minimal_medium_for-
matted.txt"

find_metabolic_support_index(intermediate_dir, graph_na-
me_combined,
graph_name_firstorg, graph_name_secondorg, seed_metabo-
lite_fname)
```

This function generates a CSV file within a folder named "results" in `intermediate_dir`. It calculates the MSI values bidirectionally for both microorganisms. The MSI values are also written in the output for ready review. **Note 3** presents an analysis of the MSI values and stuck/relieved reactions corresponding to the two microorganisms across the different media considered.

```
[INFO] Loading graph files...
[INFO] Loading seed metabolite file...
[INFO] Performing metabolic scope calculations for com-
bined and individual graphs...
[INFO] Organism 1 identified as: iCN718
[INFO] Organism 2 identified as: iYL1228
[INFO] Calculating Metabolic Support Index (MSI)...
[INFO] MSI (iCN718 in presence of iYL1228): 0.0998
```

```
[INFO] MSI (iYL1228 in presence of iCN718): 0.0076
[SUCCESS] MSI values have been written to:
/Path/to/the/directory/5_Int_dir/results/iCN718_iYL1228_-
curated_minimal_medium_MSI.csv
```

4 Notes

1. *Limitations of MetQuest.* To begin with, the predictions of the algorithm are solely dictated by the input metabolic network. If the quality of the GEM is not substantial, then the results may be unreliable, as essential reactions may get missed or nonexistent reactions might be predicted. Although MSI is highly proficient in detecting positive (beneficial) metabolic interactions, it does not shed light on potential negative interactions in a community. Further, a high MSI indicates that there is scope for more metabolic pathways to get activated. However, this does not guarantee that these pathways are active in the organism and, even if active, contribute to enhancing the organism's growth.
2. *Analysis of the MSI results.* Among the four tested media (LB aerobic and anaerobic, curated minimal medium, and curated M9 medium), the MSI values were observed to be in the range of 1.10%–11.01% for *A. baumannii* (in the presence of *K. pneumoniae*) and 0.19%–0.79% for *K. pneumoniae* (in the presence of *A. baumannii*). When added to minimal media (both curated and M9), the cofactors relieved a significant amount of stuck reactions. When comparing LB aerobic to the anaerobic medium, the MSI value increased for *A. baumannii*. However, it almost remained constant for *K. pneumoniae*, implying that stress conditions in the latter, due to oxygen deprivation, induce more metabolic support. This implies that *A. baumannii* draws high metabolic benefits from *K. pneumoniae*, per the findings reported by Semenec et al. [8]. In *A. baumannii*, across all media, many stuck reactions get relieved when grown along with *K. pneumoniae*. For instance, 47 out of 471 stuck reactions were relieved in the curated minimal medium. In contrast, in the M9 medium, 47 out of 427 stuck reactions were relieved (Fig. 1). The MSI values in the corresponding media for *A. baumannii* were 10% and 11%, respectively. However, for *K. pneumoniae*, only about 0.76%–0.79% of reactions were relieved (Fig. 2).

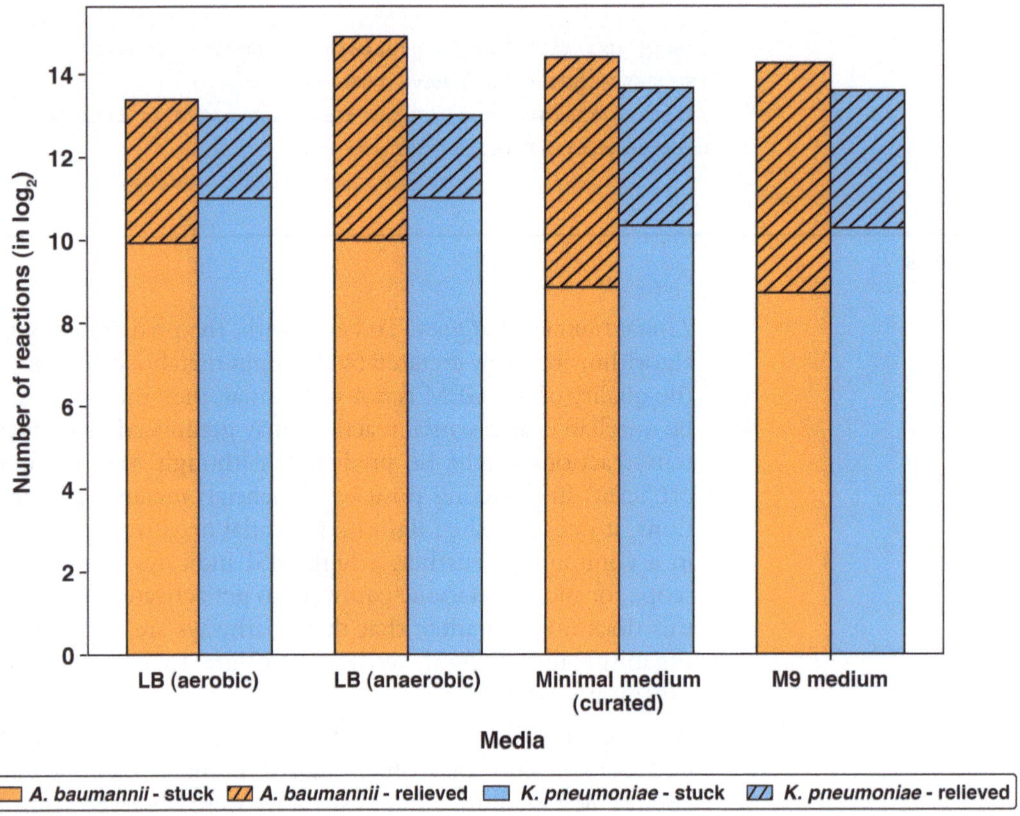

Fig. 1 Comparison of the number of stuck reactions (in $\log_2$ scale) between *A. baumannii* (orange) and *K. pneumoniae* (blue) monocultures across four different media. The dashed regions highlight the stuck reactions that get relieved in co-culture

3. *Sample MSI calculation. For A. baumannii* grown with *K. pneumoniae* in the M9 medium (with cofactors), the MSI is computed as follows:

   Number of stuck reactions in monoculture = 427

   Number of stuck reactions in co-culture = 380

$$\begin{aligned} \mathrm{MSI}\left(A.baumannii|A.baumannii, K.pneumoniae\right) &= 1 - \frac{n_{\mathrm{stuck}\,(A.baumannii|A.baumannii,K.pneumoniae)}}{n_{\mathrm{stuck}\ A.baumannii}} \\ &= 1 - \frac{380}{427} \\ &= 11.01\% \end{aligned}$$

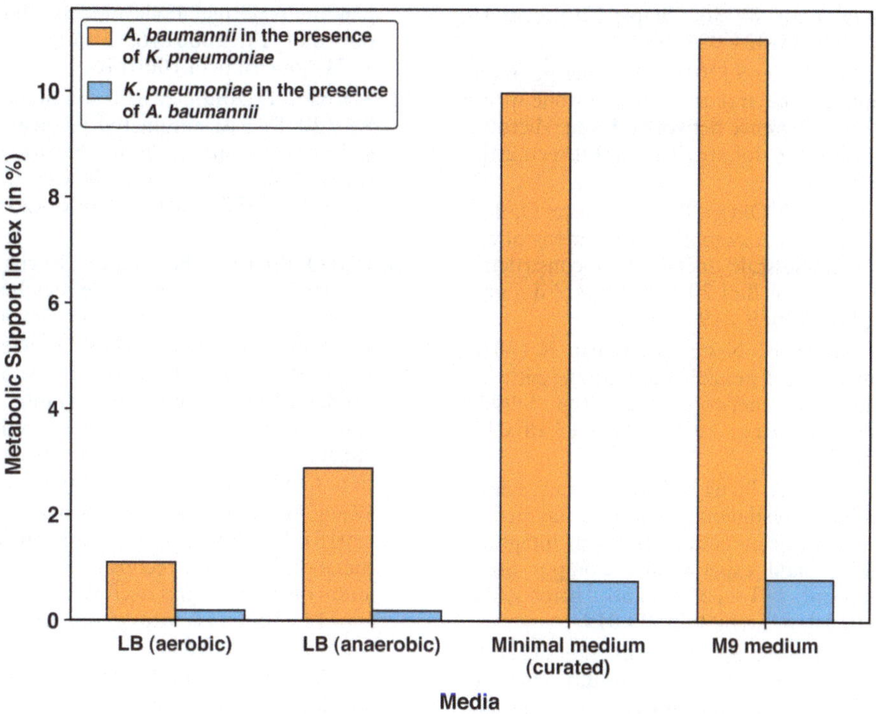

Fig. 2 Comparison of the Metabolic Support Index (MSI) values (in %) for *A. baumannii* (orange) and *K. pneumoniae* (blue) when grown together under four media conditions

Acknowledgments

P.S. acknowledges the Prime Minister's Research Fellowship (PMRF) from the Ministry of Education, Government of India. S.V. acknowledges the Half-Time Research Assistantship (HTRA) from the Ministry of Education, Government of India. K.R. acknowledges support from the Wadhwani School of Data Science and AI. The authors thank Dr. Maziya Ibrahim and Dr. Aarti Ravindran for their valuable comments.

Conflict of Interest Statement The authors have no conflicts of interest to declare that are relevant to the content of this chapter.

References

1. Lopes W, Amor DR, Gore J (2024) Cooperative growth in microbial communities is a driver of multistability. Nat Commun 15: 4709. https://doi.org/10.1038/s41467-024-48521-9
2. Kodera SM, Das P, Gilbert JA, Lutz HL (2022) Conceptual strategies for characterizing interactions in microbial communities. iScience 25:103775. https://doi.org/10.1016/j.isci.2022.103775
3. Srinivasan S, Jnana A, Murali TS (2024) Modeling microbial community networks: methods and tools for studying microbial interactions.

Microb Ecol 87:56. https://doi.org/10.1007/s00248-024-02370-7

4. Berry D, Widder S (2014) Deciphering microbial interactions and detecting keystone species with co-occurrence networks. Front Microbiol 5:219. https://doi.org/10.3389/fmicb.2014.00219
5. Mendoza SN, Olivier BG, Molenaar D, Teusink B (2019) A systematic assessment of current genome-scale metabolic reconstruction tools. Genome Biol 20:158. https://doi.org/10.1186/s13059-019-1769-1
6. Ravikrishnan A, Nasre M, Raman K (2018) Enumerating all possible biosynthetic pathways in metabolic networks. Sci Rep 8:9932. https://doi.org/10.1038/s41598-018-28007-7
7. Ravikrishnan A, Blank LM, Srivastava S, Raman K (2020) Investigating metabolic interactions in a microbial co-culture through integrated modelling and experiments. Comput Struct Biotechnol J 18:1249–1258. https://doi.org/10.1016/j.csbj.2020.03.019
8. Semenec L, Cain AK, Dawson CJ et al (2023) Cross-protection and cross-feeding between *Klebsiella pneumoniae* and *Acinetobacter baumannii* promotes their co-existence. Nat Commun 14:702. https://doi.org/10.1038/s41467-023-36252-2
9. Norsigian CJ, Pusarla N, McConn JL et al (2019) BiGG Models 2020: multi-strain genome-scale models and expansion across the phylogenetic tree. Nucleic Acids Res 48: D402–D406. https://doi.org/10.1093/nar/gkz1054
10. Glont M, Nguyen TVN, Graesslin M et al (2017) BioModels: expanding horizons to include more modelling approaches and formats. Nucleic Acids Res 46:D1248–D1253. https://doi.org/10.1093/nar/gkx1023
11. Heinken A, Hertel J, Acharya G et al (2023) Genome-scale metabolic reconstruction of 7,302 human microorganisms for personalized medicine. Nat Biotechnol 41:1320–1331. https://doi.org/10.1038/s41587-022-01628-0
12. Seaver SMD, Lerma-Ortiz C, Conrad N et al (2018) PlantSEED enables automated annotation and reconstruction of plant primary metabolism with improved compartmentalization and comparative consistency. Plant J 95: 1102–1113. https://doi.org/10.1111/tpj.14003
13. Aite M, Chevallier M, Frioux C et al (2018) Traceability, reproducibility and wiki-exploration for "à-la-carte" reconstructions of genome-scale metabolic models. PLoS Comput Biol 14:e1006146. https://doi.org/10.1371/journal.pcbi.1006146
14. Machado D, Andrejev S, Tramontano M, Patil KR (2018) Fast automated reconstruction of genome-scale metabolic models for microbial species and communities. Nucleic Acids Res 46:7542–7553. https://doi.org/10.1093/nar/gky537
15. Dias O, Rocha M, Ferreira EC, Rocha I (2015) Reconstructing genome-scale metabolic models with *merlin*. Nucleic Acids Res 43:3899–3910. https://doi.org/10.1093/nar/gkv294
16. Faria JP, Liu F, Edirisinghe JN et al (2023) ModelSEED v2: high-throughput genome-scale metabolic model reconstruction with enhanced energy biosynthesis pathway prediction. Cold Spring Harbor Laboratory
17. Wang H, Marcišauskas S, Sánchez BJ et al (2018) RAVEN 2.0: a versatile toolbox for metabolic network reconstruction and a case study on *Streptomyces coelicolor*. PLoS Comput Biol 14:e1006541. https://doi.org/10.1371/journal.pcbi.1006541
18. Gu C, Kim GB, Kim WJ et al (2019) Current status and applications of genome-scale metabolic models. Genome Biol 20:121. https://doi.org/10.1186/s13059-019-1730-3
19. Norsigian CJ, Kavvas E, Seif Y et al (2018) iCN718, an updated and improved genome-scale metabolic network reconstruction of *Acinetobacter baumannii* AYE. Front Genet 9: 121. https://doi.org/10.3389/fgene.2018.00121
20. Liao Y-C, Huang T-W, Chen F-C et al (2011) An experimentally validated genome-scale metabolic reconstruction of *Klebsiella pneumoniae* MGH 78578, iYL1228. J Bacteriol 193:1710–1717. https://doi.org/10.1128/jb.01218-10
21. Lee I, Ouk Kim Y, Park S-C, Chun J (2016) OrthoANI: an improved algorithm and software for calculating average nucleotide identity. Int J Syst Evol Microbiol 66:1100–1103. https://doi.org/10.1099/ijsem.0.000760
22. Lieven C, Beber ME, Olivier BG et al (2020) MEMOTE for standardized genome-scale metabolic model testing. Nat Biotechnol 38: 272–276. https://doi.org/10.1038/s41587-020-0446-y
23. Kumar RK, Singh NK, Balakrishnan S et al (2022) Metabolic modeling of the International Space Station microbiome reveals key microbial interactions. Microbiome 10:102. https://doi.org/10.1186/s40168-022-01279-y

Chapter 10

Constraint-Based Metabolic Modeling Approach for Microbial Communities

Satyajit Beura, Sayan Saha Roy, Amit Kumar Das, and Amit Ghosh

Abstract

Microorganisms grow in complex communities by fostering symbiotic relationships to uphold the integrity and functionality of the consortium. Deciphering the metabolic interactions within microbial communities and their impact on host environments is essential due to their association with major domains, including human health, bioremediation, and bioenergy production. However, unraveling their metabolic activity in laboratory conditions is challenging, as many microbes resist cultivation, and recreating their complex natural ecosystem with all its biological parameters presents additional hurdles. Therefore, modeling the microbial communities has become crucial for comprehending the intricate interactions within diverse microbial populations. In this chapter, we elucidate an in silico methodology for reconstructing a genome-scale metabolic model of a microbial consortium. This community modeling approach encompasses the reconstruction of microbial models, the integration of individual models into a community, and the optimization of the community model under different environmental conditions. Furthermore, a wide range of flux analysis techniques, like Flux Balance Analysis (FBA), Flux Variability Analysis (FVA), and Flux Sampling (FS), were described to investigate both the community-wide flux profile and intermicrobial interactions.

Key words Microbial consortium, Genome-scale community metabolic model, Flux balance analysis, Flux sampling, Flux variability analysis, Metabolic interaction

1 Introduction

The microbial community represents a fascinating and intricate network of diverse microorganisms coexisting and interacting within a specific environment. These communities, found in various ecosystems ranging from soil and water to the human body, play pivotal roles in shaping the overall health and functioning of their respective habitats [1–4]. Comprehending the dynamics of microbial communities has become essential in diverse fields, including ecology, environmental science, medical sciences, and biotechnology. Recent advancements in high-throughput multi-omics technologies have facilitated the exploration of the possible metabolic

Karthik Raman and Gayathri Sambamoorthy (eds.), *Modelling the Microbiome*, Methods in Molecular Biology, vol. 3006, https://doi.org/10.1007/978-1-0716-5080-6_10,

behaviors exhibited by microbial communities [5]. The inclusion of multi-omics data has enhanced both culture-dependent and culture-independent strategies, resulting in a thorough comprehension of microbial metabolism. However, the application of such methodologies to comprehend the detailed metabolic characteristics of each microorganism and the metabolic communication within a microbial community remains challenging. Consequently, various mathematical and computational modeling tools are employed to map the metabolic activities of the microbial ecosystem at both the species and community levels [6–9]. These emerging strategies of computational systems biology in community modeling promise to offer insights into metabolic interaction within microbial ecosystems.

Genome-scale metabolic modeling (GEM) serves a fundamental role in the field of computational systems biology, enabling a comprehensive understanding of the biochemical processes within individual microbes and microbial communities [10–15]. Numerous GEM databases, like BiGG Models Database [16], path2Models [17], and AGORA [18] have been developed over the past two decades (Table 1). Further, different GEM reconstruction tools, like GEMsiRV [32], GEM system [33], and COBRA Toolbox [34] have also been developed (Table 2). Initially, GEM reconstruction and analysis were introduced to explore the biochemical pathways of single organisms and translate genetic information into phenotypic characteristics. A pivotal outcome of GEM is the measurement of metabolic flux, representing the phenotypic characteristics manifested in diverse biochemical reactions of microbial species. However, traditional GEM approaches are constrained to single organisms and lack insights into community-level flux distribution. In recent times, strategies for GEM reconstruction have evolved toward genome-scale community metabolic modeling, facilitating the tracking of community-wide flux distribution patterns and metabolic interactions within complex microbial communities. Over the past two decades, a wide range of community modeling tools such as MICOM, MM Toolbox, OptCom, and others have been developed (Table 3). Further, mathematical and computational algorithms, including FBA [77], FVA [78], FS [79], and dynamic FBA (dFBA) [80], have been developed to analyze the metabolic flux profile, simulating the metabolic behavior of naturally occurring microbial consortium. Community modeling methodologies predict metabolic functions including cross-feeding, nutritional competition, and symbiotic interactions within a microbial consortium [81–83]. These modeling approaches are capable of integrating numerous microbial models to form a community model, allowing for the investigation of metabolic exchanges, as well as the consumption and production profile of the microbial community. Further, some advanced genome-scale community modeling approaches like MICOM and MM toolbox utilize

Table 1
List of GEM databases

No.	Resource	Details	Model types	References
1	BiGG Models Database	Hosts a diverse collection of 108 individual metabolic models	Various organisms (*E. coli*, Human, tubercle bacillus, etc.)	[16] [weblink]
2	MetaNetX	Features 163 models sourced from ModelSeed, Path2Models, with a consistent namespace	Various	[19] [weblink]
3	Other Organisms (The Palsson Lab)	Links to 178 existing published models	Bacteria, Eukaryota, Archaea	[20] [weblink]
4	ModelSeed	Numerous models have been generated from KEGG, MetaCyc, and SEED databases	Various	[21] [weblink]
5	Path2Models	Automatic conversion from KEGG to SBML, compatible with COBRA toolbox	Various (Eukaryota, Bacteria, Human, etc.)	[17] [weblink]
6	MEMOSys	Manages 20 models that depict a variety of 16 distinct organisms	Various (*Aspergillus niger*, *E. coli*, Human, etc.)	[22] [weblink]
7	BioMet Toolbox	Online repository with eight fungal and six bacterial models alongside various analytical tools	Fungal, Bacterial	[23] [weblink]
8	Pathway Tools	A renowned reference for metabolic reconstruction and the development of organism-specific databases	Bacteria, Eukaryota, Archaea	[24] [weblink]
9	Recon 3D	Comprehensive resource offering 3D reconstructions of human metabolism	Human	[25] [weblink]
10	BioModels Database	A platform that stores mathematical models elucidating biological phenomena, encompassing metabolic models as well	Various (Eukaryota, Bacteria, Human, etc.)	[26] [weblink]
12	Metabolic Atlas	Resource focusing on detailed human metabolic models with annotations and visualizations	Human	[27] [weblink]
13	AGORA	Curated database of gut microbes' metabolic models, aiding in understanding the human microbiome	Microbes	[18] [weblink]
14	Virtual Metabolic Human	Offers reconstructions of human metabolic networks for various tissues and cell types	Human gut microbes	[28] [weblink]

(continued)

Table 1
(continued)

No.	Resource	Details	Model types	References
15	Saccharomyces Genome Database (SGD)	Network-based resource for *Saccharomyces cerevisiae*, integrating diverse biological data for analysis	Yeast	[29] [weblink]
16	EcoCyc	Collection of Escherichia coli (*E. coli*) metabolic models	E. coli	[30] [weblink]
17	AraCyc	A database centered on the reconstruction and examination of the metabolic network of *Arabidopsis thaliana*	Arabidopsis	[31] [weblink]

metagenomic samples to retrieve microbial diversity and abundance, mapping them against the available GEM repositories. Subsequently, these mapped GEMs can be simulated together in a specified in silico environment, providing insights into the intermicrobial interactions and the metabolic flux profile on a community level. Various microbial consortia have been examined through community metabolic modeling, encompassing soil-based model communities [84, 85], biofilm-based communities [86, 87], industrial model communities [88–90], and human microbiomes [90–92]. A community metabolic model was developed by incorporating pivotal members of the rumen microbiome along with their associated viral phages [93]. Notably, this model unveiled novel interactions, including the exchange of vitamins, fatty acids, amino acids, sugars, and coenzymes within the community members, shedding light on previously unexplored metabolic dynamics. Another community modeling approach is introduced, focusing on reconstructing the microbiota of termite guts and fungal combs to elucidate the lignocellulose degradation potential within the microbiome [94]. These investigations reveal intricate metabolic exchanges involving amino acids, short-chain fatty acids (SCFAs), and simple sugars within the microbial community. Examining these patterns of interaction between different species holds promise for refining the composition of synthetic communities to enhance the carbohydrate utilization efficiency. Hence, employing community modeling helps to gain insights into the intricacies of a complex microbial consortium.

Within this chapter, we outlined an in silico approach for reconstructing a community metabolic model for predicting the metabolic activities of a microbial consortium. At its core, community metabolic modeling involved constructing genome-scale metabolic models (GEMs) for each individual entity within a microbial consortium. Subsequently, the reconstructed GEMs were

Table 2
List of tools for GEM reconstruction

Sl. no.	Tool	Input data	Reference databases	GUI available	Language	Integrated gap filling	Application and features
1	GEMsiRV [32]	Genome-scale models	MetaCyc, BiGG, ModelSEED, KEGG	✗	Java	✔	Optimization-based reconstruction of metabolic networks
2	Pathway tools [35]	Sequences with annotations	MetaCyc, Pathway/Genome Databases (PGDBs)	✔	Lisp, Python	✔	Pathway analysis and metabolic network reconstruction
3	GEM system [33]	Sequences with or without annotations	BioCyc, KEGG	–	N/A	✗	Genome-scale metabolic modeling
4	COBRA Toolbox [34]	Genome-scale models	BiGG, KEGG, MetaCyc	✔	MATLAB	✔	Constraint-based reconstruction and analysis
5	ModelSEED [20]	RAST annotation (Web)	Model SEED database	✔	N/A	✔	Reconstruction of metabolic models from annotated genomes
6	AuReMe [36]	A Fasta (.faa) file or Genbank (.gbk)	BiGG, METACYC, KEGG	✗	Python	✔	Automated reconstruction of metabolic networks
7	Pantograph tool [37]	Genome-scale models	Various	✔	Python	✗	Systems biology, network comparison
8	CoReCo [38]	Protein sequences	KEGG	✗	Python	✔	Genome-scale metabolic modeling
9	MicrobesFlux [39]	Organisms from KEGG	KEGG (Web)	✔	Python, Google Web Toolkit technology	✗	Flux balance analysis for microbial communities
10	AutoKEEGRec [40]	KEGG organism IDs	KEGG databases	✗	MATLAB	✗	Reconstruction and analysis of KEGG pathways
11	ScrumPy [41]	Annotated genome sequence	BioCyc database, NCBI	✗	Python	✗	Genome-scale metabolic modeling

(continued)

Table 2
(continued)

Sl. no.	Tool	Input data	Reference databases	GUI available	Language	Integrated gap filling	Application and features
12	rBioNet [42]	Metabolite and reaction information	RxnDB, MetDB, KEGG	✔	Matlab	✔	Systems biology, network analysis
13	Architect [43]	Annotated sequence	UniProt/SwissProt	✔	Python	✔	Systems biology, network construction
14	CarveMe [44]	DNA sequences or protein FASTA file	BiGG database	✗	Python	✔	Genome-scale metabolic modeling
15	COBRAme [45]	Metabolic models, experimental data	KEGG, BiGG, MetaCyc	✔	Python	✔	Systems biology, metabolic engineering
16	MetaDraft [46]	GenBank format file or protein FASTA	BiGG Models	✔	Python	✗	Reconstruction of metabolic networks
17	Raven toolbox [47]	Sequences with annotation	MetaCyc, KEGG	✗	MATLAB	✔	Metabolic flux analysis and pathway visualization
18	MetExplore [48]	Metabolic pathway analysis	BiGG, KEGG, MetaCyc	✔	Java	✗	Exploration and visualization of metabolic pathways
19	merlin [49]	Annotated or unannotated sequence	MetaCyc, KEGG, UniProtKB, TCDB	✔	Java	✗	Metabolic pathway analysis
20	SuBliMinal toolbox [50]	Organisms from MetaCyc and KEGG	MetaCyc, KEGG	✗	Java	✗	Metabolic modeling and network analysis
21	FAME [51]	Organisms from KEGG databases (Web)	KEGG	✔	Python	✗	Metabolic network analysis and visualization
22	PySCeS CBMPy [52]	Metabolic models	BiGG, KEGG, MetaCyc	✗	Python	✗	Simulation and analysis of physiological systems

Table 3
List of genome-scale community modeling tools

Sl. no.	Tool name	Approaches	Dynamics	Capable to accommodate large communities?	Platforms used	Description
1	DMMM [53]	Temporal, linear	Dynamic	✔	MATLAB	Introduces the first multispecies dynamic FBA approach
2	Integrated evolutionary game theory [12]	Evolutionary, linear	Dynamic	✔	Python	Captures the balance and transfer of expensive metabolites
3	SteadyCom [7]	Compartmentalized GEM, nonlinear, nonconvex	Steady-state	✔	MATLAB	Seeks to optimize the community's growth rate
4	d-optCom [6]	Temporal, bilevel, nonlinear, nonconvex	Dynamic	✗	UNIX/ LINUX	Maximizes community biomass concentration
5	MMinte [54]	Compartmentalized GEM, nonlinear	Steady-state	✔	–	Assesses microbial metabolic interactions and their impact on growth rates using 16S rRNA data
6	Brunner and Chia [55]	Temporal, linear	Dynamic	✔	Python	Substantially decreases the occurrence of issues related to FBA
7	cFBA [56]	Compartmentalized GEM, nonlinear, nonconvex	Steady-state	✗	Python	Focuses on maximizing the community's specific growth rate
8	DOLMN [57]	Compartmentalized GEM, nonlinear	Steady-state	✔	MATLAB	Utilizes MILP for optimizing labor division in constrained communities
9	FLYCOP [58]	Evolutionary, spatiotemporal, nonlinear, nonconvex	Dynamic	✔	Python	A system using COMETS and a local search algorithm selects the best consortium for a goal

(continued)

Table 3 (continued)

Sl. no.	Tool name	Approaches	Dynamics	Capable to accommodate large communities?	Platforms used	Description
10	MICOM [9]	Compartmentalized GEM, nonlinear, convex	Steady-state	✔	Python	First maximizes weighted sum of biomass reaction fluxes
11	DFBALab [59]	Temporal, linear	Dynamic	✔	MATLAB	Handles FBA infeasibility and degeneracy
12	MetaNetX [60]	Compartmentalized GEM, nonlinear, convex	Steady-state	✔	Java	Provide access to data and tools for importing GEMs, aligning them with MNXref reconciliation, and exporting the resultant GSMNs
13	Koch et al. [61]	Compartmentalized GEM, FBA, nonlinear, nonconvex	Steady-state	✔	MATLAB	Optimizes the community's expansion rate first, followed by maximizing biomass yield
14	NECom [62]	Bi-level optimization, nonlinear, nonconvex	Steady-state	✗	MATLAB	Maximizes the biomass flux of taxa as the primary internal objective
15	Henry et al. [63]	Mixed-bag GEM, linear	Steady-state	✔	N/A	Optimizes the reaction flux of biomass for all taxa to its maximum potential
16	optCom [64]	Bi-level optimization, nonlinear, nonconvex	Steady-state	✗	UNIX/LINUX	Optimizes the biomass flux and community production for each taxon
17	ACBM [65]	Spatiotemporal, linear	Dynamic	✔	MATLAB, Java	Utilizes agent-based modeling
18	CODY [66]	Spatiotemporal	Dynamic	✔	MATLAB	Incorporates EMA, enzyme kinetics, and the distribution of resources
19	COMETS [67]	Evolutionary, spatiotemporal, linear	Dynamic	✔	UNIX/LINUX	Incorporates biophysical models for simulating diffusion
20	µbialSim [68]	Temporal, linear	Dynamic	✔	MATLAB	Utilizes an augmented Euler method

21	MicrobiomeAnalyst 2.0 [69]	Evolutionary, temporal, nonlinear, convex	Dynamic	✔		N/A	Multifactor differential analysis
22	RedCOM [8]	Nested, EFVA, N/A	Steady-state	✔		MATLAB	Constructs reduced community models
23	CASINO [70]	Compartmentalized GEM, nonlinear	Steady-state	✔		MATLAB	A toolkit for studying interactions between diet and the microbiota
24	Taffs et al. [71]	Mixed bag GEM, compartmentalized GEM	Steady-state	✔		MATLAB	Offers enhanced scalability but demands increased user involvement with multiple rounds of pathway analysis
25	IndiMeSH [72]	Spatiotemporal, linear	Dynamic	✔		MATLAB	A hybrid model integrating dFBA and an individual-based approach in an angular pore network was used to simulate soil aggregates, considering habitat geometry and hydration conditions
26	EvoFBA [73]	Evolutionary, linear	Dynamic	✔		MATLAB	Modeling the progression of spontaneous mutations
27	ORKA [74]	Temporal, linear, multiple languages (Python, Perl)	Dynamic	✔		MATLAB	Utilizes the Runge–Kutta method for the solution of ODEs and alleviates underlying assumptions
28	BacArena [75]	Spatiotemporal, linear	Dynamic	✗		R	Utilizes agent-based modeling
29	Microbiome Modeling Toolbox [76]	Compartmentalized GEM, linear	Steady-state	✔		MATLAB	Aims to maximize the weighted sum of biomass reaction fluxes

integrated in a compartmentalized method to form a comprehensive genome-scale community model. This chapter also explored the application of various constraints, including media composition and relative abundance, to optimize the community model in specific environmental conditions. Furthermore, various in silico analyses, including FBA, FVA, and FS, were explored in order to examine intermicrobial interactions using community-wide flux profiles.

2 Materials

2.1 Software and Tools

COBRA Toolbox stands out as a leading choice for the reconstruction and interpretation of metabolic networks. It offers versions tailored for MATLAB, referred to as "COBRA Toolbox," for Python, known as "COBRApy," and for Julia, termed "COBRA.jl". These are integral components of the openCOBRA project (opencobra.github.io). For our purposes, we concentrate solely on utilizing COBRApy in Python for employing in silico metabolic modeling and analyzing genome-scale as well as community-scale metabolic models. Primarily, we have installed Anaconda Navigator from Anaconda website. It is a part of Anaconda distribution, encompassing Python and an extensive array of pre-installed libraries and tools frequently employed in data science, machine learning, and scientific computing. Further, we utilized Anaconda Navigator to launch both Spyder and Jupyter Notebook, enabling us to write scripts for metabolic modeling. Moreover, COBRApy was accessed from https://opencobra.github.io/cobrapy/, with installation instructions provided online.

Instruction To install COBRApy via command line, use the following: for pip installation, enter `'$ pip install cobra'`, or opt for conda installation from conda-forge with `'$ conda install -c conda-forge cobra'`.

2.2 Solvers

A constraint-based reconstruction and analysis model for biological systems represents a specific type of discrete optimization problem, often addressed through techniques like linear, mixed integer, or quadratic programming. COBRApy does not directly provide algorithms to solve these problems. Instead, it offers a biologically inspired abstraction of these techniques, facilitating conceptualization of metabolic system functioning without delving deeply into optimization problem formulations. The solving aspect is handled by external tools like the freely available software GLPK or commercial solutions such as Gurobi and CPLEX. The default configuration of COBRApy includes the GLPK solver. We have installed the Gurobi solver to optimize the model and conduct various analyses, including FBA and FVA. However, in the case of FS analysis, we utilized IBM CPLEX.

Instruction After defining the model, one can switch the solver backend by directly assigning to the model.solver property.

```
# COBRApy is by default equipped with the GLPK
model.solver = 'glpk'
# or if you have cplex installed
model.solver = 'cplex'
# or if you have gurobi installed
model.solver = 'gurobi'
```

3 Methods

In order to model microbial consortium effectively, it is crucial to first identify the composition of the community. The composition of the community can be determined by analyzing metagenomic data using various tools such as QIIME 2 [95], VITCOMIC2 [96], or through a comprehensive review of existing literature. Each constituent of the community can then be modeled individually using the genome-scale metabolic modeling approach. These models are subsequently integrated into a community to create an encompassing genome-scale community metabolic model. Following model integration, the community model can undergo various in silico analysis processes, such as FBA, FVA, and FS, to investigate the metabolic interactions between the individual microbes and flux profiles of the microbial consortium. Detailed methodologies, along with the corresponding scripts, are elaborated upon in the subsequent sections.

3.1 Reconstruction of GEMs

The whole-genome sequences of individual microbes were acquired from the NCBI. The microbe-specific annotated genome (FASTA sequence), template model, and media composition were used to construct the draft GEM using the ModelSEED web server (Fig. 1) [20].

The draft model contains several unbounded, unbalanced, and blocked reactions, orphan/dead-end metabolites, infeasible loops, and reaction gaps. To enhance its quality and predictive capability, the model requires manual curation. The unbound reactions involve thermodynamically infeasible cycles (TIC) and do not result in the consumption or production of any metabolites (Fig. 2) [93]. In order to identify these TICs in our GEMs, we disabled all nutrient uptake reactions for the microbes and performed FVA. This estimates the flux ranges for individual reactions, and those

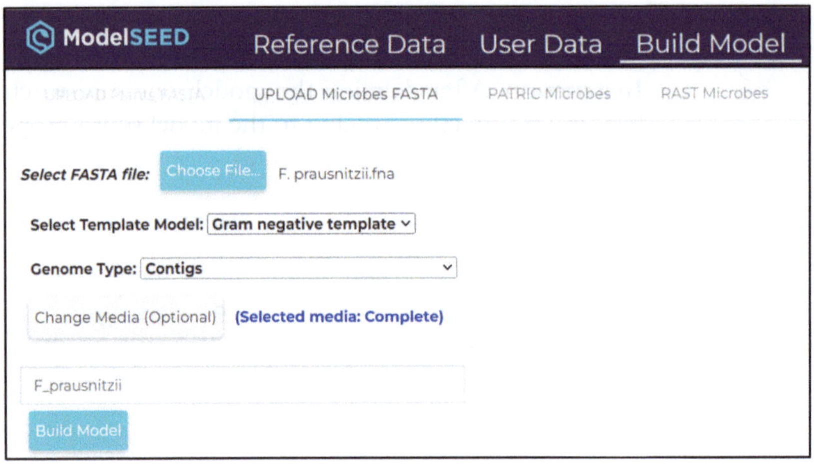

Fig. 1 Generation of draft model in ModelSEED web server

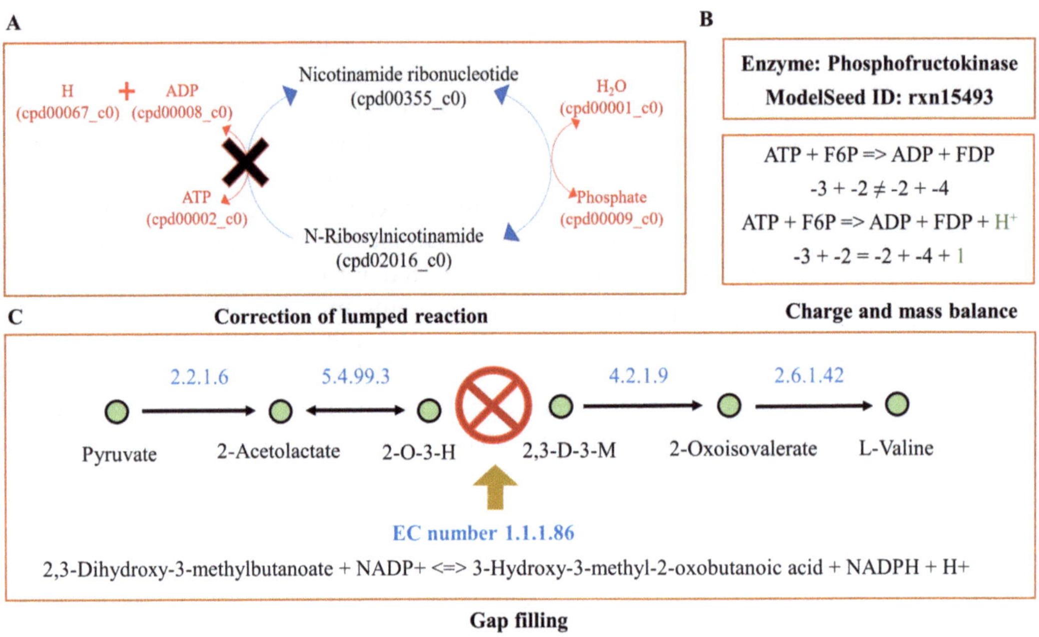

Fig. 2 Manual curation of the Draft Genome-scale model. (**A**) Panel A shows a reaction involving Nicotinamide ribonucleotide and N-Ribosylnicotinamide with associated compounds and a marked error. (**B**) Panel B details an enzyme, Phosphofructokinase, with ModelSeed ID: rxn15493, showing charge and mass balance equations for ATP, F6P, ADP, FDP, and H+. (**C**) Panel C illustrates the correction of a lumped reaction pathway from Pyruvate to L-Valine, highlighting an error at 2-O-3-H with EC number 1.1.1.86. The reaction involves 2,3-Dihydroxy-3-methylbutanoate and NADP+ converting to 3-Hydroxy-3-methyl-2-oxobutanoic acid, NADPH, and H+

hitting either lower or upper bound are defined as unbound reactions [97]. To remove the unbound reactions, the approach involved removing duplicate reactions, deactivating lumped reactions, and selectively turning on or off the reactions based on available cofactor specificity information obtained from MetaCyc [98] and ModelSEED reaction databases [21]. In addition to unbounded reactions, the draft GEMs also included unbalanced reactions. Maintaining charge and mass balance in reactions is crucial for enhancing the predictive accuracy of GEMs (Fig. 2) [99]. In order to uphold both elemental and charge equilibrium in biochemical reactions, the precise chemical formulas and charges for each metabolite were assigned by referencing CHEBI (Chemical Entities of Biological Interest) [100], PubChem, and BiGG databases [101]. Furthermore, the CHEBI and BiGG databases were utilized to correct reactions with proton imbalances by modifying the proton count on either the reactant or product side (Fig. 2). The reaction stoichiometry was adjusted for the remaining imbalanced reactions to guarantee that the number of atoms on both sides of the reaction was equal. The model also included orphan and dead-end metabolites lacking exact metabolic pathways across the GEMs. Some of these metabolites were omitted from the models, while others were included in the relevant metabolic pathway to improve the model's quality. Additionally, correcting the reversibility of all essential biochemical reactions is necessary to enhance the accuracy of the metabolic network. The appropriate direction was assigned by adjusting the upper and lower boundaries for each reaction, with reference to the MetaCyc and ModelSEED databases. Further, the model exhibited a significant proportion of blocked reactions across the metabolic network, primarily due to the absence of certain reactions in the central metabolic network. The gaps in the respective metabolic networks were addressed by incorporating the missing biochemical reactions, guided by the KEGG pathway database (Fig. 2). Curation of the metabolic networks enhances the accuracy, predictive capability, and biological relevance of these models. These curated GEMs were optimized through FBA, a mathematical method for analyzing the flux of biochemical reactions within a metabolic network. Biomass production has been considered as the objective function while optimizing the individual microbial models using FBA. The biomass equation of the GEMs represents the combination of precursor molecules, which ideally reflect the molecular makeup of the microorganism. Further, the models were optimized in minimal media to achieve the in silico growth rate. The model can be verified by comparing the predicted in silico growth of the individual models with the experimental growth rates.

Instruction The draft model can be imported into the COBRApy platform to optimize the model. Leveraging the capabilities of COBRApy, users can fine-tune and optimize these models, allowing for detailed analysis and simulation of biological systems.

```
# In [1]: # import libraries
          import cobra
          import sys
          import os
# In [2]: # import model
           model=cobra.io.read_sbml_model("C:/Users/DELL/
Desktop/Model/F_prausnitzii.xml")
# In [3]: # optimize model
          model.slim_optimize()
# Out[3]: 0.33100615775008974
```

Print model summary:

```
# In [4]: # print model summary
          model.summary()
# Out[4]: Objective
          1.0 bio1 = 0.33100615775008796
```

Metabolite uptake:

Metabolite	Reaction
cpd00009_b	EX_cpd00009_b
cpd00011_b	EX_cpd00011_b
cpd00013_b	EX_cpd00013_b
cpd00027_b	EX_cpd00027_b
cpd00030_b	EX_cpd00030_b
cpd00034_b	EX_cpd00034_b
cpd00048_b	EX_cpd00048_b
cpd00058_b	EX_cpd00058_b
cpd00063_b	EX_cpd00063_b
cpd00099_b	EX_cpd00099_b
cpd00149_b	EX_cpd00149_b
cpd00205_b	EX_cpd00205_b
cpd00254_b	EX_cpd00254_b
cpd17041_c0	rxn13782_c0
cpd17042_c0	rxn13783_c0

Metabolite secretion:

Metabolite	Reaction	Flux	C-Number	C-Flux
cpd00001_b	EX_cpd00001_b	-41.64	0	0.00%
cpd00007_b	EX_cpd00007_b	-0.1316	0	0.00%
cpd00023_b	EX_cpd00023_b	-1.778	0	0.00%
cpd00036_b	EX_cpd00036_b	-0.005256	0	0.00%
cpd00106_b	EX_cpd00106_b	-4.164	0	0.00%
cpd11416_b	EX_cpd11416_b	-0.331	0	0.00%

Model characteristics:

```
# In [5]: # model characteristics
          print(len(model.genes), len(model.reac-
tions), len(model.metabolites))
        1108 1111 1212
```

3.2 Construction of a Community Model

Consider a hypothetical scenario involving four distinct microbes, denoted as m1, m2, m3, and m4, for the purpose of community modeling. To construct a comprehensive community model, the GEMs of these individual microbes were integrated using a compartmentalized approach (Fig. 3). Each model's genes, reactions,

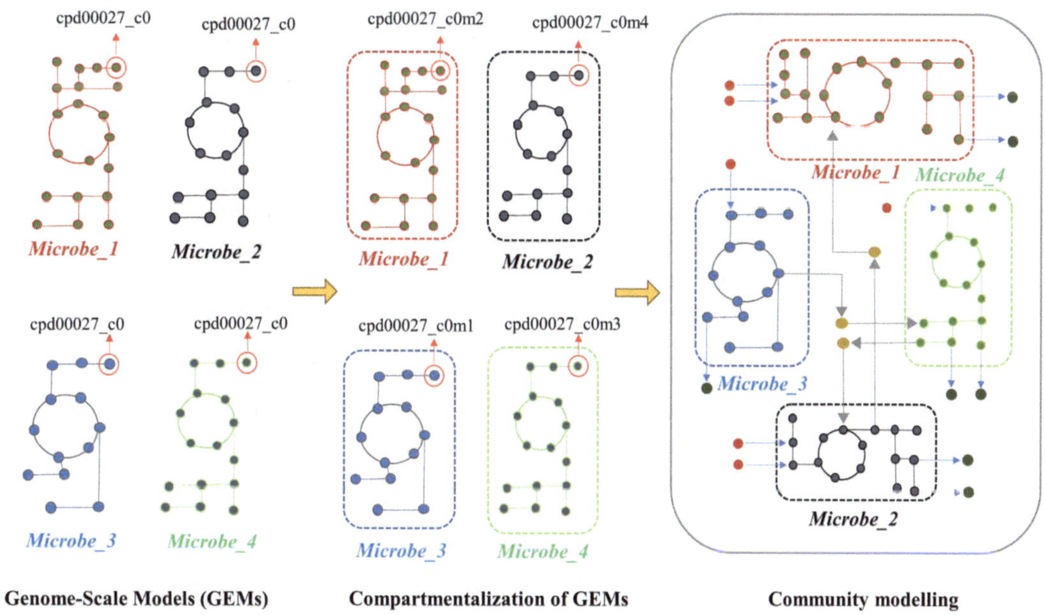

Fig. 3 Integration of individual GEMs into a Genome-scale community metabolic model

and metabolites were allocated to unique compartments and annotated accordingly during the integration process. For instance, within the model of m1, all reactions and metabolites were tagged with an extension of "m1," a practice upheld across all models (Fig. 3). Consequently, each GEM represents its own distinct compartment within the multispecies community model while sharing a common extracellular space. This arrangement facilitates metabolic exchange among the individual microbial species, as all microbes interact within the same spatial domain. Notably, metabolic compounds involved in extracellular interactions, whether consumed or produced by the microbes, were annotated with the extension "e0." This systematic compartmentalization and annotation strategy enable precise modeling of intermicrobial metabolic interactions and exchanges within the community setting.

To define the community biomass, biomass objective functions from all individual species models were integrated by summing their stoichiometric coefficients using the following optimization formula [102]:

$$Z = \left(v_{\text{biomass}}^{\text{community}}\right),$$

$$\text{Community biomass} = \sum\nolimits_{k=1}^{N} a_k \times v_{\text{biomass}}^{k}$$

Here, "Z" is the objective function, "$v_{\text{biomass}}^{\text{community}}$" represents the community biomass reaction, and "a_k" is the abundance of individual microbes in the host environments.

Instruction: By adopting a compartmentalization strategy, researchers can integrate numerous individual metabolic models into a comprehensive community metabolic model, leveraging their metagenomics abundance data. This approach allows for a holistic representation of the metabolic capabilities of the entire microbial community. The biomass reaction serves as the objective function of the community model, representing the amalgamation of individual microbial biomass constituents within the community. One crucial aspect of this integration involves updating the biomass coefficient of the GEMs to align with the latest metagenomics data. This adjustment ensures that the community model accurately reflects the prevailing environmental conditions, thereby enhancing its predictive power. Furthermore, the media composition can be updated to achieve a realistic prediction of the microbial community in a specific environment. With these refinements, the community model becomes capable of elucidating the intricate metabolic interactions among different microbial species. Furthermore, under quasi-steady-state conditions, the community metabolic model can provide valuable insights into the overall metabolic flux profile, shedding light on the dynamic interplay of metabolic activities within the microbial community. This integrated approach offers a powerful framework for studying complex microbial ecosystems and their responses to various environmental stimuli.

Step 1: Import the GEMs

```
# import libraries in python
import cobra
import sys
import os
# import microbial models
model1 = cobra.io.read_sbml_model(r"C:/Users/DELL/Desk-
top/Model/m1.xml")
model2 = cobra.io.read_sbml_model(r"C:/Users/DELL/Desk-
top/Model/m2.xml")
model3 = cobra.io.read_sbml_model(r"C:/Users/DELL/Desk-
top/Model/m3.xml")
model4 = cobra.io.read_sbml_model(r"C:/Users/DELL/Desk-
top/Model/m4.xml")
```

Step 2: Compartmentalization (It will repeat for all the individual microbes)

```
# adding extension to the genes for microbe m1 (Repeat it
for all microbial models)
count=0
for i in model1.genes:
    #if(str(i.id)=="EX_cpd11416_c0"):continue
    if(str(i.id)[0]=="6"):
        j=str(i.id)+"m1"
        i.id=j
        count=count+1
print(count)
467
count=0
for i in model1.genes:
    #if(str(i.id)=="EX_cpd11416_c0"):continue
    if(str(i.id)[0]=="f"):
        j=str(i.id)+"m1"
        i.id=j
        count=count+1
print(count)
468
# adding extension to the metabolites for microbe m1
(Repeat it for all microbial models)
for i in model1.metabolites:
    if(str(i)[-2]=="c"):
```

```
        j=str(i)+"m1"
        i.id=j
        i.compartment="c0m1"
# adding extension to the reactions for microbe m1
(Repeat it for all microbial models)
count=0
for i in model1.reactions:
    if(str(i.id)[0]=="r"):
        j=str(i.id)+"m1"
        i.id=j
        count=count+1
print(count)
1041
# adding extension to the biomass reaction for microbe m1
(Repeat it for all microbial models)
count=0
for i in model1.reactions:
    if(str(i.id)=="bio1"):
        j=str(i.id)+"m1"
        i.id=j
        count=count+1
print(count)
1
```

Step 3: Integration of individual models into a community

```
# integration of individual models into a community
reaction = Reaction('Bio1')
reaction.name = ('Community_biomass')
reaction.lower_bound = 0
reaction.upper_bound = 1000
reaction.add_metabolites({model1.metabolites.get_by_id
("cpd11416_c0m1"): -0.25,
 model2.metabolites.get_by_id("cpd11416_c0m2"): -0.25,
 model3.metabolites.get_by_id("cpd11416_c0m3"): -0.25,
 model4.metabolites.get_by_id("cpd11416_c0m4"): -0.25,
 cpd11416_c0: 1
 })

reaction.reaction
model.add_reactions([reaction])
print(count)
model.reactions.Bio1.reaction
```

0.25 cpd11416_c0m1 + 0.25 cpd11416_c0m1 + 0.25 cpd11416_c0m1 + 0.25 cpd11416_c0m1 → cpd11416_c0cm

Here, m1, m2, m3, and m4 represent the individual microbes, and cm represents the community model.

Step 4: Adding or deleting metabolites from the media to optimize the community model in a specific environment

```
# Adding media components
model.reactions.EX_cpd00001_b.bounds = (-1000, 1000)
model.reactions.EX_cpd00002_b.bounds = (-1000, 1000)
# Deleting media components
model.reactions.EX_cpd00001_b.bounds = (0, 0)
model.reactions.EX_cpd00002_b.bounds = (0, 0)
```

3.3 Community Model Analysis

3.3.1 Flux Balance Analysis

FBA is a mathematical modeling technique used in systems biology to analyze and predict the flow of metabolites through a network of biochemical reactions [77]. It operates based on the mass balance principle, ensuring that the rates of metabolite production and consumption are balanced to attain a steady-state condition. This technique was performed to estimate the flux profile of the community model. The algorithm for performing FBA is as follows:

$$\text{Maximize } v_{\text{biomass}} \text{ subject to}$$

$$\sum_{j \in J} S_{ij} V_j = 0 \quad \forall_i \in I \tag{1}$$

$$\text{LB}_j \leq V_j \leq \text{UB}_j \quad \forall_j \in I \tag{2}$$

where i = metabolite, j = reaction, S = stoichiometric coefficients, V = fluxes, LB = lower bound of reaction, UB = upper bound of reaction.

Instruction Follow the script to perform the FBA.

```
# Perform FBA Analysis
for rxn in model.reactions:
    print(rxn.id, rxn.name, rxn, rxn.flux)

        EX_cpd11416_b  EX_cpd11416_b  EX_cpd11416_b:
cpd11416_b <=> 0.3218680299371824
        EX_cpd00067_b  EX_cpd00067_b  EX_cpd00067_b:
cpd00067_b --> 0.0
        EX_cpd00106_b  EX_cpd00106_b  EX_cpd00106_b:
cpd00106_b <=> 0.0
        EX_cpd00036_b  EX_cpd00036_b  EX_cpd00036_b:
```

```
cpd00036_b <=> 78.23965134580074
          EX_cpd11576_b  EX_cpd11576_b  EX_cpd11576_b:
cpd11576_b <=> 0.0
          EX_cpd04097_b  EX_cpd04097_b  EX_cpd04097_b:
cpd04097_b <=> 0.0
          EX_cpd01030_b  EX_cpd01030_b  EX_cpd01030_b:
cpd01030_b <=> 0.0
          EX_cpd01012_b  EX_cpd01012_b  EX_cpd01012_b:
cpd01012_b <=> 0.0
          EX_cpd00276_b  EX_cpd00276_b  EX_cpd00276_b:
cpd00276_b --> 0.0
          EX_cpd00082_b  EX_cpd00082_b  EX_cpd00082_b:
cpd00082_b --> 0.0
          EX_cpd00001_b  EX_cpd00001_b  EX_cpd00001_b:
cpd00001_b <=> 715.8691640809429
          EX_cpd00023_b  EX_cpd00023_b  EX_cpd00023_b:
cpd00023_b <=> -3.8809594092326716
          EX_cpd00971_b  EX_cpd00971_b  EX_cpd00971_b:
cpd00971_b <=> -4.086357988097866e-15
          EX_cpd00058_b  EX_cpd00058_b  EX_cpd00058_b:
cpd00058_b <=> -0.05284020176393198
          EX_cpd00314_b  EX_cpd00314_b  EX_cpd00314_b:
cpd00314_b <=> 0.0
```

3.3.2 Flux Variability Analysis

The flux space of a biochemical reaction from an individual entity of a community model can significantly change with the alteration in the microbial and media composition of the community. To understand the flux space and altered flux space with respect to different conditions, a flux variability analysis is performed (FVA). FVA [78] generates the flux space for all reaction fluxes v_i in the community model by solving linear programming problems in a given optimal state (Sopt):

$$\text{maximize or minimize } v_i$$
$$\text{Subjected to} : S = \text{Sopt}$$
$$S.v = 0$$
$$\text{and } v_i, \text{ min} \leq v_i \leq v, \text{ max, for } i = 1, 2, 3, \ldots, n$$

FVA assists in identifying the expansion of metabolic flux within distinct biochemical pathways in separate compartments of the community.

Instruction Follow the script to perform the FVA.

```
# Perform FVA Analysis
from cobra.flux_analysis import flux_variability_analysis
FVA = cobra.flux_analysis.flux_variability_analysis(model,
 model.reactions[:],
 fraction_of_optimum=0.9)

# Print FVA data
FVA
```

	minimum	maximum
EX_cpd08636_b	0.000000	0.674400
EX_cpd15302_b	0.000000	0.000000
EX_cpd11416_b	0.297906	0.331006
EX_cpd00067_b	0.000000	0.000000
EX_cpd00106_b	0.376037	4.164486
...	...	...
EX_cpd00205_e0	-0.002628	-0.002365
EX_cpd00099_e0	-0.002628	-0.002365
EX_cpd00048_e0	-0.072189	-0.064970
EX_cpd00034_e0	-0.002628	-0.002365
EX_cpd00063_e0	-0.002628	-0.002365

1111 rows × 2 columns

3.3.3 Flux Sampling

Flux sampling is a computational technique crucial in analyzing the flux profile of genome-scale metabolic models as well as the community-scale metabolic models [79]. By randomly sampling the solution space, flux sampling generates sets of feasible flux distributions that adhere to model constraints and objectives. This technique is valuable for several reasons: First, it enables uncertainty assessment, revealing the variability and robustness of metabolic networks. Second, flux sampling identifies alternative optima, showcasing different metabolic strategies or trade-offs within the model. Third, it allows for the exploration of diverse metabolic states beyond the optimal solution, aiding in understanding adaptations under different conditions. Lastly, flux sampling assists in metabolic engineering applications by identifying targets or strategies for optimizing metabolic objectives. Various computational methods, including Monte Carlo sampling and

Markov Chain Monte Carlo methods, facilitate flux sampling by ensuring sampled distributions adhere to model constraints. Overall, flux sampling is a powerful tool for gaining insights into genome-scale metabolic networks, facilitating their manipulation for biotechnological and biomedical purposes.

Our methodology employed Artificial Coordinate Hit and Run (ACHR) sampler for sampling fluxes of a community model. To ensure sparsity and uncorrelatedness within the fluxes, a thinning factor of 100 was applied, facilitating comprehensive coverage of the flux space. The optimization process guiding ACHR flux sampling was conducted using IBM CPLEX. Subsequently, for each model, 10,000 sample points were generated using the sampling object. The flux sampling script has been coded in Python 3.7.

Instruction Follow the script to perform the FS.

```
from cobra.sampling import ACHRSampler
import cobra
model=cobra.io.read_sbml_model('C:/Users/DELL/Desktop/
Model/CM.xml')
achr = ACHRSampler(model, thinning=100)
samples=achr.sample(10000)
samples.to_csv("C:/Users/DELL/Desktop/Output/FS.csv")
samples
```

Flux sampling output:

10000 rows * column sizes (Total reactions in the model)

	rxn02201_c01r	rxn00836_c01r	rxn00390_c01r	rxn00062_c01r	rxn11567_c01r	rxn00364_c01r	rxn03408_c01r	rxn02177_c01r	rxn05440_c01r
0	2.982733e-25	0.313339	0.0	0.006934	0.0	-0.023916	0.000102	0.0	0.000967
1	4.998590e-27	0.312228	0.0	0.003520	0.0	-0.109746	0.000102	0.0	0.000966
2	5.649133e-26	0.312650	0.0	0.003527	0.0	-1.939280	0.000099	0.0	0.000938
3	7.499939e-27	0.319783	0.0	0.014451	0.0	-4.447857	0.000100	0.0	0.000946
4	2.116518e-25	0.321644	0.0	0.014453	0.0	-5.011569	0.000100	0.0	0.000953
...	...	...	...	...	...	...	...	...	...
9995	1.018486e-18	0.227407	0.0	2.685964	0.0	-48.256519	0.000125	0.0	0.001193
9996	1.023783e-18	0.204254	0.0	2.879856	0.0	-46.657304	0.000121	0.0	0.001154
9997	1.015652e-18	0.204060	0.0	2.582267	0.0	-46.011434	0.000118	0.0	0.001118
9998	1.016322e-18	0.197494	0.0	2.207859	0.0	-44.298727	0.000112	0.0	0.001067
9999	1.007179e-18	0.199832	0.0	1.873339	0.0	-44.111047	0.000113	0.0	0.001074

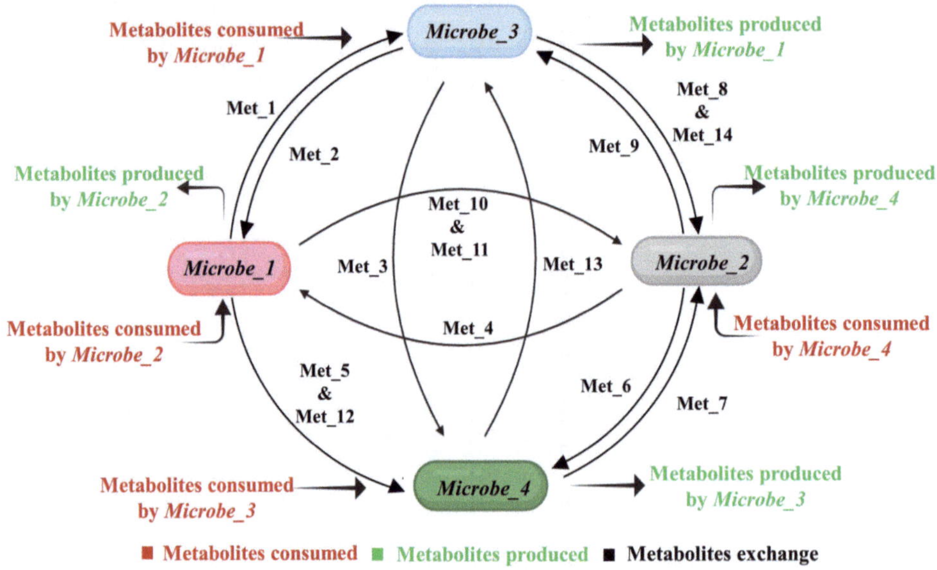

Fig. 4 Metabolic flux profile of the community model

3.3.4 Metabolic Flux Profile of the Microbial Community

The intricate network of microbes within a community relies heavily on intercommunication, achieved through the exchange of metabolites. This communication is crucial for understanding the dynamics of the entire microbial community and its effects on the host organism. To gain insights into this communication process, we employed the FBA method. It is conducted with intracellular metabolic reactions and offers a glimpse into the flux profile of individual microbes within the community. Additionally, FBA utilizing exchange reactions sheds light on the interplay between different microorganisms, facilitating comprehension of microbial cross-talk (Fig. 4). By examining flux values associated with exchange reactions, we can effectively map out the metabolic exchange patterns, thereby unraveling the intricate interactions that characterize microbial partnerships within a community. This approach provides valuable insights into the cooperative and competitive relationships that shape microbial ecosystems and their impact on host organisms.

Instruction Calculate the flux value for the exchange reactions using FBA and map it to generate the community metabolic interaction network.

Acknowledgments

The authors are thankful to the Department of Science and Technology-GoI (Grant No. CRG/2020/002080), Department of Biotechnology-GoI (Grant No. BT/PR37958/GET/119/297/2020), and Scheme for Promotion of Academic and Research Collaboration (SPARC), MHRD-GoI (Grant No. SPARC/2019–2020/P1991/SL). S.B thanks the PMRF Scheme for the fellowship. S.S.R thanks TCS RSP for the fellowship.

Conflict of Interest Statement The authors have no conflicts of interest to declare that are relevant to the content of this chapter.

References

1. Ren D, Madsen JS, de la Cruz-Perera CI et al (2014) High-throughput screening of multi-species biofilm formation and quantitative PCR-based assessment of individual species proportions, useful for exploring interspecific bacterial interactions. Microb Ecol 68:146–154. https://doi.org/10.1007/s00248-013-0315-z
2. Abreu A, Bourgois E, Gristwood A et al (2022) Priorities for ocean microbiome research. Nat Microbiol 7(7):937–947. https://doi.org/10.1038/s41564-022-01145-5
3. Hou K, Wu ZX, Chen XY et al (2022) Microbiota in health and diseases. Signal Transduct Target Ther 7(1):1–28. https://doi.org/10.1038/s41392-022-00974-4
4. Wang M, Garrido-Sanz D, Sansegundo-Lobato P et al (2021) Soil microbiome structure and function in Ecopiles used to remediate petroleum-contaminated soil. Front Environ Sci 9:624070. https://doi.org/10.3389/FENVS.2021.624070/BIBTEX
5. O'Donnell ST, Ross RP, Stanton C (2020) The progress of multi-omics technologies: determining function in lactic acid bacteria using a systems level approach. Front Microbiol 10:1–17. https://doi.org/10.3389/fmicb.2019.03084
6. Zomorrodi AR, Islam MM, Maranas CD (2014) D-OptCom: dynamic multi-level and multi-objective metabolic modeling of microbial communities. ACS Synth Biol 3:247–257. https://doi.org/10.1021/sb4001307
7. Chan SHJ, Simons MN, Maranas CD (2017) SteadyCom: predicting microbial abundances while ensuring community stability. PLoS Comput Biol 13:1–25. https://doi.org/10.1371/journal.pcbi.1005539
8. Koch S, Kohrs F, Lahmann P et al (2019) Redcom: a strategy for reduced metabolic modeling of complex microbial communities and its application for analyzing experimental datasets from anaerobic digestion. PLoS Comput Biol 15:e1006759
9. Diener C, Gibbons SM, Resendis-Antonio O (2020) MICOM: metagenome-scale modeling to infer metabolic interactions in the gut microbiota. mSystems 5:e00606-19. https://doi.org/10.1128/msystems.00606-19
10. Kundu P, Beura S, Mondal S et al (2024) Machine learning for the advancement of genome-scale metabolic modeling. Biotechnol Adv 74:108400. https://doi.org/10.1016/J.BIOTECHADV.2024.108400
11. Zuñiga C, Li CT, Yu G et al (2019) Environmental stimuli drive a transition from cooperation to competition in synthetic phototrophic communities. Nat Microbiol 4:2184–2191. https://doi.org/10.1038/s41564-019-0567-6
12. Zomorrodi AR, Segrè D (2017) Genome-driven evolutionary game theory helps understand the rise of metabolic interdependencies in microbial communities. Nat Commun 8:1–11. https://doi.org/10.1038/s41467-017-01407-5
13. Mazumdar V, Snitkin ES, Amar S, Segrè D (2009) Metabolic network model of a human oral pathogen. J Bacteriol 91:74–90. https://doi.org/10.1128/JB.01123-08
14. Chowdhury S, Fong SS (2020) Computational modeling of the human microbiome. Microorganisms 8:1–28. https://doi.org/10.3390/microorganisms8020197

15. Hastings J, Mains A, Virk B et al (2019) Multi-omics and genome-scale modeling reveal a metabolic shift during C. elegans aging. Front Mol Biosci 6:1–18. https://doi.org/10.3389/fmolb.2019.00002
16. King ZA, Lu J, Dräger A et al (2016) BiGG Models: a platform for integrating, standardizing and sharing genome-scale models. Nucleic Acids Res 44:D515–D522. https://doi.org/10.1093/NAR/GKV1049
17. Büchel F, Rodriguez N, Swainston N et al (2013) Path2Models: large-scale generation of computational models from biochemical pathway maps. BMC Syst Biol 7:116. https://doi.org/10.1186/1752-0509-7-116
18. Magnúsdóttir S, Heinken A, Kutt L et al (2017) Generation of genome-scale metabolic reconstructions for 773 members of the human gut microbiota. Nat Biotechnol 35:81–89. https://doi.org/10.1038/nbt.3703
19. Feist AM, Herrgård MJ, Thiele I et al (2009) Reconstruction of biochemical networks in microorganisms. Nat Rev Microbiol 7:129–143. https://doi.org/10.1038/NRMICRO1949
20. Henry CS, Dejongh M, Best AA et al (2010) High-throughput generation, optimization and analysis of genome-scale metabolic models. Nat Biotechnol 28:977–982. https://doi.org/10.1038/nbt.1672
21. Seaver SMD, Liu F, Zhang Q et al (2021) The ModelSEED Biochemistry Database for the integration of metabolic annotations and the reconstruction, comparison and analysis of metabolic models for plants, fungi and microbes. Nucleic Acids Res 49:D575–D588. https://doi.org/10.1093/NAR/GKAA746
22. Pabinger S, Snajder R, Hardiman T et al (2014) MEMOSys 2.0: an update of the bioinformatics database for genome-scale models and genomic data. Database (Oxford) 2014:bau004. https://doi.org/10.1093/DATABASE/BAU004
23. Garcia-Albornoz M, Thankaswamy-Kosalai S, Nilsson A et al (2014) BioMet Toolbox 2.0: genome-wide analysis of metabolism and omics data. Nucleic Acids Res 42:W175. https://doi.org/10.1093/NAR/GKU371
24. Karp PD, Paley SM, Krummenacker M et al (2010) Pathway Tools version 13.0: integrated software for pathway/genome informatics and systems biology. Brief Bioinform 11:40. https://doi.org/10.1093/BIB/BBP043
25. Brunk E, Sahoo S, Zielinski DC et al (2018) Recon3D enables a three-dimensional view of gene variation in human metabolism. Nat Biotechnol 36:272–281. https://doi.org/10.1038/nbt.4072
26. Chelliah V, Laibe C, Le Novère N (2013) BioModels Database: a repository of mathematical models of biological processes. Methods Mol Biol 1021:189–199. https://doi.org/10.1007/978-1-62703-450-0_10
27. Pornputtapong N, Nookaew I, Nielsen J (2015) Human metabolic atlas: an online resource for human metabolism. Database (Oxford) 2015:bav068. https://doi.org/10.1093/DATABASE/BAV068
28. Noronha A, Modamio J, Jarosz Y et al (2019) The Virtual Metabolic Human database: integrating human and gut microbiome metabolism with nutrition and disease. Nucleic Acids Res 47:D614–D624. https://doi.org/10.1093/nar/gky992
29. Hirschman JE, Balakrishnan R, Christie KR et al (2006) Genome Snapshot: a new resource at the Saccharomyces Genome Database (SGD) presenting an overview of the Saccharomyces cerevisiae genome. Nucleic Acids Res 34:D442. https://doi.org/10.1093/NAR/GKJ117
30. Karp PD, Riley M, Saier M et al (2002) The EcoCyc Database. Nucleic Acids Res 30:56–58. https://doi.org/10.1093/NAR/30.1.56
31. Mueller LA, Zhang P, Rhee SY (2003) AraCyc: a biochemical pathway database for Arabidopsis. Plant Physiol 132:453. https://doi.org/10.1104/PP.102.017236
32. Liao YC, Tsai MH, Chen FC, Hsiung CA (2012) GEMSiRV: a software platform for GEnome-scale metabolic model simulation, reconstruction and visualization. Bioinformatics 28:1752–1758. https://doi.org/10.1093/BIOINFORMATICS/BTS267
33. Arakawa K, Yamada Y, Shinoda K et al (2006) GEM system: automatic prototyping of cell-wide metabolic pathway models from genomes. BMC Bioinformatics 7:1–11. https://doi.org/10.1186/1471-2105-7-168
34. Heirendt L, Arreckx S, Pfau T et al (2019) Creation and analysis of biochemical constraint-based models using the COBRA Toolbox v.3.0. Nat Protoc 14:639–702. https://doi.org/10.1038/S41596-018-0098-2
35. Karp PD, Paley S, Romero P (2002) The pathway tools software. Bioinformatics 18:225–232. https://doi.org/10.1093/bioinformatics/18.suppl_1.S225

36. Aite M, Chevallier M, Frioux C et al (2018) Traceability, reproducibility and wiki-exploration for "à-la-carte" reconstructions of genome-scale metabolic models. PLoS Comput Biol 14:1–25. https://doi.org/10.1371/journal.pcbi.1006146
37. Loira N, Zhukova A, Sherman DJ (2015) Pantograph: a template-based method for genome-scale metabolic model reconstruction. J Bioinforma Comput Biol 13: 1550006. https://doi.org/10.1142/S0219720015500067
38. Pitkänen E, Jouhten P, Hou J et al (2014) Comparative genome-scale reconstruction of gapless metabolic networks for present and ancestral species. PLoS Comput Biol 10: e1003465. https://doi.org/10.1371/journal.pcbi.1003465
39. Feng X, Xu Y, Chen Y, Tang YJ (2012) MicrobesFlux: a web platform for drafting metabolic models from the KEGG database. BMC Syst Biol 6:94. https://doi.org/10.1186/1752-0509-6-94
40. Karlsen E, Schulz C, Almaas E (2018) Automated generation of genome-scale metabolic draft reconstructions based on KEGG. BMC Bioinformatics 19:1–11. https://doi.org/10.1186/s12859-018-2472-z
41. Poolman MG (2006) Scrumpy: metabolic modelling with Python. Syst Biol (Stevenage) 153:375–378
42. Thorleifsson SG, Thiele I (2011) rBioNet: a COBRA toolbox extension for reconstructing high-quality biochemical networks. Bioinformatics 27:2009–2010. https://doi.org/10.1093/BIOINFORMATICS/BTR308
43. Nursimulu N, Moses AM, Parkinson J (2022) Architect: a tool for aiding the reconstruction of high-quality metabolic models through improved enzyme annotation. PLoS Comput Biol 18:e1010452. https://doi.org/10.1371/JOURNAL.PCBI.1010452
44. Machado D, Andrejev S, Tramontano M, Patil KR (2018) Fast automated reconstruction of genome-scale metabolic models for microbial species and communities. Nucleic Acids Res 46:7542–7553. https://doi.org/10.1093/nar/gky537
45. Lloyd CJ, Ebrahim A, Yang L et al (2018) COBRAme: a computational framework for genome-scale models of metabolism and gene expression. PLoS Comput Biol 14: e1006302. https://doi.org/10.1371/JOURNAL.PCBI.1006302
46. Hanemaaijer M, Olivier BG, Röling WFM et al (2017) Model-based quantification of metabolic interactions from dynamic microbial-community data. PLoS One 12: e0173183. https://doi.org/10.1371/JOURNAL.PONE.0173183
47. Agren R, Liu L, Shoaie S et al (2013) The RAVEN Toolbox and its use for generating a genome-scale metabolic model for Penicillium chrysogenum. PLoS Comput Biol 9: e1002980. https://doi.org/10.1371/journal.pcbi.1002980
48. Cottret L, Wildridge D, Vinson F et al (2010) MetExplore: a web server to link metabolomic experiments and genome-scale metabolic networks. Nucleic Acids Res 38:W132–W137. https://doi.org/10.1093/NAR/GKQ312
49. Dias O, Rocha M, Ferreira EC, Rocha I (2015) Reconstructing genome-scale metabolic models with merlin. Nucleic Acids Res 43:3899–3910. https://doi.org/10.1093/nar/gkv294
50. Swainston N, Smallbone K, Mendes P et al (2011) The SuBliMinaL Toolbox: automating steps in the reconstruction of metabolic networks. J Integr Bioinform 8:186. https://doi.org/10.1515/jib-2011-186
51. Boele J, Olivier BG (2012) FAME, the flux analysis and modeling environment. BMC Syst Biol 6:8
52. Olivier B, Gottstein W, Molenaar D, Teusink B (2021) CBMPy release 0.8.2. Zenodo. https://doi.org/10.5281/ZENODO.5546608
53. Zhuang K, Izallalen M, Mouser P et al (2011) Genome-scale dynamic modeling of the competition between Rhodoferax and Geobacter in anoxic subsurface environments. ISME J 5: 305–316. https://doi.org/10.1038/ismej.2010.117
54. Mendes-Soares H, Mundy M, Soares LM, Chia N (2016) MMinte: an application for predicting metabolic interactions among the microbial species in a community. BMC Bioinformatics 17:1–10. https://doi.org/10.1186/s12859-016-1230-3
55. Brunner JD, Chia N (2020) Minimizing the number of optimizations for efficient community dynamic flux balance analysis. PLoS Comput Biol 16:1–20. https://doi.org/10.1371/journal.pcbi.1007786
56. Khandelwal RA, Olivier BG, Röling WFM et al (2013) Community flux balance analysis for microbial consortia at balanced growth. PLoS One 8:e64567. https://doi.org/10.1371/journal.pone.0064567
57. Thommes M, Wang T, Zhao Q et al (2019) Designing metabolic division of labor in microbial communities. mSystems 4:263–

281. https://doi.org/10.1128/MSYSTEMS.00263-18/SUPPL_FILE/MSYSTEMS.00263-18-SF010.TIF

58. García-Jiménez B, García JL, Nogales J (2018) FLYCOP: metabolic modeling-based analysis and engineering microbial communities. Bioinformatics 34:i954–i963. https://doi.org/10.1093/BIOINFORMATICS/BTY561

59. Gomez JA, Barton PI (2018) Dynamic flux balance analysis using DFBAlab. Methods Mol Biol 1716:353–370. https://doi.org/10.1007/978-1-4939-7528-0_16

60. Moretti S, Martin O, Van Du Tran T et al (2016) MetaNetX/MNXref--reconciliation of metabolites and biochemical reactions to bring together genome-scale metabolic networks. Nucleic Acids Res 44:D523–D526. https://doi.org/10.1093/NAR/GKV1117

61. Koch S, Benndorf D, Fronk K et al (2016) Predicting compositions of microbial communities from stoichiometric models with applications for the biogas process. Biotechnol Biofuels 9:1–16. https://doi.org/10.1186/S13068-016-0429-X/FIGURES/5

62. Cai J, Tan T, Chan SHJ (2020) Predicting Nash equilibria for microbial metabolic interactions. Bioinformatics 36:5649–5655. https://doi.org/10.1093/bioinformatics/btaa1014

63. Henry CS, Bernstein HC, Weisenhorn P et al (2016) Microbial community metabolic modeling: a community data-driven network reconstruction. J Cell Physiol 231:2339–2345. https://doi.org/10.1002/jcp.25428

64. Zomorrodi AR, Maranas CD (2012) OptCom: a multi-level optimization framework for the metabolic modeling and analysis of microbial communities. PLoS Comput Biol 8:e1002363. https://doi.org/10.1371/journal.pcbi.1002363

65. Karimian E, Motamedian E (2020) ACBM: an integrated agent and constraint based modeling framework for simulation of microbial communities. Sci Rep 10(1):1–10. https://doi.org/10.1038/s41598-020-65659-w

66. Geng J, Ji B, Li G et al (2021) CODY enables quantitatively spatiotemporal predictions on in vivo gut microbial variability induced by diet intervention. Proc Natl Acad Sci USA 118:e2019336118. https://doi.org/10.1073/pnas.2019336118

67. Harcombe WR, Riehl WJ, Dukovski I et al (2014) Metabolic resource allocation in individual microbes determines ecosystem interactions and spatial dynamics. Cell Rep 7:1104–1115. https://doi.org/10.1016/j.celrep.2014.03.070

68. Popp D, Centler F (2020) μBialSim: constraint-based dynamic simulation of complex microbiomes. Front Bioeng Biotechnol 8:574. https://doi.org/10.3389/fbioe.2020.00574

69. Lu Y, Zhou G, Ewald J et al (2023) MicrobiomeAnalyst 2.0: comprehensive statistical, functional and integrative analysis of microbiome data. Nucleic Acids Res 51:W310–W318. https://doi.org/10.1093/NAR/GKAD407

70. Shoaie S, Ghaffari P, Kovatcheva-Datchary P et al (2015) Quantifying diet-induced metabolic changes of the human gut microbiome. Cell Metab 22:320–331. https://doi.org/10.1016/J.CMET.2015.07.001

71. Taffs R, Aston JE, Brileya K et al (2009) In silico approaches to study mass and energy flows in microbial consortia: a syntrophic case study. BMC Syst Biol 3:114. https://doi.org/10.1186/1752-0509-3-114

72. Borer B, Ataman M, Hatzimanikatis V, Or D (2019) Modeling metabolic networks of individual bacterial agents in heterogeneous and dynamic soil habitats (IndiMeSH). PLoS Comput Biol 15:e1007127. https://doi.org/10.1371/JOURNAL.PCBI.1007127

73. Großkopf T, Consuegra J, Gaffé J et al (2016) Metabolic modelling in a dynamic evolutionary framework predicts adaptive diversification of bacteria in a long-term evolution experiment. BMC Evol Biol 16:1–15. https://doi.org/10.1186/s12862-016-0733-x

74. Schroeder WL, Saha R (2020) Introducing an optimization- and explicit Runge-Kutta-based approach to perform dynamic flux balance analysis. Sci Rep 10:1–28. https://doi.org/10.1038/s41598-020-65457-4

75. Bauer E, Zimmermann J, Baldini F et al (2017) BacArena: individual-based metabolic modeling of heterogeneous microbes in complex communities. PLoS Comput Biol 13:1–22. https://doi.org/10.1371/journal.pcbi.1005544

76. Baldini F, Heinken A, Heirendt L et al (2019) The microbiome modeling toolbox: from microbial interactions to personalized microbial communities. Bioinformatics 35:2332–2334. https://doi.org/10.1093/bioinformatics/bty941

77. Orth JD, Thiele I, Palsson BO (2010) What is flux balance analysis? Nat Biotechnol 28(3):245–248. https://doi.org/10.1038/nbt.1614

78. Gudmundsson S, Thiele I (2010) Computationally efficient flux variability analysis. BMC Bioinformatics 11:489. https://doi.org/10.1186/1471-2105-11-489
79. Herrmann HA, Dyson BC, Vass L et al (2019) Flux sampling is a powerful tool to study metabolism under changing environmental conditions. npj Syst Biol Appl 5:1–8. https://doi.org/10.1038/s41540-019-0109-0
80. Henson MA, Hanly TJ (2014) Dynamic flux balance analysis for synthetic microbial communities. IET Syst Biol 8:214–229. https://doi.org/10.1049/IET-SYB.2013.0021
81. Kundu P, Ghosh A (2023) Genome-scale community modeling for deciphering the inter-microbial metabolic interactions in fungus-farming termite gut microbiome. Comput Biol Med 154:106600. https://doi.org/10.1016/j.compbiomed.2023.106600
82. Beura S, Kundu P, Das AK, Ghosh A (2022) Metagenome-scale community metabolic modelling for understanding the role of gut microbiota in human health. Comput Biol Med 149:105997. https://doi.org/10.1016/J.COMPBIOMED.2022.105997
83. García-Jiménez B, Torres-Bacete J, Nogales J (2021) Metabolic modelling approaches for describing and engineering microbial communities. Comput Struct Biotechnol J 19: 226–246. https://doi.org/10.1016/j.csbj.2020.12.003
84. Hyun JK, Boedicker JQ, Jang WC, Ismagilov RF (2008) Defined spatial structure stabilizes a synthetic multispecies bacterial community. Proc Natl Acad Sci USA 105:18188–18193. https://doi.org/10.1073/pnas.0807935105
85. Lozano GL, Bravo JI, Garavito Diago MF et al (2019) Introducing THOR, a model microbiome for genetic dissection of community behavior. mBio 10:e02846-18. https://doi.org/10.1128/mBio.02846-18
86. Lee KWK, Periasamy S, Mukherjee M et al (2014) Biofilm development and enhanced stress resistance of a model, mixed-species community biofilm. ISME J 8:894–907. https://doi.org/10.1038/ismej.2013.194
87. Villa F, Pitts B, Lauchnor E et al (2015) Development of a laboratory model of a phototroph-heterotroph mixed-species biofilm at the stone/air interface. Front Microbiol 6:1–14. https://doi.org/10.3389/fmicb.2015.01251
88. Wolfe BE, Button JE, Santarelli M, Dutton RJ (2014) Cheese rind communities provide tractable systems for in situ and in vitro studies of microbial diversity. Cell 158:422–433. https://doi.org/10.1016/j.cell.2014.05.041
89. Kosina SM, Danielewicz MA, Mohammed M et al (2016) Exometabolomics assisted design and validation of synthetic obligate mutualism. ACS Synth Biol 5:569–576. https://doi.org/10.1021/acssynbio.5b00236
90. Venturelli OS, Carr AV, Fisher G et al (2018) Deciphering microbial interactions in synthetic human gut microbiome communities. Mol Syst Biol 14:1–19. https://doi.org/10.15252/msb.20178157
91. Filkins LM, Graber JA, Olson DG et al (2015) Coculture of Staphylococcus aureus with Pseudomonas aeruginosa drives S. aureus towards fermentative metabolism and reduced viability in a cystic fibrosis model. J Bacteriol 197:2252–2264. https://doi.org/10.1128/JB.00059-15
92. Gutiérrez N, Garrido D (2019) Species deletions from microbiome consortia reveal key metabolic interactions between gut microbes. mSystems 4:e00185-19. https://doi.org/10.1128/msystems.00185-19
93. Islam MM, Fernando SC, Saha R (2019) Metabolic modeling elucidates the transactions in the rumen microbiome and the shifts upon virome interactions. Front Microbiol 10:2412. https://doi.org/10.3389/FMICB.2019.02412
94. Beura S, Kundu P, Das AK, Ghosh A (2024) Genome-scale community modelling elucidates the metabolic interaction in Indian type-2 diabetic gut microbiota. Sci Rep 14(1):1–12. https://doi.org/10.1038/s41598-024-63718-0
95. Bolyen E, Rideout JR, Dillon MR et al (2019) Reproducible, interactive, scalable and extensible microbiome data science using QIIME 2. Nat Biotechnol 37(8):852–857. https://doi.org/10.1038/s41587-019-0209-9
96. Mori H, Maruyama T, Yano M et al (2018) VITCOMIC2: visualization tool for the phylogenetic composition of microbial communities based on 16S rRNA gene amplicons and metagenomic shotgun sequencing. BMC Syst Biol 12:47–58. https://doi.org/10.1186/S12918-018-0545-2/FIGURES/6
97. Mahadevan R, Schilling CH (2003) The effects of alternate optimal solutions in constraint-based genome-scale metabolic models. Metab Eng 5:264–276. https://doi.org/10.1016/j.ymben.2003.09.002
98. Caspi R, Altman T, Billington R et al (2014) The MetaCyc database of metabolic pathways and enzymes and the BioCyc collection of

pathway/genome databases. Nucleic Acids Res 42:459–471. https://doi.org/10.1093/nar/gkt1103

99. Thiele I, Palsson BØ (2010) A protocol for generating a high-quality genome-scale metabolic reconstruction. Nat Protoc 5:93–121. https://doi.org/10.1038/nprot.2009.203

100. de Matos P, Alcántara R, Dekker A et al (2010) Chemical Entities of Biological Interest: an update. Nucleic Acids Res 38:D249. https://doi.org/10.1093/NAR/GKP886

101. Schellenberger J, Park JO, Conrad TM, Palsson BT (2010) BiGG: a Biochemical Genetic and Genomic knowledgebase of large scale metabolic reconstructions. BMC Bioinformatics 11:213–213. https://doi.org/10.1186/1471-2105-11-213

102. Perez-Garcia O, Lear G, Singhal N (2016) Metabolic network modeling of microbial interactions in natural and engineered environmental systems. Front Microbiol 7:673. https://doi.org/10.3389/fmicb.2016.00673

Chapter 11

Constraint-Based Modeling of Microbial Communities for Metabolite Production

Maziya Ibrahim and Karthik Raman

Abstract

In this chapter, we describe an in silico approach called CAMP (Co-Culture/Community Analyses for Metabolite Production) to predict two-species microbial communities best suited for producing a desired metabolite. Here, we use genome-scale metabolic models (GEMs) to build microbial communities and constraint-based modeling methods such as flux balance analysis (FBA) to assess and identify suitable communities. Flux variability analysis (FVA) detects the maximum product flux in the communities. The interaction behavior between community members, i.e., mutualism, commensalism, parasitism, and competition, can be deduced based on the variations in the predicted growth rates of the species as monocultures and in co-cultures. In silico community optimization strategies to predict reaction knockouts that improve product flux have also been implemented. CAMP source codes are available from https://github.com/RamanLab/CAMP/tree/master/Protocol.

Key words Genome-scale metabolic models, Cross-feeding, Reaction knockouts, Biological production, Biobased economy, Sustainable bioprocesses

1 Introduction

Microbial communities are ubiquitous in nature, with some widely studied communities found in the human gut and environments such as the soil. In metabolic engineering, microbial co-cultures or communities are useful for producing high-value industrial chemicals [1]. There are tremendous benefits to engineering such microbial communities, such as improved substrate utilization, modularity, and robustness to perturbation, compared to monocultures [2]. Metabolic interactions are the underlying backbone of microbial communities. Therefore, understanding the interactions between community members is essential to using such communities for biotechnological processes. Among the various methods

Karthik Raman and Gayathri Sambamoorthy (eds.), *Modelling the Microbiome*, Methods in Molecular Biology, vol. 3006,
https://doi.org/10.1007/978-1-0716-5080-6_11,

that model microbial metabolism, strategies based on constraint-based modeling are popular due to their ease of implementation. The primary tool that facilitates such modeling is a genome-scale metabolic model (GEM). Annotation of genomic data provides the input to begin a draft reconstruction of a metabolic model. The models capture the metabolic genes and reactions in a species mathematically through stoichiometric matrices. Further, gene–protein–reaction (GPR) associations help connect enzymes that catalyze reactions in different pathways. There are several automated pipelines available for the reconstruction of metabolic models given a genome sequence. Examples include KBase [3], CarveMe [4], AurReMe [5], etc.

A community GEM is a multicompartment model where a community compartment denoted as "[u]" is present to exchange metabolites between the species. The stoichiometric matrix of a community model of two species can be written as $S_{m \times n}$. The columns in the matrix represent the total number of reactions n ($n_1 + n_2 + n_e$) from species 1 (n_1), species 2 (n_2), and the community exchange reactions (n_e). The rows of the S matrix represent the total metabolites m ($m_1 + m_2 + m_e$) from species 1 (m_1), species 2(m_2), and the metabolites that are part of the community exchange reactions (m_e) [6].

The first attempts to model microbial co-cultures were carried out by Stolyar and colleagues, who investigated the co-culture of *Desulfovibrio vulgaris* and *Methanococcus maripaludis*, capturing growth parameters [7]. Ye et al. designed an artificial community of *Ketogulonicigenium vulgare* and *Bacillus megaterium* to produce vitamin C [8]. Recent algorithms such as SteadyCom and MICOM [9] can simulate community metabolic models at a large scale with many species using steady-state FBA. SteadyCom assumes a perfect diagonal growth trajectory through the flux cone. The trajectory implies that taxa grow independently of one another, and the steady state is maintained by a linear dilution that acts the same on all species. Limitations for SteadyCom include the phenomenon of "forced altruism" that forces individual species to produce metabolites for other community members if this can maximize the community objective. Dynamic FBA methods can simulate the full trajectory from initial conditions to steady state and back to the stationary phase when nutrients are limited. However, such dynamic methods are computationally expensive. Each modeling approach has advantages and can be chosen based on the research objective [10]. All strategies will necessitate experimental validation to rate their potential (*see* **Note 1**). In the CAMP [11] method, we have employed SteadyCom to generate microbial community models and perform FBA simulations for co-cultures.

2 Materials

2.1 Software

2.1.1 COBRA Toolbox

Constraint-based reconstruction and analysis (COBRA) Toolbox [12] is an open-source software suite that runs in the MATLAB environment for analyzing metabolic models. A Python-equivalent is also available (https://opencobra.github.io/cobrapy/). In this chapter, we make use of the MATLAB version. To perform computations, we use COBRA Toolbox v.3.0. COBRA Toolbox is downloaded and installed with MATLAB by following the instructions at https://opencobra.github.io/cobratoolbox/stable/installation.html. To install MATLAB, we refer to the MathWorks® documentation (https://in.mathworks.com/help/install/install-products.html). We use MATLAB2018b or higher for easy implementation of the CAMP method.

2.1.2 SteadyCom

SteadyCom is a constraint-based optimization framework to predict steady-state flux distributions of microbial communities built with two or more metabolic models. SteadyCom is compatible with flux variability analysis. The computational time required to solve the linear programming problems is independent of the number of organisms considered. The objective function is the maximization of the community growth rate. SteadyCom contains specialized codes for flux variability analysis that utilizes the IBM ILOG CPLEX solver. Install the solver separately from IBM after signing up for an account (note that it is free for academic use). Use CPLEX 12.8 for implementing the CAMP method. Functions *createCommModel, SteadyComCplex*, and *SteadyComFVACplex* from the SteadyCom algorithm are part of the CAMP pipeline.

2.2 Microbial GEM Resources

Genome-scale metabolic models for microbes are available in various resources such as the BiGG database (http://bigg.ucsd.edu/) [13]. A recently updated resource of >7000 semi-curated metabolic models of various microbial species is also available at (https://www.vmh.life/files/reconstructions/AGORA2/) [14]. Older versions of microbial models from the AGORA resource, i.e., AGORA v1.03, are available at https://vmh.life/#downloadview. For the CAMP analysis, use models from the AGORA v.1.03 dataset.

Metabolic models can also be generated from genome sequence information through the KBase platform (https://narrative.kbase.us/legacy/catalog/apps) [3], where many tools are available to reconstruct prokaryotic metabolic models.

Metabolic models obtained through any such resource can be used in the CAMP pipeline. Ensure that these models have a feasible flux solution when the objective function is set as the biomass reaction of the metabolic model. Curate or gap-fill the metabolic model based on the metabolite of interest. Different reconstruction methods, such as KBase [3], gapseq [15], and CarveMe [4], while based on the same input genome sequence, can result in metabolic models with varying numbers of reactions and genes. For a more accurate assessment of the functional potential of microbes in communities, consider using a consensus model, as recent findings suggest this approach improves model reliability [16].

3 Methods

In this section, we will describe the CAMP pipeline in detail and stepwise. We use the AGORA v.1.03 models, as mentioned in Subheading 2.2. As a case study, we can select six models from the *Lactobacillus* genus as we are interested in identifying a two-species community that can produce the metabolite lactate in more significant amounts than when each of the models is considered as isolate, i.e., monoculture production capabilities. Access all required MATLAB scripts and metabolic model files for executing the following steps at: https://github.com/RamanLab/CAMP/tree/master/MiMBChapter

3.1 Monoculture Growth of Each Microbe

The "*monoculturegrowth*" function computes the monoculture growth rates of single microbial models and gives the fluxes from FBA and FVA as outputs for the particular metabolite reaction of interest.

The inputs to this function are as follows:

(i) Folder name (*foldername*)—Specify the folder containing all the metabolic model files saved in "*.mat*" extension, i.e., MATLAB readable files.

(ii) Reaction name (*RxnName*)—Provide the reaction name of the metabolite of interest in the model to evaluate its production capabilities.

(iii) Substrate reactions (*substrateRxns*)—Define the names of the substrate reactions in the model, typically the major carbon source of nutrients for the organism of interest.

(iv) Reaction constraints (*reactionsAndBounds*)—Set constraints for different reactions in the model under which the FBA analysis should be performed—these constraints include defined media concentrations of the metabolites.

Outputs of this function are as follows:

(i) Growth(*Growth*)—The growth rate (h^{-1}), representing the flux through the biomass reaction of the model.

(ii) Flux values (*fbaflux*)—The flux value of the product/metabolite reaction in the model.

(iii) Maximum flux (*maxflux*)—The maximum flux value calculated through the flux variability analysis of the product/metabolite reaction in the model.

(iv) Substrate uptake flux (*substrate_uptakeflux*)—The flux of the substrate reaction(s) in the model.

(v) Product yield (*maxproductyield_mono*)—The maximum product flux based on the substrate uptake flux.

```
foldername = 'models';
RxnName = 'EX_lac(e)';
substrateRxns ='EX_glc_D(e)';
reactionsAndBounds =
106×2 cell array
    {'EX_thr_L(e)'    }    {[      -1]}
    {'EX_trp_L(e)'    }    {[      -1]}
    {'EX_tyr_L(e)'    }    {[      -1]}
...
    {'EX_val_L(e)'    }    {[      -1]}
    {'EX_glc_D(e)'    }    {[      -1]}
>>  [Growth,fbaflux,maxflux,substrate_uptakeflux,maxproduc-
tyield_mono]=  monoculturegrowth(foldername,RxnName,sub-
strateRxns,reactionsAndBounds)
```

The output is stored as an Excel file named "monoculture_-results.xlsx". This file contains the computed growth rates for each model, along with the product and substrate uptake fluxes in monoculture.

Proceed by constructing two-species community models to simulate co-cultures and identify the community growth rates.

3.2 Co-Culture Simulations

The "*create_community_models*" function generates all possible two-species community models.

Inputs:

1. Folder path (input folder)—Specify the folder containing the metabolic model files in ".mat" format.
2. Folder path (output folder)—Create an empty folder to store the generated community models and provide its name/path as input.

Output:

The function creates community models by combining two individual model files. For example, if the input folder contains "*microbeA.mat*" and "*microbeB.mat*", the resulting community model will be named "*microbeA&microbeB.mat*" and stored in the specified output folder.

```
>> create_community_models(input_folder, output_folder)
     1
 Compile metabolites and reactions...
Add community metabolites and exchange reactions...
Merge gene-reaction rules...
```

The "*create_community_models*" function generates all possible 15 two-species community models when provided with six metabolic models as input. After creating the community models, perform the FBA simulations using SteadyCom. For this simulation, use the "*communitygrowth*" function.

Step 1: Identify Cross-Fed Metabolites

The function "*communitygrowth*" simulates community growth using FBA and identifies cross-fed metabolites among community members. If a metabolite is secreted into the community compartment [u] by one organism, it will have a positive flux value in the FBA solution. If the other organism takes up the same metabolite, it will have a negative flux value in the FBA solution. This metabolite is termed as "cross-fed" between the two species. A threshold is applied to identify such cross-fed metabolites. The default threshold in CAMP is 0.0001.

The inputs to this function are as follows:

(i) Folder name (*folder_name*)—Specify the folder containing all community models. This is the output from the "*create_community_models*" function.

(ii) Exchange reaction identifier (*RxnName*)—Provides the exchange reaction for the product/metabolite of interest, such as "EX_lac_D(u)".

(iii) Substrate reaction(s) (*substrateRxns*)—List the substrate reactions as a cell array, specifying the carbon substrates consumed by the community, such as "EX_glc_D(e)".

(iv) Reaction constraints (*reactionsAndBounds*)—Define constraints for various reactions in the community model under which the FBA analysis should be performed. These constraints include defined media concentrations of the metabolites. Note: The exchange reactions in the community model are denoted as "EX_metabolite(u)".

(v) Growth rate threshold (*growthrate_threshold*)—Set the minimum growth rate required for microbes in the community to be considered viable.

(vi) Cross-fed threshold (*cross_fed_threshold*)—Define the minimum flux at which a metabolite is to be considered cross-fed between the two microbes.

```
>> communitygrowth(folder_name, RxnName, substrateRxns,
reactionsAndBounds,                    growthrate_threshold,
cross_fed_threshold)
```

The results are stored in an Excel file named "*communityresults.xlsx*", which contains two sheets. The first sheet in the file records the growth rates, species abundances, and product flux values for each community model. The second sheet lists the different cross-fed metabolites in each community.

3.3 Community Assessment

Step 1: Prediction of Interaction Behavior

The type of interaction between community members is defined by comparing the growth rates of a microbe in co-culture and monoculture. Interactions can be positive or negative, i.e., it could lead to an increase or decrease in growth rate compared to monoculture growth rates. A difference of 10% in predicted growth rates of the microbes in monoculture versus co-culture has been previously established to be significant in categorizing community interactions [17]. Based on these comparisons, viable communities can be classified as follows:

(i) Amensal communities, i.e., one microbe grows less in the community while the other microbe's growth rate is unaffected.

(ii) Competitive communities, i.e., both microbes' growth rates are less than their monoculture rates.

(iii) Parasitic communities, i.e., one microbe grows faster in the community, while the other grows slower.

(iv) Neutral communities, where neither microbe's growth rate was substantially affected upon being paired with the other.

(v) Commensal communities, i.e., one microbe has an increase in growth rate, while the other remains unaffected.

(vi) Mutualistic communities where both microbes in the pair show a substantial increase in the growth rates compared to their monoculture rates.

Use the "*retrieveparameters*" function to extract relevant data recorded in the "*monoculture_results.xlsx*" and "*communityresults.xlsx*" files generated in **steps 1** and **2**. Provide file paths to "*retrieveparameters*". The parameters, including microbial growth rates, including microbe growth rates in monoculture and co-culture, substrate uptake, rates and product yield, can then be input in the "*ComputeCommunityInteraction*" function.

```
>> [orglist_mono, growth_mono, maxproductyield_mono,
org1, org2, g1, g2, s1, s2, comm_FVAmaxflux] = retrieve-
parameters(monoculture_results,communityresults)
```

ComputeCommunityInteraction calculates the observed metabolite yield

$$M_{o_i} \text{ observed metabolite yield} = \frac{\text{maximum metabolite flux in coculture}}{\text{Total substrate uptake}}$$

and the expected metabolite yield, $\boldsymbol{M_{Ei}}$, which depends on the substrate uptake by each species in co-culture. This is adapted from the ConYE model by Medlock et al. [18].

$$\boldsymbol{M_{Ei}} = (s_1 \times y_{1i}) + (s_2 \times y_{2i})$$

M_{Ei} expected metabolite yield

$s1$ total substrate uptake of species 1 in co-culture

$s2$ total substrate uptake of species 2 in co-culture

y_{1i} the maximum yield of metabolite i in species 1 in monoculture

$$= \frac{\text{maximum metabolite flux of species 1}}{\text{substrate uptake of species 1}}$$

y_{2i} the maximum yield of metabolite i in species 2 in monoculture

$$= \frac{\text{maximum metabolite flux of species 2}}{\text{substrate uptake of species 2}}$$

Candidate pairs are defined as those with a higher observed metabolite yield than expected. The default fold change threshold is set to 2. This function identifies candidate pairs for each interaction type, if present, allowing for the selection of microbial communities with enhanced metabolite production.

```
>> [candidate_Amensal, candidate_Parasitic, candidate_-
Competition, candidate_Neutral, candidate_Commensal,
candidate_Mutualistic, candidate] = ComputeCommunityIn-
teraction(orglist_mono, growth_mono, maxproductyield_-
mono, org1, org2, g1, g2, s1, s2, comm_FVAmaxflux,
foldchange)
```

One of the candidate pairs exhibiting neutral interaction behavior is *Lactobacillus animalis KCTC 3501* and *Lactobacillus plantarum WCFS1*.

3.4 Community Optimization for Product Maximization

Step 1: Reaction Knockouts that Increase Product Flux in the Communities

Strain optimization methods in metabolic engineering, such as gene knockouts, can be simulated using the community metabolic models. Here, we aim to identify reaction knockouts that would positively impact metabolite production in the community. We have implemented the FSEOF (Flux Scanning based on Enforced Objective Flux) [19] approach for this. To ensure the viability of the community model, a minimum growth rate of 0.01 h^{-1} for both species is required. The function "*FSEOFKnock*" can be used here. The candidate community model can be given as input, along with the biomass reaction identifiers of each species in the community and the reaction of interest for which the flux scanning will be performed.

```
>> [FSEOFRxnList, KOtable] = FSEOFKnock(model,RxnName,
biomassRxn)
```

The output of this function is the list of reactions (*FSEOFRxnList*) that can be deleted to achieve maximum product formation in the community. The results are stored in "KOtable", which provides details on flux values and the growth rates of each species after the deletion.

```
FESOFRxnList =
  17×1 cell array
    {'ASNN_org2'      }
    {'ASNt2_org2'     }
    {'FORt2_org2'     }
    {'FORt2r_org1'    }
    {'GLCt2r_org2'    }
    {'HEX1_org2'      }
    {'HSDy_org1'      }
    {'LDH_L_org2'     }
    {'L_LACt2r_org2'}
    {'NADH5_org1'     }
    {'NH4DISex_org2'}
    {'OBTFL_org1'     }
    {'PFL_org2'       }
    {'PPAKr_org1'     }
```

```
      {'PPAt2_org1'    }
      {'PTA2_org1'     }
      {'THRD_L_org1'   }
KOtable =
  47×6 table
          Rxn1                     Rxn2              ProductFlux
CommunityGrowth     GrowthOrg1     GrowthOrg2

      'FORt2_org2'           'FORt2r_org1'             8.7484
0.42251          0.34006        0.082451
      'FORt2r_org1'          'PFL_org2'                8.7484
0.42251          0.34006        0.082451
...
```

As seen in the above example with the community model of *Lactobacillus animalis KCTC 3501* and *Lactobacillus plantarum WCFS1*, with lactate being the desired product, the analysis suggests knocking out pathways leading to formate production. This is consistent with existing experimental evidence in other microorganisms, such as *E. coli*, where the inactivation of pyruvate formate lyase is known to increase the yield of D-lactate from glycerol [20].

4 Notes

1. *Thresholds.* Results can vary depending on the thresholds used in the simulations. Therefore, interpretation of results should be made in the context of these parameters. In addition, an extensive literature compilation of the organisms of interest will help in assisting the interpretation of the best two-species community. The predictions from the CAMP workflow serve as an initial screening tool for microbial communities and should be used to guide experimental investigations.

References

1. Ko YS, Kim JW, Lee SY et al (2020) Tools and strategies of systems metabolic engineering for the development of microbial cell factories for chemical production. Chem Soc Rev. https://doi.org/10.1039/d0cs00155d
2. Zhou K, Qiao K, Edgar S, Stephanopoulos G (2015) Distributing a metabolic pathway among a microbial consortium enhances production of natural products. Nat Biotechnol 33:377–383. https://doi.org/10.1038/nbt.3095
3. Arkin AP, Cottingham RW, Henry CS et al (2018) KBase: the United States department of energy systems biology knowledgebase. Nat Biotechnol. https://doi.org/10.1038/nbt.4163
4. Machado D, Andrejev S, Tramontano M, Patil KR (2018) Fast automated reconstruction of

genome-scale metabolic models for microbial species and communities. Nucleic Acids Res 46:7542–7553. https://doi.org/10.1093/nar/gky537

5. Aite M, Chevallier M, Frioux C et al (2018) Traceability, reproducibility and wiki-exploration for "à-la-carte" reconstructions of genome-scale metabolic models. PLoS Comput Biol 14:e1006146. https://doi.org/10.1371/journal.pcbi.1006146
6. Ibrahim M, Raajaraam L, Raman K (2021) Modelling microbial communities: harnessing consortia for biotechnological applications. Comput Struct Biotechnol J. https://doi.org/10.1016/j.csbj.2021.06.048
7. Stolyar S, Van Dien S, Hillesland KL et al (2007) Metabolic modeling of a mutualistic microbial community. Mol Syst Biol 3:1–14. https://doi.org/10.1038/msb4100131
8. Ye C, Zou W, Xu N, Liu L (2014) Metabolic model reconstruction and analysis of an artificial microbial ecosystem for vitamin C production. J Biotechnol 182–183:61–67. https://doi.org/10.1016/j.jbiotec.2014.04.027
9. Diener C, Gibbons SM, Resendis-Antonio O (2020) MICOM: metagenome-scale modeling to infer metabolic interactions in the gut microbiota. mSystems 5:e00606–e00619. https://doi.org/10.1128/msystems.00606-19
10. Diener C, Gibbons SM (2023) More is different: metabolic modeling of diverse microbial communities. mSystems 8:e01270–e01222. https://doi.org/10.1128/msystems.01270-22
11. Ibrahim M, Raman K (2021) Two-species community design of lactic acid bacteria for optimal production of lactate. Comput Struct Biotechnol J 19:6039–6049. https://doi.org/10.1016/j.csbj.2021.11.009
12. Heirendt L, Arreckx S, Pfau T et al (2019) Creation and analysis of biochemical constraint-based models using the COBRA Toolbox v.3.0. Nat Protoc 14:639–702. https://doi.org/10.1038/nprot.2007.99
13. Norsigian CJ, Pusarla N, McConn JL et al (2020) BiGG Models 2020: multi-strain genome-scale models and expansion across the phylogenetic tree. Nucleic Acids Res 48: D402–D406. https://doi.org/10.1093/nar/gkz1054
14. Heinken A, Hertel J, Acharya G et al (2023) Genome-scale metabolic reconstruction of 7,302 human microorganisms for personalized medicine. Nat Biotechnol 41:1320–1331. https://doi.org/10.1038/s41587-022-01628-0
15. Zimmermann J, Kaleta C, Waschina S (2021) gapseq: informed prediction of bacterial metabolic pathways and reconstruction of accurate metabolic models. Genome Biol 22:295. https://doi.org/10.1186/s13059-021-02295-1
16. Hsieh YE, Tandon K, Verbruggen H, Nikoloski Z (2024) Comparative analysis of metabolic models of microbial communities reconstructed from automated tools and consensus approaches. NPJ Syst Biol Appl 10:384. https://doi.org/10.1038/s41540-024-00384-y
17. Heinken A, Thiele I (2015) Anoxic conditions promote species-specific mutualism between gut microbes in silico. Appl Environ Microbiol 81:4049–4061. https://doi.org/10.1128/AEM.00101-15
18. Medlock GL, Carey MA, McDuffie DG et al (2018) Inferring metabolic mechanisms of interaction within a defined gut microbiota. Cell Syst 7:245–257.e7. https://doi.org/10.1016/j.cels.2018.08.003
19. Choi HS, Lee SY, Kim TY, Woo HM (2010) In silico identification of gene amplification targets for improvement of lycopene production. Appl Environ Microbiol 76:3097–3105. https://doi.org/10.1128/AEM.00115-10
20. Mazumdar S, Clomburg JM, Gonzalez R (2010) Escherichia coli strains engineered for homofermentative production of D lactic acid from glycerol. Appl Environ Microbiol 76: 4327–4336. https://doi.org/10.1128/AEM.00664-10

Chapter 12

Personalized Constraint-Based Modeling of Microbial Communities from Metagenomic Data

Jordi Roma Pi and Almut Heinken

Abstract

High-throughput metagenomic sequencing techniques such as 16S rRNA and shotgun sequencing have enabled an unprecedented understanding of the structure and function of microbiome communities such as the human gut microbiome. Tailored dietary or therapeutic interventions targeting the microbiome could advance personalized medicine; however, predicting such interventions requires predictive systems biology methods. Constraint-Based Reconstruction and Analysis (COBRA) is a mechanistic systems biology approach that relies on detailed genome-scale reconstructions of a target organism's metabolism. A resource of genome-scale reconstructions of human microbes, AGORA, and its expansion in size and scope, AGORA2, have been developed through a semi-automated refinement pipeline, DEMETER. A user-friendly analysis pipeline, mgPipe, allows building and interrogating personalized models of microbiome communities from AGORA and AGORA2. Through sample-specific simulations, mgPipe can stratify patients and controls by the distinct metabolic capabilities of their microbiomes, starting from the processed metagenomic sequencing data. Building on this functionality, the protocol provides a comprehensive workflow for the contextualization of metagenomics data through personalized, mechanistic modeling. Comprehensive tutorials for the DEMETER and mgPipe workflows are presented, which will enable both systems biologists and microbiome scientists to contextualize metagenomic data and perform mechanistic simulations of diet–microbiome–host interactions.

Key words Metagenomics, Constraint-Based Reconstruction and Analysis, Systems biology, Microbiome, Metabolism, Personalized medicine

1 Introduction

The advent of high-throughput sequencing methods has resulted in the availability of increasing amounts of omics data on public databases. Metagenomics sequencing techniques such as 16S rRNA and shotgun sequencing have enabled culture-independent analyses of microbiomes [1]. Large cohort studies such as MetaHIT

Supplementary Information The online version contains supplementary material available at https://doi.org/10.1007/978-1-0716-5080-6_12.

Karthik Raman and Gayathri Sambamoorthy (eds.), *Modelling the Microbiome*, Methods in Molecular Biology, vol. 3006, https://doi.org/10.1007/978-1-0716-5080-6_12,

and the Integrated Human Microbiome Project have expanded our understanding of the diversity, composition, and structure of the human microbiome, providing thousands of metagenomic samples from multiple body sites [2, 3]. Through such studies, alterations in gut microbiome composition have been shown to be associated with multifactorial diseases including inflammatory bowel disease, obesity, type 2 diabetes, colorectal cancer, and neurodegenerative disorders [4–7]. Identifying functional metabolic differences underlying these compositional changes may enable the prediction of personalized dietary, prebiotic, or probiotic interventions [8]. Computational systems biology methods integrate diverse omics data into comprehensive, predictive models that could identify these metabolic differences, offering a means to translate microbiome variation into personalized medicine strategies [9].

Constraint-Based Reconstruction and Analysis (COBRA) is a commonly used systems biology approach. Through techniques such as flux balance analysis (FBA), COBRA allows for the prediction of metabolic fluxes that are feasible in a given organism and environment [10]. The prerequisite is the availability of a genome-scale reconstruction built from a target organism's annotated genome sequence and manually curated against the available bibliome [10]. Typically, genome-scale reconstructions cover the scope of an organism's metabolic pathways [10]. Through the implementation of constraints such as nutrient availability or gene expression, genome-scale reconstructions can give rise to condition-specific models that can predict individual-specific fluxes or dietary and therapeutic interventions that could favorably influence metabolic pathways in patients [9].

While initially constraint-based modeling of microbial metabolism focused on single organisms, microbes in natural environments never exist in isolation. This motivated the development of constraint-based modeling approaches for the simulation of multi-species interactions [11]. Compared with simulating the metabolic fluxes in a single organism, integrating metabolic networks of multiple organisms drastically increases the complexity and scale of the performed simulation as well as the size of the feasible solution space [12]. Different strategies have been implemented to address the challenge of multi-species interactions, which are reviewed in detail elsewhere, and each has its own strengths and weaknesses [12–14]. One commonly used strategy to integrate genome-scale reconstructions of multiple species is the compartmentalized approach, which was first pioneered for pairs of seven microbes in every combination [15]. In the compartmentalized approach, two or more genome-scale reconstructions are placed into a shared in silico environment in which each species has their own set of exchange reactions through which metabolites can be exchanged with the environment and with each other, making the computed metabolic interactions tractable [16]. Recent applications have combined 100 or more genome-scale reconstructions into compartmentalized models, demonstrating the scalability of the approach [17]. To model the simultaneous growth of microbes coupled in a shared environment, multiple strategies have been implemented. One option is to simultaneously optimize the biomass objective function of each coupled microbe; however, this method is dependent on the biomass formulation and may result in biologically implausible predictions [18]. Moreover, due to the steady-state assumption of constraint-based modeling, many alternate tradeoffs between microbes are possible [12]. This challenge can be overcome by computing all possible tradeoffs between microbes through multi-objective optimization [19] or by introducing a bilevel optimization problem in which a community-level objective is additionally solved [20]. Instead of predicting interactions in microbial communities without any prior assumptions, interspecies interactions can also be inferred by mapping metagenomic sequencing data, namely 16S rRNA or shotgun sequencing data, onto microbiome models with different levels of taxonomic and functional resolution [16].

Besides suitable tools for the integration of multiple species, the increasing size and scope of constraint-based multi-species modeling applications [16] was also facilitated by the increasing number of available microbial genome-scale reconstructions. Genome-scale reconstruction is laborious [21], and several automated and semi-automated genome-scale reconstruction tools have been developed to address this challenge, including Model SEED, CarveMe, merlin, and gapseq [22–25]. The semi-automated reconstruction tool DEMETER specializes in the genome-scale reconstruction of human microbes [26]. DEMETER relies on the availability of manually assembled experimental data and high-quality gene annotations that had been performed in their genomic context [26]. DEMETER previously led to the development of AGORA, a resource of 773 genome-scale reconstructions of the most common strains in the human gut microbiome [27]. A recently published expansion of AGORA, deemed AGORA2, accounts for 7302 strains found in human gut, skin, oral cavity, and vagina and captures drug transformations by human microbes [28]. Additionally, AGORA2 accounts for 127 mouse gut-associated taxa [28]. AGORA and AGORA2 share the same namespace and are fully compatible with each other; hence, they are referred to as AGORA* in the following text.

With the development of AGORA*, it has become possible to build microbiome-wide community models of microbiomes by mapping taxon-level abundances onto the AGORA* namespace. To this end, a microbiome modeling pipeline, mgPipe, was developed [29]. The aim of mgPipe is to provide a user-friendly pipeline for the creation and interrogation of microbiome community models starting from metagenomic sequencing data [29]. By relying on the scalability and tractability of the compartmentalized modeling approach, the definition of a community biomass objective function, and the AGORA* resource, mgPipe allows for the creation and interrogation of sample-specific models of the human microbiome [29]. While mgPipe was initially designed to represent microbiome communities from the human gut [29], in principle, relative abundances from any source or environment can serve as inputs. As mgPipe is most compatible with AGORA*, however, it is best suited for modeling mammalian microbiomes.

The mgPipe workflow consists of three main steps: (1) Creation of sample-specific microbiome community models, (2) prediction of sample-specific pathways and metabolite secretion capabilities, and (3) visualization of the output, followed by statistical analysis, if stratification criteria are provided. The main outputs of mgPipe include predicted sample-specific reaction and pathway abundances as well as net uptake and secretion capabilities for all metabolites exchangeable by at least one microbe in any sample. These predicted functions allow one to stratify samples by their metabolic potential, elucidating functional differences in the microbiomes of, for example, patients and controls. Simulations performed with mgPipe have provided insights into the altered metabolic capabilities in, for example, inflammatory bowel disease, Parkinson's disease, Alzheimer's disease, Type 1 diabetes, and preterm birth [17, 30–33]. Notably, the metabolic capabilities reflect the theoretical potential for each sample and are directly driven by each community's relative taxon abundances; hence, mgPipe generates unique simulation results rather than one of many possible flux distributions. This overcomes the challenge of uncertainty regarding growth ratios and alternative solutions when modeling microbiome communities [12]. Moreover, due to the use of the compartmentalized approach, mgPipe is scalable, enabling the generation of community models with >100 members, and tractable as microbes consuming and secreting metabolites in each community can be pinpointed exactly [34].

The mgPipe software was initially released as part of the Microbiome Modeling Toolbox together with a module on modeling microbe–microbe interactions [29]. The Microbiome Modeling Toolbox is written in MATLAB and has been implemented in the COBRA Toolbox [35]. An updated version with improved computational efficiency and additional visualization and analysis functions, deemed Microbiome Modeling Toolbox 2.0, has been released [34]. The Microbiome Modeling Toolbox 2.0 has been optimized to reduce computation times and takes full advantage of parallel computing [34]. In fact, the updated implementation of mgPipe enabled the construction of over 14,000 microbiome community models [36], demonstrating its efficiency and scalability. An implementation of mgPipe in Python, deemed pymgpipe, is also available [37]. The pymgpipe implementation offers functionality in the open-source software Python akin to the mgPipe implementation in MATLAB [37], allowing users to create microbiome community models as described above.

This protocol provides a step-by-step guide for executing the mgPipe workflow, covering software installation, environment setup, input data preparation, workflow execution, and subsequent analyses. It offers a comprehensive overview of sample-specific community metabolic modeling beginning with processed metagenomic sequencing data. The steps include preparing input files, reconstructing taxa not captured by AGORA* through the DEMETER workflow, generating sample-specific community models, predicting community-level fluxes, visualizing the results, and performing statistical analyses.

2 Materials (Table 1)

3 Methods

3.1 Software Installation

DEMETER and the Microbiome Modeling Toolbox 2.0 have been implemented in the MATLAB-based COBRA Toolbox [35]. The installation requirements are as follows:

- MATLAB (Mathworks, Inc.), available at https://www.mathworks.com

Table 1
Materials required for the pipeline

Tool/ environment	Description	Reference
MATLAB	Numerical computing environment widely used in academia and industry	Mathworks, Inc.
IBM CPLEX	Optimization solver for linear programming problems	IBM, Inc.
AGORA	Genome-scale reconstructions of 773 human gut microbes	[27]
AGORA2	Expansion of AGORA accounting for 7302 human microbes	[28]
GitHub	Environment for open-source software sharing. Hosts the COBRA Toolbox	https://github.com/
COBRA Toolbox	Software library for Constraint-based Reconstruction and Analysis in MATLAB. Includes DEMETER and mgPipe.	[36]

- Parallel Computing Toolbox, Bioinformatics Toolbox, and Statistics and Machine Learning Toolbox in MATLAB
- COBRA Toolbox, available at https://github.com/opencobra/cobratoolbox
- IBM CPLEX (IBM, Inc.) linear programming solver, available at https://www.ibm.com/products/ilog-cplex-optimization-studio (optional but recommended)

Note The installation of the COBRA Toolbox relies on GitHub. *See* https://git-scm.com/book/en/v2/Getting-Started-Installing-Git for further instructions on installing GitHub. *See* https://opencobra.github.io/ for guidelines on installing the COBRA Toolbox and the MATLAB interface of IBM CPLEX.

To initialize the COBRA Toolbox and linear programming solver, run the code:

```
initCobraToolbox
solverOK = changeCobraSolver('ibm_cplex','LP');
```

3.2 Step 1: Preparation of Input Data

3.2.1 Relative Abundance Data

The prerequisite for the application of the mgPipe workflow is the availability of processed metagenomic sequencing data in the form of taxon- or Operational Taxonomic Unit (OTU)-level relative abundances. Methods for sequencing fecal samples and processing metagenomic sequencing data are outside the scope of this protocol and have been reviewed elsewhere [1, 38]. Both 16S rRNA and shotgun sequencing data can serve as input data for mgPipe. For taxon abundances on the strain level, the corresponding AGORA* reconstructions can directly serve as the input. For abundances on the species level and higher, pan-models built from AGORA* can be used. While genus-level abundances generated by 16S rRNA sequencing data are suitable as inputs, it is recommended to use strain- or species-level abundances as they will result in more accurate predictions of taxon-specific metabolic pathways. The use of abundances on family level or higher is not advised.

To convert taxon or OTU abundance data tables, for example, those generated by MetaPhlAn [39], into an input suitable for use in mgPipe, it is necessary to match taxon nomenclature with the namespace of AGORA*. Relative abundances on the strain level need to be mapped manually to the corresponding AGORA* reconstructions. For relative abundance data on species level or higher, the function translateMetagenome2AGORA in the Microbiome Modeling Toolbox performs this. The required input *abunFilePath* contains the path to the table containing relative abundances to be mapped onto AGORA*. The desired sequencing depth (e.g., species, genus) is specified in the input *sequencingDepth*. The input *reconstructionResource* specifies whether AGORA or AGORA2 is used.

```
[translatedAbundances,normalizedAbundances,unmappedRows]
=translateMetagenome2AGORA(MetagenomeAbundancePath,
sequencingDepth,reconstructionResource)
```

Note

translateMetagenome2AGORA was designed for MetaPhlAn outputs and may not function for abundance files from other sources. The outputs need to be inspected manually. Check the output unmappedRows for taxa that could not be mapped. A user-friendly tool deemed MARS that automates the mapping from relative abundances to AGORA* nomenclature is also available [48].

Hint For unmapped taxa, verify the current nomenclature for the taxon and compare with the information in the reference tables on AGORA* strains available at https://github.com/opencobra/COBRA.papers/tree/master/2021_demeter/input. Consult NCBI Taxonomy (https://www.ncbi.nlm.nih.gov/taxonomy) for taxa that were recently reassigned.

To prevent infeasibly high growth rates, it is necessary to normalize relative abundances such that they sum up to 1 or less for each sample. The function *normalizeCoverage* in the Microbiome Modeling Toolbox performs this. The required input *abunFilePath* contains the path to the table containing relative abundances to be normalized. The input cutoff denotes the cutoff below which taxa should be removed (default: 0.0001).

```
[normalizedCoverage,normalizedCoveragePath] = normalize-
Coverage(abunFilePath,cutoff)
```

Note

It is recommended to remove unmapped taxa after normalization to avoid inflating taxon contributions to overall abundance.

Supplementary File 1 shows an example abundance table mapped to AGORA. The example contains relative abundances in gut microbiomes of pediatric Crohn's Disease patients and healthy controls [40] that had been applied to modeling the gut microbiome in a previous study [41]. For simplicity, the dataset has been reduced to the 20 most abundant strains in five Crohn's Disease and five control samples. The complete mapped dataset is available at https://github.com/ThieleLab/CodeBase/tree/master/Simulations_Heinken_Microbiome_2019/InputFiles.

3.2.2 Stratification Criteria

If stratification criteria (e.g., cases and controls) are available, they can be entered to inform the analysis. They will not affect the simulations themselves but allow stratifying groups by the metabolic capabilities of their microbiomes. The Microbiome Modeling Toolbox contains functions for plotting and statistical analyses to this effect. Stratification criteria should be prepared as a table with sample names in the first column and stratification criteria in the second column. Optionally, more criteria (e.g., sex, diet regime) can be added.

Supplementary File 2 contains a table with stratification criteria for the above examples of five Crohn's Disease and five control samples.

3.2.3 Diet

Dietary regimes can be entered into mgPipe through the *dietFilePath* input, which denotes the name of the file with dietary constraints. A variety of pre-made diets have been formulated (e.g., Average European, Vegan, Gluten-free) and are available at https://github.com/opencobra/COBRA.papers/tree/master/2018_microbiomeModelingToolbox/input. Custom diets can be formulated using the Diet Designer tool on the Virtual Metabolic Human database [42]. Additionally, the MATLAB implementation of mgPipe offers using an individual diet for each microbiome sample. In that case, the file specified in *dietFilePath* needs to contain an entry specifying the file with dietary constraints for each sample.

3.3 Reconstruction of Additional Strains with DEMETER

3.3.1 Strain Selection and Genome Retrieval

After having identified the taxa that are present in the metagenomic sequencing data but not captured by AGORA*, these taxa can be reconstructed through the DEMETER tool. The resulting semi-automatically refined reconstructions will be fully compatible with AGORA*. The prerequisite is the identification of unreconstructed taxa on at least the species level and the availability of a genome sequence in FASTA format. Both reference genomes and metagenome-assembled genomes are suitable as inputs. Taxa that had only been identified on the genus level or higher (unless the genus only contains a single species), or for which no genome sequence is available, cannot be reconstructed. Classified but yet unnamed species with a sequenced genome, however, can be reconstructed. Reference genomes can be retrieved, for example, from NCBI Assembly (https://www.ncbi.nlm.nih.gov/assembly/). If multiple assemblies are available, the most complete genome sequence should be selected. If multiple strains had been sequenced for the target species, the reference strain and/or the most complete genome sequence should be selected.

3.3.2 Generation of Draft Reconstructions in KBase

DEMETER requires an automatically generated draft genome-scale reconstruction as the input, which can be generated through the online platform KBase [43]. Workflows in KBase are facilitated through the Narrative interface, in which apps for data import, processing, and export can be run. Draft reconstructions can be built in batch for multiple genomes. A dedicated template narrative for the generation of draft genome-scale reconstructions is available at https://narrative.kbase.us/narrative/ws.82195.obj.8. The reconstruction workflow in KBase consists of the following steps:

- Import of genome sequences into the Narrative for sequences in FASTA format. This can be done by
  - Drag and drop into the Staging Area and subsequent import as Assembly
  - Direct import from KBase (if available) from the Add Data → Public tab
- Create a data object containing all genomes to annotate
  - If in Assembly format: "Batch Create Assembly Set" app
  - If in Genome format: "Batch Create Genome Set" app
- Annotation of the sequences
  - If in Assembly format: "Annotate Multiple Microbial Assemblies" app (using the Assembly Set as input)
  - If in Genome format: "Annotate Multiple Microbial Genomes" app (using the Genome Set as input)
- Generation of draft reconstructions
  - "Build Multiple Metabolic Models" using the annotatedGenomes variable created in the previous step as input
- Export of draft reconstructions
  - "Bulk Download Modeling Objects" app

After export from KBase, draft reconstructions need to be placed in a single folder that should not contain other files. An example set of ten draft reconstructions is available at https://github.com/opencobra/COBRA.papers/tree/master/2021_demeter/exampleDraftReconstructions.

3.3.3 Preparation of Taxonomic Information

To perform taxon-specific refinement of the draft reconstructions, DEMETER requires information on the taxonomic assignment of each organism. This information can be retrieved from NCBI Taxonomy (https://www.ncbi.nlm.nih.gov/taxonomy/). The information needs to be prepared as a table in text or spreadsheet format in the same form as the taxonomic information table for all AGORA2 strains found at https://github.com/opencobra/COBRA.papers/blob/master/2021_demeter/input/AGORA2_infoFile.xlsx.

Troubleshooting The IDs in the first column need to match the names of the resulting refined reconstructions exactly. During the pipeline run, the names of draft reconstruction are adapted. To retrieve the resulting adapted names of the refined reconstructions, run the function:

```
refinedModelIDs = printRefinedModelIDs(draftFolder);
```

3.3.4 Propagation of Experimental Data

Previously, experimental data for >1500 microbial species for carbon sources, fermentation pathways, growth requirements, consumed metabolites, and secreted metabolites had been collected [28]. These data can be propagated to newly reconstructed strains of species/genera already present in AGORA2 by running the following command:

```
[infoFilePath,inputDataFolder] = prepareInputData
(infoFilePath);
```

Where the input *infoFilePath* is the path and file name of the table with taxonomic information prepared in the previous step. For the example set of ten draft reconstructions, a table with taxonomic information is available at https://github.com/opencobra/COBRA.papers/blob/master/2021_demeter/example_infoFile.xlsx.

3.3.5 Addition of Manually Collected Experimental Data

Propagated experimental data is saved in the folder *inputDataFolder*. In each table, reconstruction IDs are listed in rows and functions that can be refined and are listed in columns. It is recommended to manually inspect the propagated experimental data for our organism(s) at this point to verify it as organisms may differ at the strain level. Moreover, it is recommended that we perform additional literature searches to inform the refinement of our organism(s), especially for species not yet present in AGORA2. Experimental data for organisms of interest can be found in peer-reviewed articles, microbial reference textbooks, as well as phenotype databases such as BacDive [44].

Experimental data is entered as follows: "1" indicates that a pathway or function is present, and "0" indicates it is not present or no information is available. Reconstructions can be refined for carbon sources, fermentation pathways, metabolite uptake, and metabolite secretion by accordingly modifying the respective tables in the correct rows and columns.

Hint Carefully check the information in peer-reviewed literature and databases and ensure to enter information in the correct row and column.

Growth requirements are defined by entering "-1" for compounds that are nonessential and "1" for compounds known to be essential for the taxon. If no information is available, "0" is entered.

Hint Defined growth media reported for a species can be used to exclude compounds that are not present in the medium as nonessential.

To be recognized by the pipeline, the manually edited files need to be present in the folder defined in the *inputDataFolder* variable.

Attention If prepareInputData is run again, the files are overwritten. Make sure to back up manually collected data elsewhere for records.

3.3.6 Refinement Pipeline

To refine the draft reconstructions, the command *runPipeline* is executed. The input parameter *draftFolder*, which defines the path to the folder containing the draft reconstructions, is required. Optional inputs are as follows:

- *infoFilePath* defines the path to the table with taxonomic information (optional but recommended).
- *inputDataFolder* is the path where the files with propagated and/or manually collected experimental data located.
- *numWorkers* indicates the number of cores used for parallel computing.
- *reconVersion* defines a name for the generated reconstruction resource (default: "Reconstructions").
- *translateModels* indicates whether translated versions of draft reconstructions without any other modifications should be exported additionally (true = yes, false = no). This will allow comparing refined with draft reconstructions in the test suite.
- *createSBML* indicates whether refined reconstructions should be saved in SBML format in addition to MAT format (true = yes, false = no).
- Additionally, the paths where results are saved can be defined manually through the *refinedFolder*, *translatedDraftsFolder*, and *summaryFolder* inputs; otherwise, default path names are used.

The pipeline is then run as follows:

```
[reconVersion,refinedFolder,translatedDraftsFolder,sum-
maryFolder] = runPipeline(draftFolder, 'infoFilePath',
infoFilePath, 'inputDataFolder', inputDataFolder, 'num-
Workers', numWorkers, 'reconVersion', reconVersion,
'translateModels', true, 'createSBML', true);
```

The results are saved in the folders output in the *refinedFolder*, *translatedDraftsFolder*, and *summaryFolder* variables (default folder names will be used if not manually defined).

- *refinedFolder* contains the refined reconstructions in MAT format.
- *refinedFolder* + suffix _SBML contains the refined reconstructions in SBML format (optional).
- *translatedDraftsFolder* contains the draft reconstructions adapted to Virtual Metabolic Human (https://www.vmh.life/) nomenclature.
- *summaryFolder* contains files listing performed modifications, reaction deletions, and reaction additions to each draft reconstruction during the refinement pipeline.

3.3.7 Quality Control Tests

DEMETER includes a comprehensive test suite that checks the generated refined reconstructions for agreement with known metabolic traits where available, capability to produce biomass, growth on defined media, thermodynamic consistency (i.e., production of reasonable amounts of ATP), and correct reconstruction structure (i.e., mass and charge balance).

The command runTestSuiteTools executes the test suite. Required inputs include the location on the refined reconstructions (*refinedFolder*), the path to the table with taxonomic information (*infoFilePath*), the location of the propagated and collected experimental data (*inputDataFolder*), and the name of the reconstruction resource (*reconVersion*). Optional inputs include the location of draft reconstructions translated to VMH nomenclature (*translatedDraftsFolder*), which will enable comparison between refined and draft reconstructions, and the number of cores dedicated to parallelization (*numWorkers*). All inputs had already been defined in previous steps of the pipeline.

The test suite is run as follows:

```
[testResultsFolder,curationReport] = runTestSuiteTools
(refinedFolder, infoFilePath, inputDataFolder, reconVer-
sion, 'translatedDraftsFolder', translatedDraftsFolder,
'numWorkers', numWorkers);
```

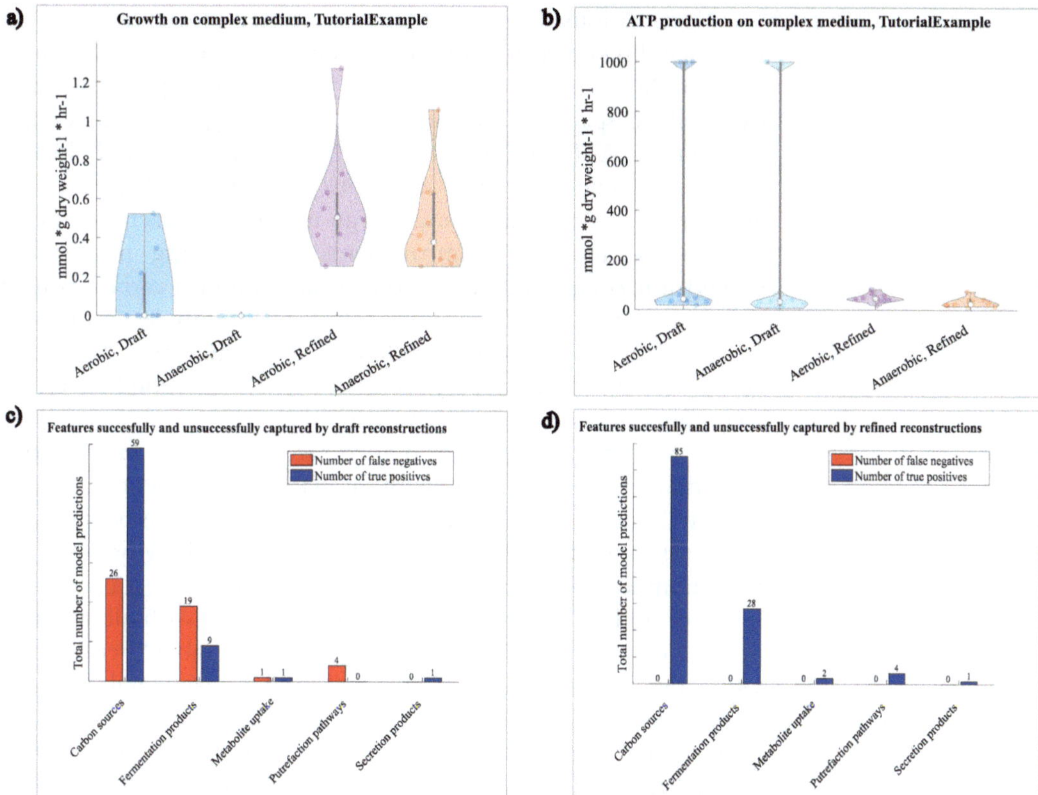

Fig. 1 Characteristics of reconstructions plotted in DEMETER shown for draft and refined reconstructions. (**a**) Aerobic and anaerobic growth rates on complex medium, (**b**) aerobic and anaerobic ATP production potential on complex medium, (**c**) false negatives and true positives against input experimental data for draft reconstructions, and (**d**) false negatives and true positives against input experimental data for refined reconstructions. The plots can be reproduced following the tutorial available at https:/github.com/opencobra/COBRA.tutorials/tree/master/reconstruction/demeter

Results from quality control tests will be found in the path *testResultsFolder*. Draft and refined reconstruction results will be in two separate subfolders.

As a first quality check, the ability of the strains to produce biomass on rich medium (consisting of every metabolite the model can consume) and a complex medium (consisting of a more realistic subset of available nutrients) is tested. Growth rates are determined aerobically and anaerobically (Fig. 1a). Thermodynamic consistency is checked through the computation of ATP production potential on the complex medium aerobically and anaerobically (Fig. 1b). ATP production of up to 250 mmol*g dry weight^{-1}*h^{-1} aerobically and 200 mmol*g dry weight^{-1}*h^{-1} anaerobically is considered in a realistic range. A significantly higher ATP production indicates the existence of futile cycles [21], which will be addressed in the debugging step (see below).

Next, various model properties are tested in each reconstruction including correctly formulated gene–protein–reaction associations (GPRs), correct compartments (only cytosol, periplasm, and extracellular space are allowed), mass and charge balance, leaking metabolites, ATP production without a carbon source, and blocked reactions. While most reconstructions will have blocked reactions and a small subset of mass- and charge-unbalanced reactions, incorrect GPRs, incorrect compartments, and leaking metabolites should not occur, and ATP production without a carbon source should be 0 mmol*g dry weight^{-1}*h^{-1}. It is highly recommended to check the generated files and verify this is the case. Note that certain tables will only be generated if the corresponding errors were found in at least one reconstruction.

Finally, the agreement of refined reconstructions and, optionally, draft reconstructions with available experimental data is exported and visualized. An overview of true positives and false negatives for each type of data is plotted (Fig. 1c, d). Moreover, the total number of strains with available data in the reconstruction resource and the percentage of strains agreeing with all available positive data for each data type are given in the table "PercentagesAgreement". Note that since the collected experimental data currently does not account for negative data (i.e., known metabolic incapabilities), it is currently not possible to compute the accuracy of predictions through the test suite. For the calculation of the accuracy of AGORA2 predictions against three independent datasets, the reader is directed to [28].

3.3.8 Debugging Suite

Unlike other semi-automated refinement tools, for example, CarveMe [23], DEMETER does not start from a curated universal reconstruction but from draft reconstructions generated through Model SEED. As the Model SEED pipeline is not specific for human microbes and has been updated recently [45], DEMETER is not adapted for all reaction content contained in draft reconstructions, especially for less studied and/or non-human microbes. Moreover, automated gap-filling is avoided in DEMETER, if possible, to reduce the number of reactions without genetic or biochemical evidence in the resulting refined reconstructions. As a result, refined reconstructions generated through DEMETER may still have errors such as unrealistically high ATP production or an inability to execute known metabolic functions. As a first step in further iterative refinement, a debugging suite can be performed by running the command *runDebuggingTools*. The debugging suite double-checks all reconstructions that had failed a test in the test suite. Additional gap-filling and curation are performed automatically to eliminate failed predictions, enable biomass production, and/or remove futile cycles.

The inputs for the debugging suite are the same as that for the test suite. To run the debugging tools, run the command:

```
[debuggingReport,fixedModels,failedModels]=runDebugging-
Tools(refinedFolder,testResultsFolder,inputDataFolder,
infoFilePath, reconVersion,'numWorkers',numWorkers);
```

The output *debuggingReport* gives an overview of performing additional curation. *fixedModels* lists the names of reconstructions that passed the test suite after the additional curation, while *failedModels* lists the reconstructions that still did not pass the test suite. The results in *testResultsFolder* will also be replaced. Check the results described above located in *testResultsFolder* to identify the specific failed predictions in the reconstructions listed in *failedModels*. Manual curation and gap-filling will be necessary to fix reconstructions that, for example, do not capture known carbon sources or fermentation products or produce unreasonable amounts of ATP. For instructions on performing manual curation, the reader is directed to a standard operating procedure on the manual generation of high-quality genome-scale reconstructions [21].

3.3.9 Prediction and Visualization of Model Properties

Besides reconstruction refinement, the DEMETER pipeline also provides functions for the convenient calculation of reconstruction features and properties in batches. Predicted properties include reconstruction statistics (i.e., reaction, metabolite, and gene numbers) and stoichiometric and flux consistency as defined in Ref. [46]. Additionally, the reaction and metabolite presence/absence, metabolite uptake and secretion potential, and the capacity to synthesize metabolites in cytosol are computed for each strain. If taxonomic information on the reconstructed strains is provided, the latter data enable clustering by taxonomic relatedness through t-distributed stochastic neighbor embedding (tSNE). The function *computeModelProperties* performs the calculation of model statistics, prediction of model properties and metabolic capabilities, and subsequent clustering by computed properties via tSNE for a set of reconstructions. Figure 2 shows an example of clustering genome-scale reconstructions based on reaction presence/absence profiles while annotating the strains with family-level taxonomy. This visualization highlights functional similarities among taxa and helps identify groups of organisms with shared metabolic capabilities.

The inputs for *computeModelProperties* are as follows:

- *modelFolder* is the folder with a set of reconstructions to analyze (i.e., AGORA* or a subset thereof).
- *infoFilePath* defines the path to the table with taxonomic information for the reconstructions to analyze (required to enable clustering).

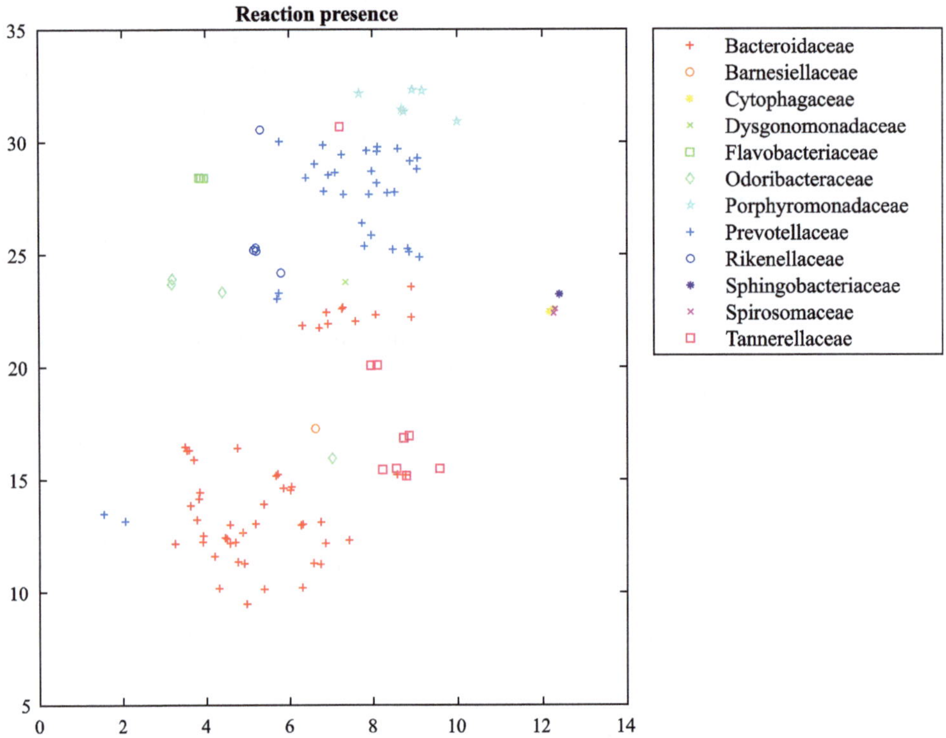

Fig. 2 tSNE clustering of Bacteroidetes strains in AGORA based on the presence and absence of reactions. The plot can be reproduced following the tutorial available at https:/github.com/opencobra/COBRA.tutorials/tree/master/reconstruction/demeter

- *reconVersion* defines a name for the generated reconstruction resource (required in this case).
- *numWorkers* indicates the number of cores used for parallel computing.

To run the prediction and visualization of model properties:

```
computeModelProperties(modelFolder, infoFilePath, recon-
Version, 'numWorkers', numWorkers);
```

3.4 Creation and Interrogation of Personalized Microbiome Community Models with mgPipe

The Microbiome Modeling Toolbox [29] module mgPipe enables the construction, interrogation, and analysis of personalized microbiome models. To initiate the modeling process, mgPipe requires specific input data, including details on diet, microbial abundances, AGORA database models, and optionally DEMETER models if needed (Fig. 3). These inputs set the stage for personalized microbiome modeling. For the modeling phase, mgPipe adopts a compartmentalized approach, strategically delineating microbial strains within community models. Each strain, characterized by its unique

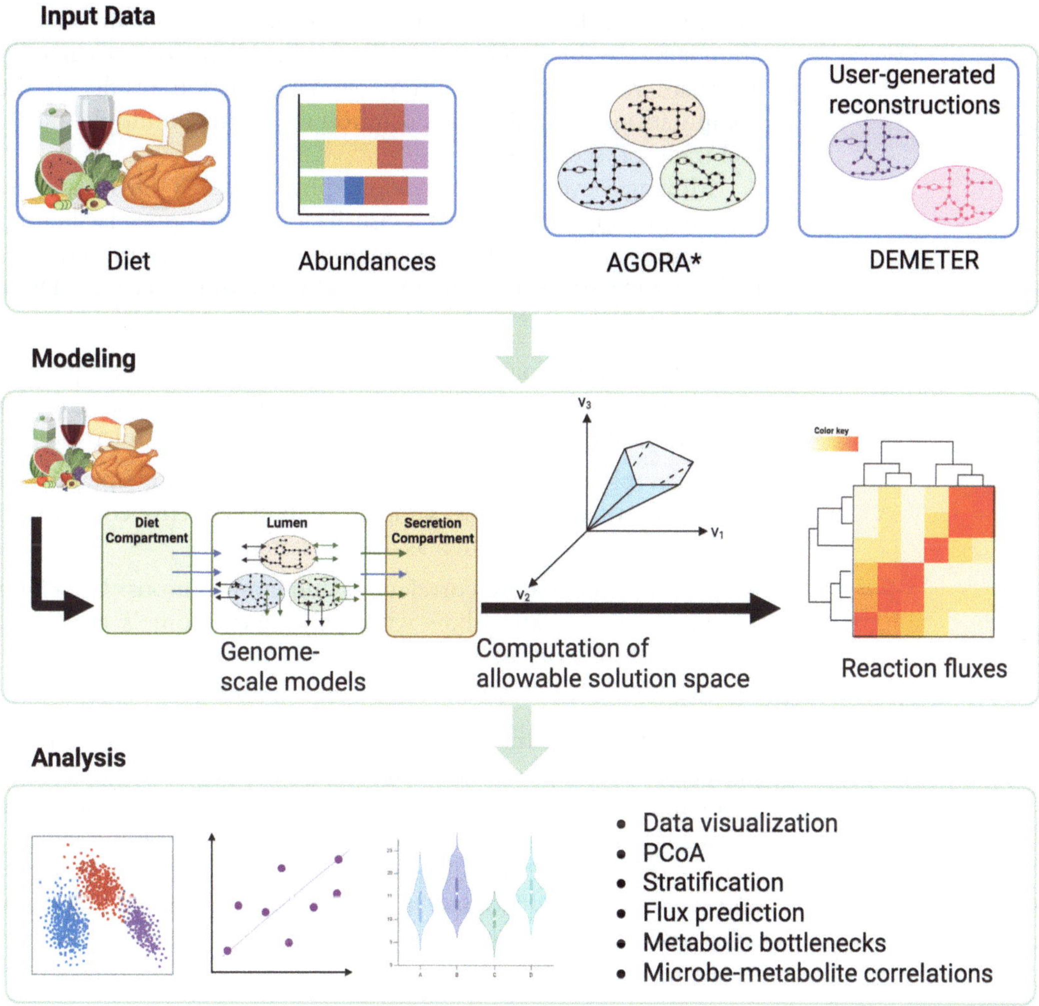

Fig. 3 Overview of the general workflow for the creation and interrogation of personalized microbiome community models using mgPipe. (Created through BioRender)

metabolic activities, contributes to the metabolism of the microbiome community. This compartmentalization is illustrated in Fig. 3, offering a visual depiction of discrete components and their interplay in the modeling process. The Flux Variability Analysis (FVA) [47] method is employed to systematically optimize reaction fluxes across all microbial strains. This step provides a quantitative assessment of the metabolic activities, unfolding the dynamic interplay of microbial strains within the community. Due to the assumptions of FVA, the theoretical flux potential in any possible solution rather than one of many flux distributions is predicted. Hence, mgPipe offers the advantage of yielding a single solution, reducing the complexity associated with constraint-based modeling of microbiome communities [12]. An additional component of the workflow is the community biomass function, which captures the collective growth of all microbial strains using relative abundances as the stoichiometry. By forcing all microbes to grow in a certain ratio, the community biomass reaction results in the computation of a unified flux solution rather than multiple microbial trade-off scenarios.

Moving beyond the modeling phase, Fig. 3 illustrates the subsequent application of data visualization techniques facilitating the stratification of individual microbiomes (e.g., cases and controls) and mechanistically linking metabolites with microbes. These analyses include:

- Principal coordinates analysis (PCoA): A robust technique distilling complex data into essential components, aiding in the identification of patterns and variations within the microbiome.
- Violin plots: Visualize the total net metabolite uptake and secretion capacity across samples, enabling effective comparison of different groups, such as disease states.
- Microbe–metabolite correlations and contributions: Unveiling quantitative relationships within the microbiome's metabolic network, providing insights into connections between microbial communities and their produced metabolites.

In conclusion, this protocol, systematically illustrated in Fig. 3, provides a comprehensive framework for mechanistic metabolic modeling of microbiome communities. It not only guides the construction and analysis of personalized microbiome models but also emphasizes the importance of a compartmentalized approach and the community biomass function in unraveling the intricacies of microbiome dynamics.

3.4.1 Creation of Pan-Models

To run mgPipe, a table with relative abundances mapped to the AGORA* nomenclature is required, with rows containing the names of the models to join and columns containing the names of the samples. To prepare the input file, the taxonomy information tables for AGORA* can be consulted, which are available in the COBRA.papers folder (AGORA_infoFile.xlsx and AGORA2_infoFile.xlsx for AGORA and AGORA2, respectively). Relative abundances for each sample must sum up to 1 or less total. If the relative abundance data is on the species or genus level, pan-models must be built from AGORA* through the *createPanModels* function. The namespace of the input file must then be adapted to the created pan-models (e.g., "panEscherichia_coli").

```
createPanModels(agoraPath,panPath,taxonLevel)
```

3.4.2 Initialization of the Pipeline

To initialize the basic mgPipe workflow, the function *initMgPipe* is run. The function has three required inputs:

- *modPath*: Path containing the folder containing AGORA* models or pan-models
- *abunFilePath*: Path to the table with relative abundance information
- *computeProfiles*: Defines whether full net secretion profiles are computed (if set to false, only microbiome models are generated and net secretion fluxes are not computed)

Additionally, several optional inputs allow the customization of performed simulations:

- *dietFilePath*: Name of the file with the simulated dietary regime that will be implemented on the personalized models (default: "Average European" diet).
- *infoFilePath*: Path to a table with subject classification criteria (e.g., cases, controls).
- *resPath*: Path to the folder where results will be saved (default path will be used otherwise).
- *numWorkers*: Number of cores dedicated to parallel computing.
- *lowerBMBound*: Defines the lower bound enforcing community biomass (default = 0.4). This can be customized for non-gut microbiomes or microbiome models that are infeasible when optimized with the default constraints.
- *upperBMBound*: Defines the upper bound enforcing a limit on community biomass (default = 1).
- *includeHumanMets*: Indicates if human-derived metabolites present in the gut should be provided to the models (default: true). This parameter should be changed for non-gut microbiomes.
- *adaptMedium*: Indicates if the medium should be adapted through the adaptVMHDietToAGORA function or used as is (default = true).
- *pruneModels*: Indicates whether reactions that do not carry flux on the input diet should be removed from the microbe models. Recommended for large datasets (default: false).

To run the basic mgPipe workflow, enter the code:

```
[init, netSecretionFluxes, netUptakeFluxes, Y, model-
Stats, summary] = initMgPipe(modPath, abunFilePath, com-
puteProfiles,          'dietFilePath',          dietFilePath,
```

```
'infoFilePath', infoFilePath, 'resPath', resPath, 'num-
Workers', numWorkers);
```

Troubleshooting

If you get an error regarding the number of workers requested for the parallel pool, you can change the number of default workers by going to Settings → Parallel Computing Toolbox → Cluster Profile Manager → Edit in MATLAB and set the number for the option "Number of workers to start on your local machine" to 4 (recommended number of cores for mgPipe).

The main computed outputs if *computeProfiles* was set to true are the net uptake and secretion fluxes of the total microbiome community for each sample, which can be found in the folder specified in *resPath*. A Principal Coordinates Analysis (PCoA) of the predicted net secretion profiles is also created (Fig. 4a). If stratification criteria were provided, samples will be color-coded by the stratification group in the plot. The reaction abundance summarized for all microbes in each sample is also computed (ReactionAbundance.csv), as well as the absolute presence or

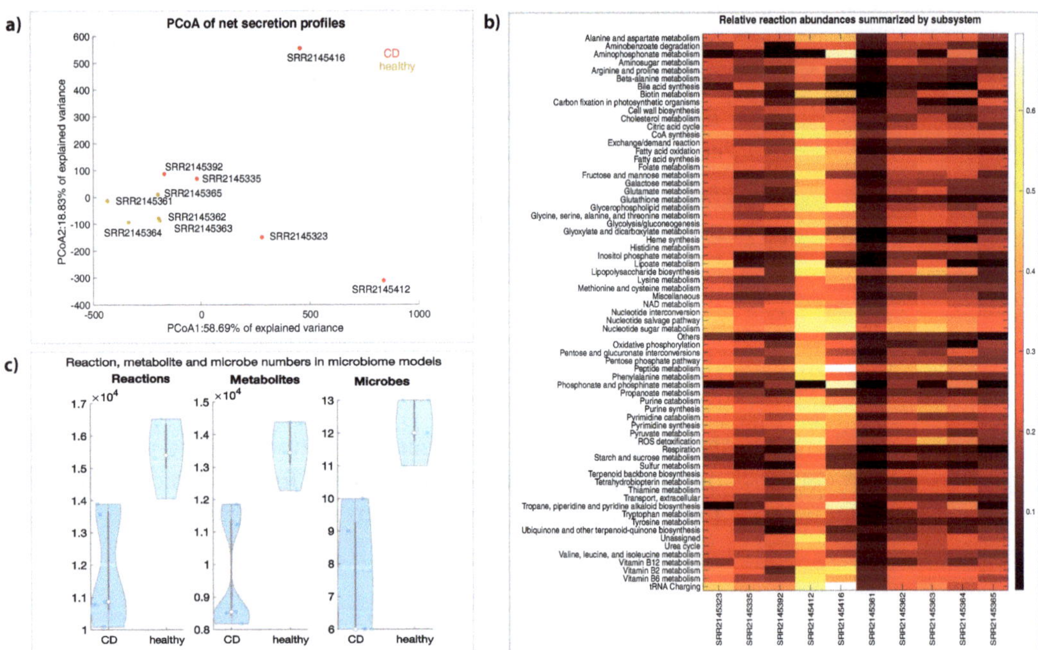

Fig. 4 Features of microbiome community models plotted in a standard mgPipe run in MATLAB. Shown is the analysis of an example dataset containing five microbiomes of Crohn's Disease (CD) patients and five healthy controls. As stratification criteria were provided, communities of CD and healthy microbiomes are plotted separately. (**a**) PCoA of net secretion profiles, (**b**) relative abundances on subsystem level, and (**c**) number of reactions, metabolites, and microbes in each personalized community model. The plots can be reproduced using a provided MATLAB LiveScript (Supplementary File 3)

absence of each reaction in each sample (ReactionPresence.csv). The subsystem abundance, which is the reaction abundance summarized by the metabolic subsystem, is also calculated (SubsystemAbundance.csv) and plotted (Fig. 4b). Finally, the number of reactions, metabolites, and organisms in each sample is calculated (ModelStatistics.csv) and plotted (Fig. 4c). The sample-specific microbiome models parameterized with dietary constraints will be available in the "Diet" subfolder and can be used for further custom analyses if desired.

3.4.3 Calculation of Correlations Between Taxon Abundances and Fluxes

For an overview of metabolite–taxa relationships, the computed uptake and secretion profiles for each metabolite can be correlated with taxon abundances on different levels (species, genus, etc.). Note that a correlation may not necessarily imply that the metabolite is synthesized by a given taxon as correlations may reflect indirect associations. Plots will be automatically generated for microbe–metabolite correlations on different taxonomical levels. Figure 5a shows an example for the computed microbe–metabolite correlations on the species level.

Required inputs for the calculation of correlations include:

- *taxInfo*: The path to the file with taxonomic information on the AGORA* microbes (located at cobratoolbox/papers/

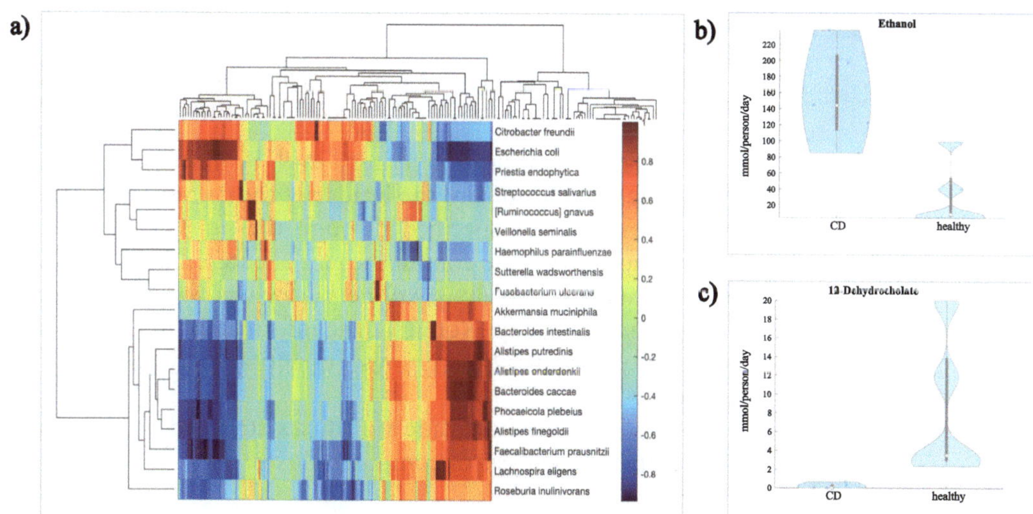

Fig. 5 Optional analyses and visualization offered in mgPipe. Shown is the analysis of an example dataset containing five microbiomes of Crohn's Disease (CD) patients and five healthy controls. As stratification criteria were provided, communities of CD and healthy microbiomes are plotted separately. (**a**) Spearman's correlations between species abundances and net secreted metabolite amounts, (**b**) Violin plots of net secretion of ethanol, (**c**) Violin plots for net secretion of 12-dehydrocholate. The plots can be reproduced using a provided MATLAB LiveScript (Supplementary File 3)

2018_microbiomeModelingToolbox/input/AGORA_infoFile.xlsx and cobratoolbox/papers/2021_demeter/input/AGORA2_infoFile.xlsx).

- *fluxPath*: Path to the file with fluxes that should be correlated (e.g., net secretion fluxes)

Optionally, the method for determining correlations (Spearman, Pearson, Kendall) can be changed through the corrMethod input.

```
[FluxCorrelations, PValues, TaxonomyInfo] = correlate-
FluxWithTaxonAbundance(abunFilePath, fluxPath, taxInfo,
corrMethod);
```

3.4.4 Visualization and Statistical Analysis

If stratification groups are available for the samples (e.g., cases and controls), further visualization and statistical analysis of the computed fluxes can be performed. This includes statistical analyses, yielding, for example, net secreted and consumed metabolites that differed between groups. Moreover, violin plots of net uptake and secretion fluxes stratified by groups are generated. The function carrying out these steps, *analyzeMgPipeResults*, has the following required inputs:

- *infoFilePath*: Path to a table with subject classification criteria (e.g., cases, controls)
- *resPath*: Path to the folder where results will be saved (needs to be same as defined for initMgPipe or otherwise default path)

By default, stratification criteria will be retrieved from the second column. Optionally, the header of the column with stratification criteria can be specified through the *sampleGroupHeaders* input. If more than one column header is entered in *sampleGroupHeaders*, statistical analysis will be performed for the criteria in each column separately. Note that the information in each column must be qualitative descriptions; quantitative values are not suitable.

To execute the visualization of fluxes stratified by a group as well as statistical analysis of computed fluxes:

```
analyzeMgPipeResults(infoFilePath,resPath, 'sampleGrou-
pHeaders', sampleGroupHeaders);
```

All violin plots will be available in the “ViolinPlots” folder afterward. Results of statistical analyses will be available in the “Statistics” folder. Alternatively, the results found in the generated spreadsheets in.csv format can be visualized and further analyzed through custom methods outside mgPipe. Figure 5b and c show the generated violin plots for computed production fluxes for ethanol and 12-dehydrocholate stratified by the group.

3.4.5 Calculation of Microbe–Metabolite Contributions

For metabolites of particular interest (e.g., for which the community-wide secretion potential was significantly different between disease cases and controls), the taxa consuming and secreting the metabolites in each sample can be computed directly rather than merely inferring these taxon–metabolite relationships through correlations. This will provide insight into the contribution of each strain to each metabolite. For metabolites for which the community-wide secretion potential did not differ, the individual strains contributing to metabolites may still be significantly different. In Ref. [17], the computation of strain–metabolite contributions had been applied for a cohort of inflammatory bowel disease patients and controls.

Note Due to the assumptions of flux variability analysis, meaning that flux spans of each reaction are calculated independently of each other and not each minimal and maximal flux can be achieved in every possible flux distribution, the summarized microbe contributions for a given metabolite may be higher than the whole community's net secretion potential for the metabolite.

One input variable for *predictMicrobeContributions* is required:

- *constrModPath*: The path to the folder with saved personalized models constrained with the simulated diet that had been created during the *initMgPipe* run

Optional inputs include:

- *metList*: A cell array with a list of Virtual Metabolic Human [42] IDs for metabolites to analyze (default: all exchanged metabolites in the models)
- *contPath*: Custom path for saving the results (default folder name will be used otherwise)
- *numWorkers*: Number of cores dedicated to parallel computing

Note This computation can be time-consuming; hence, limiting the analysis to metabolites of interest is recommended.

To run the computation of strain contributions:

```
[minFluxes,maxFluxes,fluxSpans] = predictMicrobeContributions(constrModPath, 'metList', metList, 'resultsFolder', contPath, 'numWorkers', numWorkers);
```

The output "minFluxes" shows the fluxes in the reverse direction through all internal exchange reactions that had nonzero flux for each analyzed metabolite. The output "maxFluxes" shows the corresponding forward fluxes. "fluxSpans" shows the distance between minimal and maximal fluxes for each internal exchange reaction. Note that contrary to usual conventions, in this case, negative fluxes represent secretion exchange fluxes and positive fluxes represent uptake exchange fluxes by each individual microbe. The contributions can also be found in spreadsheet format in a newly created folder called "Contributions".

Statistical analysis can also be performed on microbe–metabolite contributions by running the code:

```
analyzeMgPipeResults(infoFilePath, contPath, 'sampleGroupHeaders', sampleGroupHeaders);
```

The results of the statistical analysis will also be available in the "Statistics" folder.

3.4.6 Calculating Shadow Prices

Shadow prices are a measurement for the value of a metabolite toward the optimized objective function. In the context of microbiome community models, this enables us to identify microbes that are particularly relevant for certain metabolites produced by the community. These microbes may serve as bottlenecks for the availability of the produced metabolite. In Ref. [41], shadow price analysis had been performed to identify bottlenecks in the production of secondary bile acids.

A dedicated function, *analyseObjectiveShadowPrices*, exists for the extraction of shadow prices across many models. Note that the function can also be used to only extract the calculated optimal solutions for any objectives of interest.

Required inputs include:

- *constrModPath*: The path to the folder with saved personalized models constrained with the simulated diet that had been created during the *initMgPipe* run
- *objectiveList*: A cell array with the list of objective functions in the models to compute

Note While any objective function in any analyzed model can serve as an input, for most purposes, the most suitable objective function is the fecal exchange reaction (e.g., "EX_ac[fe]") for metabolites of interest.

Optional inputs include:

- *SPDef*: Defines whether positive, negative, or all shadow prices should be computed (default: "nonzero")
- *spPath*: To define a path where computed shadow prices will be saved (default path name will be used otherwise)
- *numWorkers*: Number of cores dedicated to parallel computing

Note The definition of the shadow price depends on the solver. To verify the shadow price definitions for each solver, run the test script *testDualRCostDefinition*.

To run the computation of shadow prices:

```
[objectives,shadowPrices]=analyseObjectiveShadowPrices
(constrModPath, objectiveList, 'SPDef', SPDef, 'results-
Folder', spPath, 'numWorkers', numWorkers);
```

Similar to the previous analyses, we can also perform statistical analysis on the computed shadow prices:

```
analyzeMgPipeResults(infoFilePath, spPath,'sampleGrou-
pHeaders', sampleGroupHeaders);
```

The results of the statistical analysis will also be available in the "Statistics" folder.

4 Supplementary Tutorial

A tutorial illustrating an application example for a reduced dataset of five CD patients and five controls is provided as a MATLAB LiveScript in Supplementary File 3. The tutorial relies on an example abundance file provided in Supplementary File 1 and an example file with sample stratification criteria provided in Supplementary File 2. To run the tutorial, unzip the three Supplemental Files and place them in the same folder. Ensure that the installation requirements described in Subheading 3 are followed. Afterward, open the LiveScript in MATLAB and execute it via "Run".

Acknowledgments

A.H. and J.R.P. received funding from the Agence Nationale de la Recherche (ANR) under the decree 2021-1710. A.H. was also supported by a Chaire Gutenberg grant awarded by the Cercle Gutenberg.

Conflict of Interest Statement The authors have no conflicts of interest to declare that are relevant to the content of this chapter.

References

1. Bharti R, Grimm DG (2021) Current challenges and best-practice protocols for microbiome analysis. Brief Bioinform 22(1): 178–193. https://doi.org/10.1093/bib/bbz155
2. Qin J et al (2010) A human gut microbial gene catalogue established by metagenomic sequencing. Nature 464(7285):59–65. https://doi.org/10.1038/nature08821
3. Integrative HMPRNC (2014) The Integrative Human Microbiome Project: dynamic analysis of microbiome-host omics profiles during periods of human health and disease. Cell Host Microbe 16(3):276–289. https://doi.org/10.1016/j.chom.2014.08.014
4. Schirmer M, Garner A, Vlamakis H, Xavier RJ (2019) Microbial genes and pathways in inflammatory bowel disease. Nat Rev Microbiol 17(8):497–511. https://doi.org/10.1038/s41579-019-0213-6
5. Delzenne NM, Rodriguez J, Olivares M, Neyrinck AM (2020) Microbiome response to diet: focus on obesity and related diseases. Rev Endocr Metab Disord 21(3):369–380. https://doi.org/10.1007/s11154-020-09572-7
6. Janney A, Powrie F, Mann EH (2020) Host-microbiota maladaptation in colorectal cancer. Nature 585(7826):509–517. https://doi.org/10.1038/s41586-020-2729-3
7. Cryan JF et al (2020) The gut microbiome in neurological disorders. Lancet Neurol 19(2): 179–194. https://doi.org/10.1016/S1474-4422(19)30356-4
8. Ratiner K, Ciocan D, Abdeen SK, Elinav E (2024) Utilization of the microbiome in personalized medicine. Nat Rev Microbiol 22(5): 291–308. https://doi.org/10.1038/s41579-023-00998-9
9. Sen P, Oresic M (2023) Integrating omics data in genome-scale metabolic modeling: a methodological perspective for precision medicine. Metabolites 13(7):855. https://doi.org/10.3390/metabo13070855
10. O'Brien EJ, Monk JM, Palsson BO (2015) Using genome-scale models to predict biological capabilities. Cell 161(5):971–987. https://doi.org/10.1016/j.cell.2015.05.019
11. Zengler K, Palsson BO (2012) A road map for the development of community systems (CoSy) biology. Nat Rev Microbiol 10(5): 366–372. https://doi.org/10.1038/nrmicro2763
12. Diener C, Gibbons SM (2023) More is different: metabolic modeling of diverse microbial communities. mSystems 8(2):e0127022. https://doi.org/10.1128/msystems.01270-22
13. Gottstein W, Olivier BG, Bruggeman FJ, Teusink B (2016) Constraint-based stoichiometric modelling from single organisms to microbial communities. J R Soc Interface 13(124): 20160627. https://doi.org/10.1098/rsif.2016.0627
14. Succurro A, Ebenhoh O (2018) Review and perspective on mathematical modeling of microbial ecosystems. Biochem Soc Trans 46(2):403–412. https://doi.org/10.1042/BST20170265
15. Klitgord N, Segre D (2010) Environments that induce synthetic microbial ecosystems. PLoS Comput Biol 6(11):e1001002. https://doi.org/10.1371/journal.pcbi.1001002
16. Heinken A, Basile A, Thiele I (2021) Advances in constraint-based modelling of microbial communities. Curr Opin Syst Biol 27: 100346. https://doi.org/10.1016/j.coisb.2021.05.007
17. Heinken A, Hertel J, Thiele I (2021) Metabolic modelling reveals broad changes in gut microbial metabolism in inflammatory bowel disease patients with dysbiosis. NPJ Syst Biol

Appl 7(1):19. https://doi.org/10.1038/s41540-021-00178-6

18. Scott WT Jr et al (2023) A structured evaluation of genome-scale constraint-based modeling tools for microbial consortia. PLoS Comput Biol 19(8):e1011363. https://doi.org/10.1371/journal.pcbi.1011363
19. Budinich M, Bourdon J, Larhlimi A, Eveillard D (2017) A multi-objective constraint-based approach for modeling genome-scale microbial ecosystems. PLoS One 12(2):e0171744. https://doi.org/10.1371/journal.pone.0171744
20. Zomorrodi AR, Maranas CD (2012) OptCom: a multi-level optimization framework for the metabolic modeling and analysis of microbial communities. PLoS Comput Biol 8(2):e1002363. https://doi.org/10.1371/journal.pcbi.1002363
21. Thiele I, Palsson BO (2010) A protocol for generating a high-quality genome-scale metabolic reconstruction. Nat Protoc 5(1):93–121. https://doi.org/10.1038/nprot.2009.203
22. Henry CS et al (2010) High-throughput generation, optimization and analysis of genome-scale metabolic models. Nat Biotechnol 28(9):977–982. https://doi.org/10.1038/nbt.1672
23. Machado D, Andrejev S, Tramontano M, Patil KR (2018) Fast automated reconstruction of genome-scale metabolic models for microbial species and communities. Nucleic Acids Res 46(15):7542–7553. https://doi.org/10.1093/nar/gky537
24. Capela J et al (2022) merlin, an improved framework for the reconstruction of high-quality genome-scale metabolic models. Nucleic Acids Res 50(11):6052–6066. https://doi.org/10.1093/nar/gkac459
25. Zimmermann J, Kaleta C, Waschina S (2021) gapseq: informed prediction of bacterial metabolic pathways and reconstruction of accurate metabolic models. Genome Biol 22(1):81. https://doi.org/10.1186/s13059-021-02295-1
26. Heinken A, Magnusdottir S, Fleming RMT, Thiele I (2021) DEMETER: efficient simultaneous curation of genome-scale reconstructions guided by experimental data and refined gene annotations. Bioinformatics 37:3974–3975. https://doi.org/10.1093/bioinformatics/btab622
27. Magnusdottir S et al (2017) Generation of genome-scale metabolic reconstructions for 773 members of the human gut microbiota. Nat Biotechnol 35(1):81–89. https://doi.org/10.1038/nbt.3703
28. Heinken A et al (2023) Genome-scale metabolic reconstruction of 7,302 human microorganisms for personalized medicine. Nat Biotechnol 41:1320–1331. https://doi.org/10.1038/s41587-022-01628-0
29. Baldini F et al (2019) The Microbiome Modeling Toolbox: from microbial interactions to personalized microbial communities. Bioinformatics 35(13):2332–2334. https://doi.org/10.1093/bioinformatics/bty941
30. Martinelli F et al (2023) Whole-body modelling reveals microbiome and genomic interactions on reduced urine formate levels in Alzheimer's disease. Res Sq. https://doi.org/10.21203/rs.3.rs-3306891/v1
31. Lamichhane S et al (2022) Dysregulation of secondary bile acid metabolism precedes islet autoimmunity and type 1 diabetes. Cell Rep Med 3:100762. https://doi.org/10.1016/j.xcrm.2022.100762
32. Kindschuh WF et al (2023) Preterm birth is associated with xenobiotics and predicted by the vaginal metabolome. Nat Microbiol 8(2):246–259. https://doi.org/10.1038/s41564-022-01293-8
33. Hertel J et al (2019) Integrated analyses of microbiome and longitudinal metabolome data reveal microbial-host interactions on sulfur metabolism in Parkinson's disease. Cell Rep 29(7):1767–1777.e1768. https://doi.org/10.1016/j.celrep.2019.10.035
34. Heinken A, Thiele I (2022) Microbiome Modelling Toolbox 2.0: efficient, tractable modelling of microbiome communities. Bioinformatics 38:2367–2368. https://doi.org/10.1093/bioinformatics/btac082
35. Heirendt L et al (2019) Creation and analysis of biochemical constraint-based models using the COBRA Toolbox v.3.0. Nat Protoc 14(3):639–702. https://doi.org/10.1038/s41596-018-0098-2
36. Heinken A et al (2025) A genome-scale metabolic reconstruction resource of 247,092 diverse human microbes spanning multiple continents, age groups, and body sites. Cell Syst 16:101196. https://doi.org/10.1016/j.cels.2025.101196
37. Meydan Y, Baldini F, Korem T (2023) pymgpipe: microbiome metabolic modeling in Python. J Open Source Softw 8(88):5545. https://doi.org/10.21105/joss.05545
38. Knight R et al (2018) Best practices for analysing microbiomes. Nat Rev Microbiol 16(7):410–422. https://doi.org/10.1038/s41579-018-0029-9
39. Manghi P et al (2023) MetaPhlAn 4 profiling of unknown species-level genome bins

improves the characterization of diet-associated microbiome changes in mice. Cell Rep 42(5):112464. https://doi.org/10.1016/j.celrep.2023.112464

40. Lewis JD et al (2015) Inflammation, antibiotics, and diet as environmental stressors of the gut microbiome in pediatric Crohn's disease. Cell Host Microbe 18(4):489–500. https://doi.org/10.1016/j.chom.2015.09.008
41. Heinken A et al (2019) Systematic assessment of secondary bile acid metabolism in gut microbes reveals distinct metabolic capabilities in inflammatory bowel disease. Microbiome 7(1):75. https://doi.org/10.1186/s40168-019-0689-3
42. Noronha A et al (2019) The Virtual Metabolic Human database: integrating human and gut microbiome metabolism with nutrition and disease. Nucleic Acids Res 47(D1):D614–D624. https://doi.org/10.1093/nar/gky992
43. Arkin AP et al (2018) KBase: the United States Department of Energy Systems Biology Knowledgebase. Nat Biotechnol 36(7):566–569. https://doi.org/10.1038/nbt.4163
44. Reimer LC et al (2019) BacDive in 2019: bacterial phenotypic data for high-throughput biodiversity analysis. Nucleic Acids Res 47(D1):D631–D636. https://doi.org/10.1093/nar/gky879
45. Seaver SMD et al (2021) The ModelSEED Biochemistry Database for the integration of metabolic annotations and the reconstruction, comparison and analysis of metabolic models for plants, fungi and microbes. Nucleic Acids Res 49(D1):D575–D588. https://doi.org/10.1093/nar/gkaa746
46. Fleming RMT, Vlassis N, Thiele I, Saunders MA (2016) Conditions for duality between fluxes and concentrations in biochemical networks. J Theor Biol 409:1–10. https://doi.org/10.1016/j.jtbi.2016.06.033
47. Gudmundsson S, Thiele I (2010) Computationally efficient flux variability analysis. BMC Bioinformatics 11:489. https://doi.org/10.1186/1471-2105-11-489
48. Hulshof T, Nap B, Martinelli F and Thiele I (2024) Microbial Abundances Retrieved from Sequencing data-automated NCBI Taxonomy (MARS): a pipeline to create relative microbial abundance data for the Microbiome Modelling Toolbox and utilizing homosynonyms for efficient mapping to resources. Bioinform Adv 4(1):vbae068. https://doi.org/10.1093/bioadv/vbae068

Chapter 13

Predicting Interspecies Metabolic Dependencies in Microbial Communities by Integrating Flux Coupling Analysis with SteadyCom

Steve Zhang, Hugh C. McCullough, and Hyun-Seob Song

Abstract

Microbial communities play a pivotal role in a wide range of ecological processes and engineered systems. The level of complexity in analyzing microbial communities necessitates the use of predictive mathematical models such as genome-scale metabolic networks. To identify metabolic interdependence among member species within microbial communities, we present a recently developed computational tool combining two existing approaches: SteadyCom and Flux Coupling Analysis (FCA). Both approaches leverage genome-scale metabolic networks as their primary inputs, albeit for distinct objectives. SteadyCom is a community modeling tool used to estimate flux distributions and metabolite exchanges within and across species, respectively, by constraining individual specific growth rates to be equal. In contrast, FCA has been used to identify the causal relationships among reactions with a primary focus on the analysis of individual metabolic networks, i.e., what reactions must be active for a target reaction to be active. The combination of these two approaches allows us to find how metabolic reactions in individual species are coordinated as an interacting community. Without any additional computations, the implementation of this algorithm also provides information on what reactions are blocked in the network. We provide all the information needed to implement these coupled computational tools in a Python Jupyter notebook.

Key words Flux coupling, Metabolic dependency, Microbial communities, Genome-scale metabolic networks, Amino acid auxotrophs

1 Introduction

The omnipresence of microbes ensures their activity affects every environment and organism on the planet, serving as the backbone of essential ecological processes, such as nutrient cycling [1, 2]. These processes are critical in shaping an ecosystem's diversity and functionality. Furthermore, microbes significantly affect human health by providing a range of benefits, including the chelation of ingested heavy metals and the prevention of diseases such as cancer and inflammatory bowel disease [3, 4], as well as being a source of potential harm, such as cases of food being contaminated

Karthik Raman and Gayathri Sambamoorthy (eds.), *Modelling the Microbiome*, Methods in Molecular Biology, vol. 3006, https://doi.org/10.1007/978-1-0716-5080-6_13,

with pathogens, which has elicited extensive research into their prevention [5, 6]. The metabolic capabilities of an individual species do not solely determine its impact on these processes because the phenotype emerges through complex interspecies interactions within a microbial community.

Genome-scale metabolic models (GEMs) serve as invaluable tools for exploring both the metabolism of individual organisms and their interactions in microbial communities [7]. GEMs collect all relevant biochemical reactions and pathways, as informed by genome annotations. To study interspecies metabolic interactions in microbial communities, individual GEMs of member species are merged through compartmentalization [8]. The resulting community network has one "community" compartment shared by all member species; each species has one "extracellular" compartment and at least one "intracellular" compartment. Metabolites are exchanged across these extracellular and intracellular compartments.

Flux Balance Analysis (FBA) [9] is the most popularly used method to estimate the distribution of metabolite flows (or fluxes) through the reactions in GEMs such that a predefined objective is maximized. Typically, FBA seeks the solution by maximizing biomass production, though it can accommodate other alternatives as well. SteadyCom extends FBA to the analysis of GEMs of microbial communities [8]. Unlike FBA, which simply maximizes a given objective, SteadyCom determines intracellular flux distributions and cross-species metabolite exchanges under the constraint of equal specific growth rates among all members and predicts how they vary over a range of specific growth rates. Imposition of the equal specific growth rate constraint enforces community stability, leading to each member's relative abundances being constant over time. SteadyCom also describes the participation of individual organisms in the community metabolism by accounting for their abundances.

Flux Coupling Analysis (FCA) is a complementary computational method used for identifying dependencies between reactions [10]. Specifically, it addresses the following critical question: For a certain reaction to be active, what other reactions must also be active? Understanding these causal relationships among reactions is useful in numerous applications. For instance, FCA can serve as an invaluable tool for metabolic engineering when targeting specific genes and reactions for enhancement or suppression. FCA also finds blocked reactions, through which no fluxes are carried. The existence of these reactions may be a sign of model incompleteness, perhaps missing crucial reactions that, if included, could activate those currently blocked.

While FCA has been traditionally applied to the analysis of individual GEMs, this chapter extends it to community metabolic networks via a combination with SteadyCom. This integrative method enables identifying metabolic interdependencies within microbial communities by predicting which reactions in one species need to be coupled with the reactions in other species for their co-growth. We illustrate the implementation of our computational method using example systems of mutant *Escherichia coli* strains that are auxotrophic for distinct essential amino acids, making them metabolically dependent on each other.

2 Materials

2.1 Genome-Scale Metabolic Models

Metabolic models are often represented in Systems Biology Markup Language (SBML) format, as is the case with models provided by BiGG [11]. From BiGG (http://bigg.ucsd.edu/models/iAF1260), we downloaded the *E. coli* metabolic model iAF1260 in SBML format as an example.

2.2 Software Tools

Python Libraries to Install: The programming language Python is needed to execute the code. Installation instructions are found at https://www.python.org/downloads/.

Python libraries may be installed with pip, a Python package installer. This can be run from the command line. Most operating systems come with a built-in terminal emulator that provides a command-line interface. Examples include Terminal on macOS and Windows Terminal on Windows. To install pip, run the following in the command line:

```
python -m ensurepip
```

To install libraries with pip, run:

```
python -m pip install library_name
```

Below is a list of Python libraries our codes require. Run the command to install them for each library by replacing `library_name` with the library's name.

1. copy
2. gurobipy
3. numpy
4. pickle
5. xmltodict

During the installation of Python, a directory containing the installed libraries will automatically be made and allow them to be used in code.

Linear Programming Solver: We use the Python API, with installation instructions provided at https://support.gurobi.com/hc/en-us/articles/360044290292-How-do-I-install-Gurobi-for-Python, of the Gurobi™ optimizer, a linear programming (LP) problem solver, available for free for academics at www.gurobi.com. Make an account at that site using an academic email to obtain a license that will enable the fullest use of Gurobi. Additional instructions are found at https://support.gurobi.com/hc/en-us/articles/12684663118993-How-do-I-obtain-a-Gurobi-license.

Installing Jupyter: We provide a Jupyter notebook file that walks through the code running SteadyCom and FCA. Jupyter must be installed to run this file. Installation instructions are provided at https://jupyter.org/install. It may also be installed through the pip installation method described in this section, with the library name being "jupyterlab".

Installing SteadyCom and Flux Coupling Analysis source code: The implementation is located at https://github.com/hyunseobsong/flux-coupling. The example files can be run after being downloaded and the prerequisite dependencies detailed above have been installed. The files also show how the code should be organized for the steps, which are roughly described in Subheading 3, to be executed in the recommended order. All analyses were run on a 64-bit MacBook Pro with a 2.2 GHz quad-core Intel Core i7 processor, which contains the resources of typical modern processors.

3 Methods

3.1 Loading the Metabolic Model

The metabolic model's information must first be extracted from the SBML file. This can be done with the parseXML function.

```
MetabolicMdl = parseXML("iAF1260.xml")
```

Inputs:

1. File name: Path to the SBML file.

Outputs:

1. A list containing the following elements:
   1. A stoichiometric matrix with dimensions m * n, where m and n are the number of metabolites and reactions. Each

row and column represents a metabolite and reaction. A negative number in the i'th row and j'th column means the i'th metabolite is a reactant in the j'th reaction. If it were positive, it would be a product. Thus, each column represents the reactions' stoichiometry.

2. Reaction flux lower bounds. These represent the extent a reaction can proceed in the reverse direction.
3. Reaction flux upper bounds. These represent the extent a reaction can proceed in the forward direction.
4. A list of metabolites. Note that some metabolites may appear multiple times but exist in different compartments. For example, the metabolite water or H_2O may appear in the list as "h2o_e" and "h2o_p", which refer to water while in the extracellular and periplasm compartments, respectively. Metabolic models of individual organisms can have multiple internal compartments (e.g., cytosolic and periplasmic) that are all enclosed in a single extracellular compartment, which represents the cell's environment from which metabolites may enter and leave from.
5. A list of reactions.

Using the pickle module, this list can be "pickled", the process of serializing a Python object and stored into a pkl file, then later "unpickled", deserializing or loading the data into a variable. Serialization is the process of converting data into a storable format, like a file, and deserialization is the process of converting the information in that format back into the original data. Only use pkl files from trustworthy sources (*see* **Note 1**). The code example below shows the pickle module being used to save and load the data.

```
import pickle
#Pickling, or saving, the metabolic model
with open("iAF1260.pkl","wb") as File:
 pickle.dump(MetabolicMdl, File)
#With the pkl file, the file can be unpickled or loaded into
a variable
with open("iAF1260.pkl","rb") as File:
 MetabolicMdl = pickle.load(File)
```

This way, the SBML file does not need to be read every time the program begins, and any changes, such as the ones described in Subheadings 3.2 and 3.3, made to the model will be conserved.

3.2 Editing the Metabolic Model

One change can be to add a reaction:

```
import numpy as np
#Adding column to represent the reaction
MetabolicMdl[0]  =  np.hstack((MetabolicMdl[0],np.zeros(
(MetabolicMdl[0].shape[0],1))))
#Filling the stoichiometry of the reaction
MetabolicMdl[0][MetabolicMdl[3].index('met__L_c')][-1] =
MetabolicMdl[0][MetabolicMdl[3].index('h_c')][-1] = -1
MetabolicMdl[0][MetabolicMdl[3].index('met__L_p')][-1] =
MetabolicMdl[0][MetabolicMdl[3].index('h_p')][-1] = 1
#Adding name of newly added reaction
MetabolicMdl[4].append('METt3pp')
MetabolicMdl[5].append('METt3pp')
#Editing the name of a reaction in the model, in this
case, the last one
MetabolicMdl[4][-1] = MetabolicMdl[5][-1] = 'R_METt3pp'
#Adding bounds for the reaction
MetabolicMdl[1].append(0)
MetabolicMdl[2].append(1000)
```

When using the "index" function, one possible error that may occur is "ValueError". This is due to the parameter, in this case the metabolite name, being absent from the list of metabolites (represented in the example as "MetabolicMdl[3]"). This can be fixed by correcting the spelling.

3.3 Simulating Auxotrophy

Auxotrophy is the inability of an organism to synthesize a compound that is essential for its growth. This can occur among microbes that have co-evolved and developed metabolic interdependencies. To simulate auxotrophy in metabolic models, we can force the lower and upper bounds of reactions essential to the synthesis of that compound to equal zero. This is known as "blocking" a reaction, which prevents that reaction from carrying flux or, in other words, from being used. Care should be taken when choosing which reaction to block a reaction, since it may disable other metabolic functionalities.

```
RxnIdx = MetabolicMdl[4].index("ARGSL")
MetabolicMdl[1][RxnIdx] = MetabolicMdl[2][RxnIdx] = 0
```

We use this method to create in silico consortia of *E. coli* with each member auxotrophic in amino acids that the other members produce. Figure 1 contains tables depicting the auxotrophy of two such consortia, one with two and the other with three members.

Binary consortium

	Arg	Lys
E. coli #1	O	X
E. coli #2	X	O

Ternary consortium

	Arg	Lys	Met
E. coli #1	O	X	X
E. coli #2	X	O	X
E. coli #3	X	X	O

Fig. 1 List of *E. coli* strains included in binary and ternary consortia. It shows which amino acid each *E. coli* member can produce and which amino acids each is auxotrophic in. An *E. coli* member with "O" to an amino acid can synthesize that amino acid. Entries with "X" means that member is auxotrophic in that amino acid. In case A, the two members are auxotrophic in one of two amino acids. In case B, the three members are auxotrophic in two of three amino acids. In both cases, each member is dependent on the other member(s) for survival. Arg = arginine, Lys = lysine, and Met = methionine

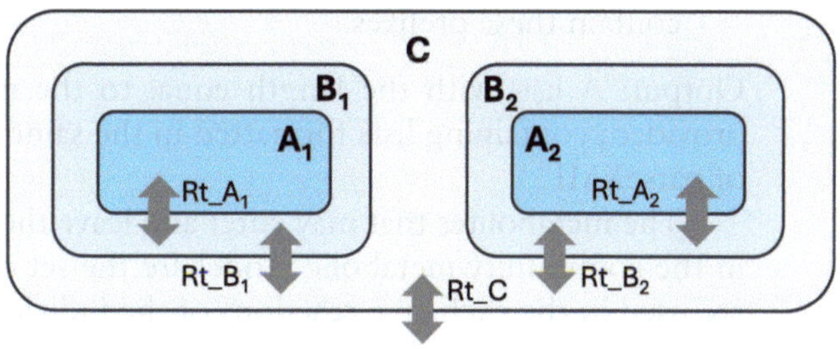

Fig. 2 A conceptual representation of community model structure. The rectangles in different colors, A (A_1 and A_2), B (B_1 and B_2), and C, respectively represent the intracellular, extracellular, and community compartments where the subscripts 1 and 2 denote species 1 and 2. Three arrows are displayed, each representing a type of transport reaction the community model contains. In this example, each has a name prefixed with "Rt". Rt_A transports metabolites between a community member's intracellular and extracellular spaces. These are often transporter reactions. Rt_B transports metabolites between a community member's extracellular space and the community space. Rt_C is responsible for metabolites entering and exiting the community space, which is analogous to the exchange reactions found in the GEMs of individual organisms

3.4 Creating the Community Metabolic Model

A community metabolic model is an aggregate of individual metabolic models. Earlier, we mentioned that each GEM contains an extracellular space in the community model. The space enclosing the extracellular spaces of each organism, and therefore connecting them to each other in the model, is called the community compartment. Figure 2 provides an example of how this community model is structured, with reactions shuttling metabolites through these compartments.

To create a community metabolic model, we use the createCommModel function.

```
TheListOfMdls = TwoComm(MetMdl) #Load auxotrophic E. coli
consortia
CommMdl = createCommModel(TheListOfMdls,"EX")
```

Inputs:

1. Models: List containing arbitrary number of metabolic models. The list will contain the members comprising the community.
2. Prefixes of exchange reactions, which transport metabolites in and out of the extracellular compartment. To prevent errors in the construction of community metabolic models, ensure that each prefix of every exchange reaction from each GEM is included here and that it is only the exchange reactions that contain these prefixes.

Output: A list, with the length equal to the number of models provided, containing lists formatted in the same way as the output of parseXML.

The metabolites that may enter and leave the community space in the community metabolic model are the set of metabolites that are used in the exchange reactions of the individual GEMs used to create the community model.

GEMs from other databases may have their own unique prefix identifying exchange reactions. Each of those prefixes may be input into the createCommModel function to ensure clear identification between the exchange reactions and nonexchange, in other words internal, reactions. The code below is an example of using the createCommModel function when all the GEMs' exchange reactions are identified by three unique prefixes:

```
createCommModel
(TheListOfMdls,"Prefix1","Prefix2","Prefix3")
```

Metabolite and reaction names of the community members will be slightly changed. Reactions and their respective metabolites will have a "_c#" appended to their names, with # being the position of that model in the list that was passed as an argument to createCommModel. Reaction names of the community uptake/export reactions and the metabolites they involve will also be added. They will follow the naming convention of the metabolic models' exchange reactions and their metabolites but be suffixed with "_u". This preserves the original reaction names in the community metabolic model, maintaining consistency and allowing users to easily identify and link metabolites and reactions in each compartment to their corresponding individual GEMs.

Below is an example of editing an exchange reaction in a specific compartment in the community metabolic model. The exchange reaction "EX_glc__D_e_c0" transports glucose in and out of the first community member's (denoted by "_c0") extracellular compartment. To edit the equivalent reaction of the second member, replace the "0" in "EX_glc__D_e_c0" with "1".

```
RxnIdx = CommMdl[4].index("EX_glc__D_e_c0")
CommMdl[1][RxnIdx] = -9 #Changing the lower bound of the
reaction to -9
```

The original lower bound of the exchange reaction was −8. By changing it to −9, this would allow for more glucose to enter the first member's extracellular space, allowing that member to consume more of it. The increased nutrient availability would lead to higher predicted growth for this member, impacting downstream metabolic analyses like Flux Balance Analysis.

3.5 Finding the Maximum Community Growth Rate

The community model can be passed as input to the FindMaxGrowthRate function, which, analogous to Flux Balance Analysis, will find the highest growth rate predicted by the model.

```
GrowthRate,TheMdl,BioIndices = FindMaxGrowthRate
(CommMdl,"BIOMASS")
```

Inputs:

1. List representing the community metabolic model: This is the model created by createCommModel described earlier.
2. Substrings exclusively found in each community member's biomass reaction name: To prevent errors in growth rate prediction, ensure that each substring from every biomass reaction of each GEM is included here and that it matches only the biomass reactions containing that substring.

Outputs:

1. Community growth rate: This growth rate is also the specific growth rate of all the community's members.
2. Linear programming model: The LP model contains all the information needed by the solver to find a solution. This information includes decision variables, constraints, and parameter values such as the method of arriving at a solution. The variables may be extracted using the model's getVars() method (e.g. MyModel.getVars()). Each variable represents a reaction and contains the flux value that led to the maximum

growth rate. Using the model's write() method, it may be stored into a file (*see* **Note 2**), which, assuming gurobipy was imported as import gurobipy as gp, can be loaded into a variable with gp.read(filename).

3. A numerical list with length equal to the number of community members, with the nth number containing the nth community member's biomass reaction name in the community model's list of reactions. This is needed in the RemoveBlockedReactions function described below.

3.6 Removing Blocked Reactions

The first step in conducting FCA is to remove blocked reactions [10]. As stated earlier, blocked reactions cannot be used in steady-state metabolic analyses; in other words, their flux would always be zero. Since they can never be active, no reaction will ever be dependent on their activity to be active themselves. Thus, their removal would allow FCA to run quicker without affecting the results. They will be removed with the function "RemoveBlockedReactions". Different methods to solve LPs may change RemoveBlockedReactions's results; *see* **Note 3**.

```
TrimmedLPMdl,TrimmedMdl,TheKFT = RemoveBlockedReactions
(LPMdl,GrowthRate, BiomassIndices)
```

Inputs:

1. Linear programming model of the community metabolic model: This is the model created by FindMaxGrowthRate described earlier.
2. Growth rate threshold: This value is the minimum predicted community growth rate that the solution the solver shall arrive at (*see* **Note 4** for advice on choosing a value). At higher values, the range of each reaction's flux becomes more limited as they become increasingly constrained toward biomass production. Thus, different thresholds may lead to different reactions being regarded as blocked.
3. BiomassIndices: Indices of each community member's biomass reactions.

Outputs:

1. Trimmed linear programming model: This is the model created by FindMaxGrowthRate described earlier, except that all the constraints solely involving variables pertaining to blocked reactions, as well as the variables themselves, have been removed.

2. List representing the trimmed model: This is a list of five elements in the same format as the list from createCommModel. The stoichiometric matrix will have any columns associated with blocked reactions, and any rows associated with metabolites that are solely involved in blocked reactions, removed. Any associated lower and upper bounds, reaction, and metabolite names of the removed reactions will be removed as well.
3. Known Flux Table (KFT): A table, with length equal to the number of unblocked reactions, whose elements are pairs. The pair's first value is the reaction's name. The second value is a list of pairs. The first element of this pair is the unblocked reaction's name, and the second element is the nonzero flux that the reaction was predicted to achieve.

With the removal of blocked reactions, the trimmed model will have fewer variables and constraints than the original model, which will also result in faster determination of flux couplings. Blocked reactions may also be a sign of the quality of the metabolic model's reconstruction; *see* **Note 5**. The KFT will further quicken FCA by avoiding the determination of couplings between reactions found to be uncoupled in RemoveBlockedReactions. This is done as follows: Upon finding a nonzero flux value to ascertain that a reaction is unblocked, it will record what subset of reactions was also found to carry flux, meaning that the newly identified unblocked reaction will be possibly coupled only to this subset of reactions, thereby reducing the number of LPs FCA needs to solve (*see* **Note 6**).

During this step, the solution may be infeasible if the specified value exceeds the maximum growth rate determined by FindMaxGrowthRate. To resolve this, the constraints causing the infeasibility must be modified or removed. In this case, to avoid the error, the growth rate threshold must be set equal to or lower than the value obtained from FindMaxGrowthRate.

3.7 Identifying Couplings Between Reactions

To find flux couplings, LPs will be solved that forces a reaction (e.g., rxnA) to be active, that is, have nonzero flux, and a second reaction (e.g., rxnB) to be inactive, that is, have zero flux [10]. If this results in an infeasible solution—meaning the model predicts it is impossible for rxnA to be active while rxnB is inactive—then rxnA is directionally coupled to rxnB; in other words, for rxnA to be active, rxnB must also be active but not necessarily the reverse. If this infeasibility occurs and it is also found that no solution exists in which rxnB has nonzero flux while rxnA has zero flux, then the two reactions are either fully or partially coupled. If a constant nonzero flux in rxnA forces rxnB to also have a constant nonzero flux, and vice versa, they are fully coupled. Otherwise, they are partially coupled, meaning that rxnA having nonzero flux causes rxnB to have a variable but nonzero flux, and vice versa. In the reaction pair of rxnA and rxnB, if a feasible solution exists where flux through one does not imply flux through another and vice versa, then they are uncoupled, meaning neither one of the reactions being active is contingent on the other being active.

All reactions' flux couplings may be acquired with the FluxCouplingAnalysis function.

```
FCTBl = FluxCouplingAnalysis(TrimmedLPMdl,GrowthRate,
TheKFT)
```

Inputs:

1. Trimmed linear programming model of the community metabolic model: This is the model created by RemoveBlockedReactions described earlier.
2. Growth rate threshold: This value is the minimum predicted community growth rate that the solution the solver shall arrive at. This value must be equal to or less than the value from FindMaxGrowthRate. At higher values, the range of each reaction's flux becomes more limited as they become increasingly constrained toward biomass production. Thus, different thresholds may lead to different reaction couplings.
3. Known flux table: This is the KFT output by RemoveBlockedReactions

Outputs:

1. Flux Coupling Table (FCT): This is a square numerical matrix, with the side length equal to the number of reactions in the linear programming model. Entries range from 0 to 4 inclusive. The entry at coordinates (x, y) holds what type of coupling the x'th reaction has with the y'th reaction. The meaning of each entry is as follows:
   0. The x'th and y'th reactions are uncoupled.
   1. The x'th and y'th reactions are fully coupled.
   2. The x'th and y'th reactions are partially coupled.
   3. The x'th reaction is directionally coupled to the y'th reaction.
   4. The y'th reaction is directionally coupled to the x'th reaction.

A reference to the trimmed model should be kept, allowing for associations to be made between the reaction information and the FCT. The FCT may be pickled together with the list representing the trimmed model. Since there should be many more unblocked

A

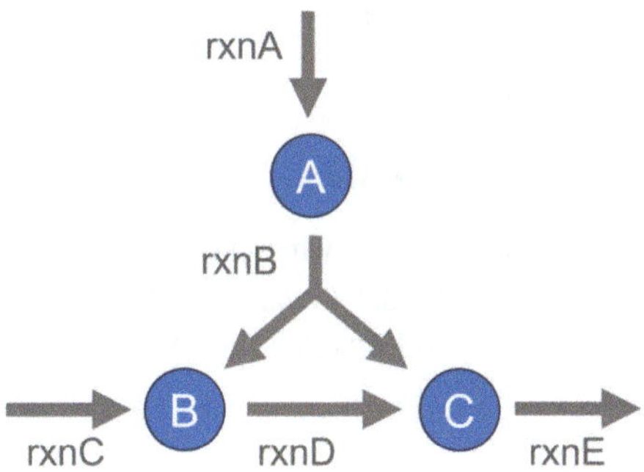

B

Reactions

Metabolites	rxnA	rxnB	rxnC	rxnD	rxnE
A	1	-1	0	0	0
B	0	1	1	-1	0
C	0	1	0	1	-1

C

rxnA	rxnB	rxnC	rxnD	rxnE
rxnB, 1	rxnA, 1	rxnD, 1	rxnC, 1	rxnC, 1
rxnD, 1	rxnD, 1	rxnE, 1	rxnE, 1	rxnD, 1
rxnE, 2	rxnE, 2			

D

	rxnA	rxnB	rxnC	rxnD	rxnE
rxnA	1	1	0	3	3
rxnB	1	1	0	3	3
rxnC	0	0	1	3	3
rxnD	4	4	4	1	2
rxnE	4	4	4	2	1

Fig. 3 Illustration of the concepts of KFT and FCT: (**a**) A toy network with three metabolites and five reactions. The circles and arrows represent metabolites and reactions, respectively. An arrow pointing to a circle means that reaction produces that metabolite. An arrow pointing away means it consumes that metabolite; (**b**) A stoichiometry table where each row and column represents a metabolite and reaction, respectively. Positive and negative numbers in the i'th row and j'th column mean the j'th reaction produces and consumes that metabolite respectively; (**c**) The KFT for the toy network in (**a**). Under each reaction, when that reaction is active, that is, has nonzero flux, there is a list of pairs, where the reaction also found to be active and its flux comprise the pair's first and second elements. In this example, when an LP was solved with rxnC's flux constrained to be at least 1, the solution also set the flux of rxnD and rxnE to 1; (**d**) The FCT for the toy network in (**a**). Each row and column represents a reaction. Each entry contains the flux coupling present between the i'th row and j'th column. In this example, the number in the third row and fourth column is 3, meaning rxnC is directionally coupled to rxnD

reactions than there are blocked, the FCT will be very large. In Fig. 3, we provided a simple example to illustrate how the KFT and FCT are generated for a given network.

4 Brief Look into Results

An excerpt of the first row of the FCT is [1, 0, 0..., 3, 3, 3, 0, 0, 3..., 0, 0, 0]. Ellipses are included to avoid printing the entirety of the large row. In the first row, we see the reaction is, as expected, fully coupled to itself. Its relationship with the other reactions is either uncoupled, signified by "0", or is directionally coupled to them, as signified by "3", meaning they are dependent on those reactions but not vice versa. The multitude of "3" in the first row means this reaction alone largely is not needed for other reactions

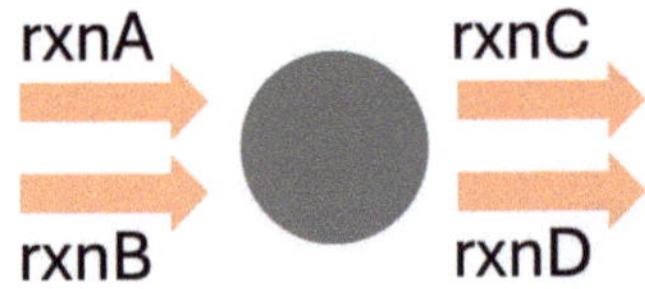

Fig. 4 A schematic of an example metabolic network with one metabolite (circle) and four reactions (arrows). Each reaction is labeled with its name above and below the representative arrow. rxnA and rxnB produce the metabolite, while rxnC and rxnD consume it

to be active and that this reaction's deletion would not significantly impact the microorganism's metabolic capabilities.

Such a deletion could lead to additional couplings between reactions. Figure 4 illustrates an example. For rxnC and rxnD to be active, either rxnA or rxnB must also be active. Consequently, there are no reactions coupled to each other. But, if rxnA were removed, then rxnC and rxnD would be directionally coupled to rxnB, since now only rxnB produces the metabolite that rxnC and rxnD consume.

5 Notes

1. Although the pickle module is widely used to serialize data, it is possible, with methods outside the scope of this chapter, to create a pkl file that executes arbitrary, malicious code upon unpickling. While we hope the methods detailed in the chapter will be widely used, widespread usage could lead to more pkl files being shared, increasing the possibility of encountering dangerous pkl files.
2. A file from the model's write() method will typically be a.mps or.lp file. The former should be used for the most accurate reproduction of the model. The latter records the model's information in a more human-readable format at the cost of reproduction accuracy. Viewing this file in a text editor can grant the user a better understanding of the constraints added to the model. Also, changes to the model's parameters are *not* recorded, so the model after being loaded will first set parameter values defined in the gurobi.env file in https://github.com/hyunseobsong/flux-coupling. Afterward, parameters not defined in the gurobi.env file will be assigned default values.
3. The method the LP solver uses may decide whether a reaction is blocked or if two reactions are coupled. The chances of this occurring rise in large metabolic models, that is, those containing many reactions. By default, gurobi sets its method parameter value to −1, which automatically selects a method. To ensure consistent results, the user may set the value to 0. This

option uses the primal simplex, which we have found leads to quick and accurate computation. This is already done in the gurobi.env file, in the line "Method 0".

4. Using a growth rate threshold value too close to the maximum value predicted by FindMaxGrowthRate may lead to an infeasible solution, even though that value was predicted earlier and therefore was predicted to create a feasible solution. This is a side effect of computers performing double-precision arithmetic with numbers stored in binary format. Small inaccuracies may arise from these operations that are imperceptible when calculations are few but may accumulate with many calculations building off the previous ones. There are many opportunities for this to transpire since FCA solves many LPs, each with different constraints. These constraints were absent in the LPs solved by FindMaxGrowthRate. To avoid this happening, the maximum value to use as the growth rate threshold value should be 95–99% of the value FindMaxGrowthRate calculated.
5. An excessive number of blocked reactions means the model is lacking many reactions, meaning the GEM would underestimate the organism's metabolic capabilities. Thus, absolute conclusions about FCA from such models should not be drawn.
6. Even after removing blocked reactions, it is common for GEMs to retain well over a thousand reactions. Since community models are aggregates of individual GEMs, the time needed to complete FCA may still be long.

Acknowledgments

This material is based upon work supported by the National Science Foundation under Grant No. 2125155. Any opinions, findings, and conclusions or recommendations expressed in this material are those of the author(s) and do not necessarily reflect the views of the National Science Foundation.

Conflict of Interest Statement The authors have no conflicts of interest to declare that are relevant to the content of this chapter.

References

1. Morris A, Meyer K, Bohannan B (2020) Linking microbial communities to ecosystem functions: what we can learn from genotype-phenotype mapping in organisms. Philos Trans R Soc B Biol Sci 375(1798). https://doi.org/10.1098/rstb.2019.0244
2. Zhou X, Chen X, Qi X, Zeng Y, Guo X, Zhuang G, Ma A (2023) Soil bacterial

communities associated with multi-nutrient cycling under long-term warming in the alpine meadow. Front Microbiol 14:1–10. https://doi.org/10.3389/fmicb.2023.1136187

3. Ogunrinola GA, Oyewale JO, Oshamika OO, Olasehinde GI (2020) The human microbiome and its impacts on health. Int J Microbiol 2020. https://doi.org/10.1155/2020/8045646
4. Song HS, Lindemann SR, Lee DY (2021) Editorial: predictive modeling of human microbiota and their role in health and disease. Front Microbiol 12:782871. https://doi.org/10.3389/fmicb.2021.782871
5. Manas P, Pagan R (2005) Microbial inactivation by new technologies of food preservation. J Appl Microbiol 98(6):1387–1399. https://doi.org/10.1111/j.1365-2672.2005.02561.x
6. Zhang S, Ahamed F, Song H-S (2022) Knowledge-informed data-driven modeling for sparse identification of governing equations for microbial inactivation processes in food. Front Food Sci Technol 2:1–12. https://doi.org/10.3389/frfst.2022.996399
7. Song H-S, Cannon W, Beliaev A, Konopka A (2014) Mathematical modeling of microbial community dynamics: a methodological review. Processes 2(4):711–752. https://doi.org/10.3390/pr2040711
8. Chan SHJ, Simons MN, Maranas CD (2017) SteadyCom: predicting microbial abundances while ensuring community stability. PLoS Comput Biol 13(5). https://doi.org/10.1371/journal.pcbi.1005539
9. Orth JD, Thiele I, Palsson BO (2010) What is flux balance analysis? Nat Biotechnol 28(3): 245–248. https://doi.org/10.1038/nbt.1614
10. David L, Marashi SA, Larhlimi A, Mieth B, Bockmayr A (2011) FFCA: a feasibility-based method for flux coupling analysis of metabolic networks. BMC Bioinformatics 12. https://doi.org/10.1186/1471-2105-12-236
11. King ZA, Lu J, Dräger A, Miller P, Federowicz S, Lerman JA, Ebrahim A, Palsson BO, Lewis NE (2016) BiGG Models: a platform for integrating, standardizing and sharing genome-scale models. Nucleic Acids Res 44 (D1):D515–D522. https://doi.org/10.1093/nar/gkv1049

Chapter 14

Describing and Designing Microbial Community Metabolic Models In Silico: A Comprehensive Protocol Utilizing FLYCOP

Ana del Ramo, David San León Granado, and Juan Nogales

Abstract

Microbial communities play crucial roles in multiple natural and engineered environments, contributing to biogeochemical cycling, waste treatment, and biotechnological processes. Much like for single organisms, genome-scale metabolic models (GEMs) have become essential in contextualizing, designing and optimizing microbiomes. Computational modeling of microbial community metabolism provides valuable insights into community dynamics, interactions, and metabolic capabilities. Here, we present a comprehensive protocol for describing and engineering microbial community metabolic models in silico, leveraging FLYCOP (FLexible sYnthetic Consortium Optimization). Starting with available individual GEMs, this protocol covers the construction of condition-specific GEMs for community members, the generation of community-based metabolic models, and the analysis of community-wide metabolic capabilities and interactions. Furthermore, we showcase the utility of FLYCOP by illustrating its application in: (i) describing a community-driven complex biological process (e.g., denitrification) and (ii) designing a synthetic community for biotechnological purposes (e.g., production of violacein).

Key words Genome-scale metabolic model, FLYCOP, Synthetic Microbial Consortia, COMETS, Flux Balance Analysis

1 Introduction

Microbial communities are essential for maintaining ecosystem stability, promoting human health, driving biotechnological innovation, and regulating Earth's biogeochemical cycles. Additionally, Synthetic Microbial Communities (SMCs) are emerging as promising multicellular biocatalysts with significant potential to surpass engineered single strains in various biotechnological applications. For example, they enhance metabolic capabilities by leveraging complementary metabolic pathways; they increase robustness and stability, as diverse microorganisms can buffer against

Ana del Ramo and David San León Granado contributed equally with all other contributors.

Karthik Raman and Gayathri Sambamoorthy (eds.), *Modelling the Microbiome*, Methods in Molecular Biology, vol. 3006, https://doi.org/10.1007/978-1-0716-5080-6_14,

environmental perturbations, ensuring more stable processes; and they enable the implementation of division of labor, where each microorganism can specialize in specific subtasks [1, 2].

Similar to their role in individual organisms, genome-scale metabolic models (GEMs) are now indispensable for comprehending, designing, and optimizing natural and synthetic microbial communities. GEMs provide a comprehensive overview of all metabolic reactions occurring in an organism and for the extension of a given community. GEMs predict how genetic and environmental perturbations affect fitness, thus enabling optimization of cell performance while optimizing bioproduction processes. These models have been extensively utilized to understand, at a systems-level, the metabolic characteristics of their respective species [3, 4]. Additionally, GEMs are powerful tools for designing more efficient cell factories by identifying crucial metabolic targets for engineering [5]. Although still underutilized, this approach can also be extended to multiple microorganisms within a consortium.

This protocol aims to tackle this challenge by providing a step-by-step procedure for understanding and optimizing microbial consortia using genome-scale metabolic models (GEMs) within the framework called FLYCOP (FLexible sYnthetic Consortium Optimization) [6]. FLYCOP allows for the selection of the optimal consortium configuration to maximize a specific objective or fitness function from a variety of configurations. This is achieved by dynamically considering changes in metabolite concentrations over time. These concentrations are determined using a dynamic Flux Balance Analysis under the framework provided by COMETS [7].

We demonstrate the utility of FLYCOP by simulating the denitrification process and optimizing the biotechnological production of violacein within microbial communities. We compare the in silico results with those obtained experimentally. The detailed algorithm and some examples can be found in the original publication of FLYCOP [6]. The Jupyter notebook with the code examples can be accessed in https://github.com/SBGlab/FLYCOP_protocol

2 Materials

In this section, we describe the required material and tools for describing and engineering microbial community metabolic models in silico.

2.1 Equipment

Operating System

Linux OS installed with Internet connection; this protocol was tested in Ubuntu 20.04 Desktop.

Minimum Hardware Requirements

2 GHz dual core processor, 4 GiB RAM (system memory).

2.2 Software Tools

1. *Conda, Python, and Pip.*

Package environmental manager tools like Conda and pip. Conda is a cross-platform package and environment manager that installs and manages Conda packages to create isolated environments. Python is a high-level, general-purpose programming language widely used for data analysis, artificial intelligence, and scientific computing, and pip is the Python Packaging Authority's recommended tool for installing packages.

To install Conda, follow this tutorial: (https://conda.io/projects/conda/en/latest/user-guide/install/index.html).

It is recommended to create a new Conda environment and install Python and pip inside. To do that, open a linux terminal and use the following commands:

```
conda create -n FLYCOP python=3.8 pip
```

Activate the environment:

```
conda activate FLYCOP
```

2. *COBRA.*

The well-known computational framework COBRA [8] for analyzing Genome-scale metabolic models will be used. Install COBRA inside the FLYCOP environment. Version tested for this protocol: 0.26.3.

```
pip install cobra
```

3. *COMETS.*

To install COMETS [4] version 0.4.2 or above, follow the instructions in (https://www.runcomets.org/installation) and the interface with Python, cometspy version 0.4.2 or above with the command.

```
pip install cometspy
```

4. *Solver.*

One critical choice with Linear Problems is the solver; due to COMETS compatibility, only GLPK and Gurobi are allowed. GLPK is included by default in COBRA, but Gurobi is preferred due to better performance. Gurobi has free licenses for the academy (https://www.gurobi.com/academia/academic-program-and-licenses/). To obtain a license, it is necessary to be registered and follow the Gurobi instructions.

With the cometspy version included in this protocol, the newest compatible version is 10.0.3.

```
pip install gurobipy=10.0.3
```

5. *SMAC3.*

SMAC3 [9] is the software used by FLYCOP to search optimal values for a specific fitness function. To install SMAC3 version 2.0.2 or above, use the following commands:

```
conda install gxx_linux-64 gcc_linux-64 swig
pip install smac
```

6. *FLYCOP.*

Install FLYCOP python package.

```
pip install FLYCOP
```

7. *Numpy.*

The multiple analysis requires the usage of large vectors and matrices. To increase the performance, the software Numpy is desired. Install Numpy version 1.23.5.

```
pip install numpy=1.23.5
```

8. *Pandas.*

Pandas is a Python package that provides flexible data structures and high-performance data analysis tools.

Install Pandas version 1.5.3.

```
pip install pandas=1.5.3
```

9. *Plotly.*

Plotly is a Python package that provides interactive graphing and data visualization capabilities, like a wide range of static, animated, and interactive charts and plots.

Install Plotly version 5.14.1.

```
pip install plotly=5.14.1
```

10. *R and RStudio.*

R is a programming language and software environment designed for statistical computing and graphics, widely used for data analysis, statistical modeling, and visualization. RStudio is an integrated development environment (IDE) for R.

Install R and RStudio following the official instructions in https://posit.co/download/rstudio-desktop/

11. *Tidyverse and GGally.*

Tidyverse (https://www.tidyverse.org/) is a collection of R packages designed for data science, providing tools for data manipulation, visualization, and analysis. GGally (https://ggobi.github.io/ggally/) is an R package that extends ggplot2, offering additional functions and plots to simplify complex data visualization tasks. After the R and RStudio installation, to install the R frameworks tidyverse and Ggally, write in the RStudio console:

```
install.packages("tidyverse")
install.packages("GGally")
```

3 Methods

3.1 Example 1: Analysis of a Community-Scale Denitrification Process

The case study is based on ref. 10. The authors show that for denitrification, community metabolite dynamics can be predicted from the genes possessed by each community member. They conclude that a consumer–resource model can predict community metabolite dynamics from single-strain phenotypes and that metabolite dynamics can be inferred from metagenomes, aiding in the design of denitrifying communities. Here, we will optimize one of these communities, demonstrating how to integrate FLYCOP into a metagenomic analysis.

Following data from the aforementioned study, we will build a consortium composed of two *Pseudomonas putida* strains with different denitrification capabilities. For this example, one strain exhibits the ability to utilize nitrate as the final electron acceptor (Nar phenotype), while the other strain employs nitrite (Nir phenotype). We will use the latest *P. putida* KT2440 GEM, *i*JN1462 [11], as the basic model for both strains. Specific models will be constructed by constraining *i*JN1462 to capture the condition-specific phenotype of each strain.

Preliminary Evaluation of the Selected Models

The first step involves reviewing and obtaining an overview of the selected models to perform the desired biological process. This step is essential for identifying target metabolites and reactions that may already be included in the model, as well as determining those that need to be added.

Table 1
Metabolites implicated in the denitrification route and their associated id's according to BiGG notation

Metabolite name	BiGG id
Nitrate	no3
Nitrite	no2
Nitric oxide	no
Nitrous oxide	n2o
Nitrogen	n2

Table 2
Reactions implicated in the denitrification route and their associated id's according to BiGG notation

Reaction name	BiGG id
Nitrate reductase	NITR
Nitrite reductase	NITR_NO
Nitric oxide reductase	NORx
Nitrous oxide reductase	N2OR

1. *Look for the correspondence between these reactions, metabolites, and BiGG notation.*

   See **Note 1** (Tables 1 and 2).

2. *Load the metabolic model.*

   See **Note 2**.

```
import cobra as cb
model= cb.io.load_model('iJN1463')
```

3. *Create dictionaries for id's of metabolites and reactions of interest.*

```
met_id={'nitrate': 'no3', 'nitrite': 'no2', 'nitric_oxide':
'no', 'nitrous_oxide': 'n2o', 'nitrogen': 'n2'}
rxn_id={'nitrate_reductase': 'NITR', 'nitrite_reductase':
'NITR_NO', 'nitric_oxide_reductase': 'NORx', 'nitrous_oxi-
de_reductase': 'N2OR'}
```

4. *Get all the metabolites' id's from the model.*

```
modelmet=model.metabolites
modelmetid=[met.id for met in modelmet]
```

5. *Remove the compartment from the id of the metabolites.*

   See **Note 3**.

```
modelmetid_wo_compartment=[met.id.split('_')[0] \
for met in modelmet]
```

6. *Retrieve all the reaction id's from the model.*

```
modelrxnid=[rxn.id for rxn in model.reactions]
```

7. *Identify the metabolites of interest that are not present in the model.*

```
mets_in_model=[]
mets_missing=[]
for (km,vm) in met_id.items():
 if vm not in modelmetid_wo_compartment:
 print("""Metabolite '%s' not found in model""" \ %(vm))
 mets_missing.append(vm)
 else:
 mets_in_model.append(str(vm)+'_')
```

8. *Identify the reactions of interest that are not present in the model.*

```
rxns_missing=[]
for (kr,vr) in rxn_id.items():
 if vr not in modelrxnid:
 print("""Reaction '%s' not found in model""" \ %(vr))
 rxns_missing.append(vr)
```

9. *Confirm whether the metabolites "no3" and "no2" are present in the extracellular compartment of the model.*

If they are included (result is True), no further action is needed. If they are not included, add them in the same manner as described in the violacein example.

```
for met in modelmetid:
 if met.startswith("no3_e") or \ met.startswith("no2_e"):
 print(True)
```

10. *Verify whether a transport reaction and an exchange reaction for the metabolites "no3" and "no2" are already included in the model.*

Since the denitrification pathway occurs in the periplasm, it is crucial to ensure that a proper transport reaction is included in the model. This is necessary to facilitate the transportation of metabolites to the compartment where the pathway takes place.

See **Note 4**.

```
mets_of_interest=['no3', 'no2']
for r in modelrxnid:
 sustrate_compartiments={met.id[:-2]:met.id[-2:] \ for met in
model.reactions.get_by_id(r).reactants}
 product_compartiments ={met.id[:-2]:met.id[-2:] \ for met in
model.reactions.get_by_id(r).products}
 shared_mets =set(sustrate_compartiments.keys()) \ & set(pro-
duct_compartiments.keys())
 for i in mets_of_interest:
 if i in shared_mets:
 if len(shared_mets) >= 1 and \ all([sustrate_compartiments
[i] != \ product_compartiments[i]]):
 print('%s is a transport reaction for \ metabolite: %s,
between the compartments: %s and %s' \ %(r,i,sustrate_compar-
timents[i],\ product_compartiments[i]))

#Check if exchange reaction exists

for m in mets_of_interest:
 if 'EX_'+m+'_e' in modelrxnid:
 print('An exchange reaction for the metabolite \ %s is in the
model' %(m))
```

Construction of Condition-Specific Models

Now that we have a grasp of what the model encompasses and its deficiencies regarding this specific pathway, we will proceed to work on each strain individually.

Construction of In Silico Nar Strain

The construction of the Nar phenotype requires allowing NO_3 uptake and blocking the flux through NTRIR2x (nitrite reductase (NADH)). Additionally, the uptake of oxygen should be limited to create a microaerobic environment.

11. *Constraint NTRIR2x, exchange the reaction for O_2 (EX_o2_e), and enable the uptake for the exchange reaction*

for NO_3 (EX_NO3_e). Also allow uptake of succinate (EX_succ_e) and constrain the uptake of glucose (EX_glc__D_e).

See **Note 5**.

```
import copy
model_Nar=copy.deepcopy(model)
model_Nar.reactions.NTRIR2x.bounds=(0,0)
model_Nar.reactions.EX_o2_e.lower_bound=-0.5
model_Nar.reactions.EX_no3_e.bounds=(-2,-2)
model_Nar.reactions.EX_succ_e.lower_bound=-5
model_Nar.reactions.EX_glc__D_e.lower_bound=0
model_Nar.summary()
```

Simulate Nar Phenotype

12. *Transform the COBRA model to COMETS model object and establish the initial layout.*

    See **Note 8**.

```
import cometspy as c

model_Nar_comets=c.model(model_Nar)
model_Nar_comets.id='Nar_strain'
layout_nar=c.layout([model_Nar_comets])
```

13. *Update the COMETS layout to establish the initial biomass for the model at 0.025 g/L and configure the culture media with 25 mM of succinate, 15 mM of ammonium, 2 mM of nitrate, and 2 mM of "pqqA_kt", a metabolite required for model growth.*

```
layout_nar.initial_pop=[[0.0,0.0,float(0.025)]]
layout_nar.add_typical_trace_metabolites()
layout_nar.set_specific_metabolite("succ_e", 25)
layout_nar.set_specific_metabolite("nh4_e", 15)
layout_nar.set_specific_metabolite("no3_e", 2)
layout_nar.set_specific_metabolite("pqqA_kt_c", 2)
```

14. *Configure the simulation parameters.*

    See **Note 9**.

```
params=c.params()
params.all_params['maxCycles']=640
params.all_params['spaceWidth']=0.05
```

```
params.all_params['allowCellOverlap']= True
params.all_params['maxSpaceBiomass']= 1000
params.all_params['MediaLogRate']=1
params.all_params['exchangestyle']='Standard FBA'
params.all_params['writeTotalBiomassLog']=True
params.all_params['writeMediaLog']=True
```

15. *Run COMETS simulation.*

```
simulation_nar=c.comets(layout_nar,params)
simulation_nar.run()
```

16. *Create a dataframe to record the concentration of nitrate and nitrite throughout the simulation.*

```
import numpy as np

media_nar = simulation_nar.media #We get the media composition
results
dataframe_nar=simulation_nar.total_biomass #We get the biomass
results
columns_nar=['cycle','Nar_strain']
dataframe_nar.columns=columns_nar
max_cycles=params.all_params['maxCycles']
for d in ['no3_e','no2_e']:
 met =media_nar[media_nar['metabolite'] == d]
 temp=np.zeros(max_cycles+1) #Create an array with all zeros
 dataframe_nar[d]=temp #We add it to the dataframe
 j=1
 while j < (max_cycles+1): #For each cycle
 if (met.cycle==j).any(): #If the row exists
  dataframe_nar.loc[j-1,d]  =  float(met[met['cycle']==j]['con-
c_mmol']) #Its dataframe value is changed
 j+=1

 if float(dataframe_nar[str(d)].iloc[-1])==0:
 if dataframe_nar[d].iloc[-2]==0:
 continue
 else:
 dataframe_nar[d].iloc[-1]=dataframe_nar[d].iloc[-2]
```

17. *Plot the results.*

The result of the Nar phenotype simulation is displayed in Fig. 1a. Since we are plotting a single figure to represent the three phenotypes in this example, the code to create the image will be included under the community-scale model simulation (in **step**

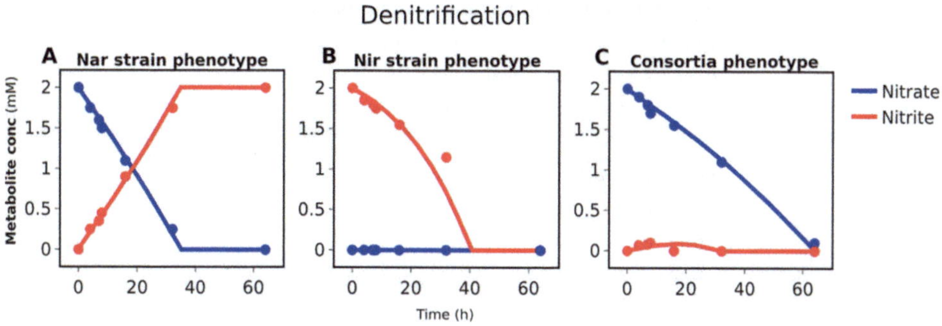

Fig. 1 Comparison between experimental data (lines) and in silico (points) values predicted with Comets for nitrate and nitrite consumption. (**a**) Nitrate and nitrite concentrations over time in the presence of the Nar strain. (**b**) Nitrate and nitrite concentrations in the presence of the Nir strain. (**c**) Nitrate and nitrite concentrations with the consortium of Nar and Nir strains

iii.f). The results demonstrate the nitrate consumption and nitrite production capabilities of the Nar strain. The in silico results perfectly match the experimental data.

Construction of In Silico Nir Strain

Since *P. putida* and subsequently its GEM lack the ability to use nitrite as the electron acceptor, the construction of the Nir phenotype requires the addition of NITR_NO, NORx, and N2OR reactions as well as the metabolites NO, N_2O, and N_2.

Additionally, it is necessary to limit oxygen uptake, allow N_2 to leave the system, and block the flux through the reaction NTRIR2x (nitrite reductase (NADH).

18. *Add the reaction "Nitrite reductase (NO-forming)" (NITR_NO) to the model.*

    See **Note 6**.

```
model_Nir=copy.deepcopy(model)

#Introduction of needed reactions
# NITR_NO
reaction = cb.Reaction('NITR_NO')
reaction.name = 'Nitrite reductase (NO-forming)'
reaction.subsystem = ''
reaction.lower_bound = 0
reaction.upper_bound = 1000 # This is the default

#metabolites associated with the reaction

focytc_c=model_Nir.metabolites.focytc_c
h_c=model_Nir.metabolites.h_c
```

```
no2_c=model_Nir.metabolites.no2_c
ficytc_c=model_Nir.metabolites.ficytc_c
h2o_c=model_Nir.metabolites.h2o_c
no_c=model_Nir.metabolites.no_c

reaction.add_metabolites({
 focytc_c: -1,
 no2_c: -1,
 h_c: -2,
 ficytc_c: 1,
 no_c:1,
 h2o_c: 1,})

reaction.gene_reaction_rule = ''
model_Nir.add_reactions([reaction])
```

19. *Add the reaction "Nitric oxide reductase" (NORx) and metabolite "Nitrous oxide" (n2o) to the model.*

```
#Introduction of needed reactions
# NORx
reaction = cb.Reaction('NORx')
reaction.name = 'Nitric oxide reductase'
reaction.subsystem = ''
reaction.lower_bound = 0 #
reaction.upper_bound = 1000 # This is the default

#metabolites associated with the reaction

focytc_c=model_Nir.metabolites.focytc_c
h_c=model_Nir.metabolites.h_c
no_c=model_Nir.metabolites.no_c
ficytc_c=model_Nir.metabolites.ficytc_c
h2o_c=model_Nir.metabolites.h2o_c
n2o_c=cb.Metabolite(
 'n2o_c',
 formula='N2O',
 name='Nitrous oxide',
 compartment='c',
 charge=0)

reaction.add_metabolites({
 focytc_c: -2,
 no_c: -2,
 h_c: -2,
 ficytc_c: 2,
 n2o_c:1,
```

```
 h2o_c: 1,})

reaction.gene_reaction_rule = ''

model_Nir.add_reactions([reaction])
```

20. *Add the reaction "Nitrous oxide reductase (nadh)" (N2OR) and metabolite "dinitrogen" (n2) to the model.*

```
#Introduction of needed reactions
# N2OR
reaction = cb.Reaction('N2OR')
reaction.name = 'Nitrous oxide reductase (nadh)'
reaction.subsystem = ''
reaction.lower_bound = 0 #
reaction.upper_bound = 1000 # This is the default

#metabolites associated with the reaction

nadh_c=model_Nir.metabolites.nadh_c
h_c=model_Nir.metabolites.h_c
n2o_c=model_Nir.metabolites.n2o_c
nad_c=model_Nir.metabolites.nad_c
h2o_c=model_Nir.metabolites.h2o_c
n2_c=cb.Metabolite(
 'n2_c',
 formula='N2',
 name='dinitrogen',
 compartment='c',
 charge=0)

reaction.add_metabolites({
 nadh_c: -1,
 n2o_c: -1,
 h_c: -1,
 nad_c: 1,
 n2_c:1,
 h2o_c: 1,})

reaction.gene_reaction_rule = ''

model_Nir.add_reactions([reaction])
```

21. *Add "Dinitrate transport" (N2TR) reaction to the model.*

```
#Introduction of needed reactions
# N2TR
```

```
reaction = cb.Reaction('N2TR')
reaction.name = 'Dinitrate transport'
reaction.subsystem = ''
reaction.lower_bound = 0 #
reaction.upper_bound = 1000 # This is the default

#metabolites associated with the reaction
n2_c=model_Nir.metabolites.n2_c
n2_e=cb.Metabolite(
 'n2_e',
 formula='N2',
 name='dinitrogen',
 compartment='e',
 charge=0)

reaction.add_metabolites({
 n2_c: -1,
 n2_e: 1})

reaction.gene_reaction_rule = ''

model_Nir.add_reactions([reaction])
```

22. *Introduction of the exchange reaction for metabolite "n2" using the "add_boundary" method from COBRApy.*

```
model_Nir.add_boundary( \
model_Nir.metabolites.get_by_id("n2_e"), \ type="exchange")
```

23. *Disable glucose (EX_glc__D_e) uptake, inactivate NTRIR2x, and constrain the oxygen (EX_o2_e) uptake. Allow the model to uptake NO_2 (EX_no2_e) and succinate (EX_succ_e).*

    See **Notes 5** and 7.

```
model_Nir.reactions.EX_glc__D_e.lower_bound=0
model_Nir.reactions.NTRIR2x.bounds=(0,0)
model_Nir.reactions.EX_o2_e.lower_bound=-0.5
model_Nir.reactions.EX_no2_e.lower_bound=-1.5
model_Nir.reactions.EX_succ_e.lower_bound=-5
model_Nir.summary()
```

Simulate Nir Phenotype

24. *Transform the COBRA model to COMETS model object and establish the initial layout.*

    See **Note 8**.

```
model_Nir_comets=c.model(model_Nir)
model_Nir_comets.id='Nir_strain'
layout_nir=c.layout([model_Nir_comets])
```

25. *Update the COMETS layout to establish the initial biomass for the model at 0.013 g/L and configure the culture media with 25 mM of succinate, 15 mM of ammonium, 2 mM of nitrite, and 2 mM of "pqqA_kt", a metabolite required for model growth.*

```
layout_nir.initial_pop=[[0.0,0.0,float(0.013)]]
layout_nir.add_typical_trace_metabolites()
layout_nir.set_specific_metabolite("succ_e", 25)
layout_nir.set_specific_metabolite("nh4_e", 15)
layout_nir.set_specific_metabolite("no2_e", 2)
layout_nir.set_specific_metabolite("pqqA_kt_c", 2)
```

26. *Run COMETS simulation.*

```
simulation_nir=c.comets(layout_nir,params)
simulation_nir.run()
```

27. *Create a dataframe to record the concentration of nitrate and nitrite throughout the simulation.*

```
media_nir = simulation_nir.media #We get the media composition
results
dataframe_nir=simulation_nir.total_biomass #We get the biomass
results
columns_nir=['cycle','Nir_strain']
dataframe_nir.columns=columns_nir
max_cycles=params.all_params['maxCycles']

for d in ['no3_e','no2_e']:
 met =media_nir[media_nir['metabolite'] == d]
 temp=np.zeros(max_cycles+1) #Create an array with all zeros
 dataframe_nir[d]=temp #We add it to the dataframe
 j=1
 while j < (max_cycles+1): #For each cycle
 if (met.cycle==j).any(): #If the row exists
  dataframe_nir.loc[j-1,d] = float(met[met['cycle']==j]['con-
c_mmol']) #Its dataframe value is changed
 j+=1

 if float(dataframe_nir[str(d)].iloc[-1])==0:
```

```
if dataframe_nir[d].iloc[-2]==0:
continue
else:
dataframe_nir[d].iloc[-1]=dataframe_nir[d].iloc[-2]
```

28. *Plot the results.*

The result of the Nir phenotype simulation is displayed in Fig. 1b. The in silico simulation exhibits a behavior very similar to that shown by the experimental data.

The code to create the image will be included under the community-scale model simulation (in **step iii.f**).

Construction of Community-Scale Model

The next step is the creation of the community-scale model to mirror the behavior observed in our reference. To achieve this, we will utilize COMETS.

29. *Establish the initial COMETS layout.*

```
layout=c.layout([model_Nir_comets,model_Nar_comets])
```

30. *Update the COMETS layout to establish the initial biomass for both models at 0.012 g/L and configure the culture media with 25 mM of succinate, 2 mM of nitrate, 15 mM of ammonium, and 2 mM of "pqqA_kt", a metabolite required for model growth.*

```
layout.initial_pop=[[0.0,0.0,float(0.012),float(0.012)]]
#Both initial biomasses are going to be 0.05g
layout.add_typical_trace_metabolites()
layout.set_specific_metabolite("succ_e", 25) layout.set_spe-
cific_metabolite("no3_e", 2) layout.set_specific_metabolite
("nh4_e", 15)
layout.set_specific_metabolite("pqqA_kt_c", 2)
```

31. *Run COMETS simulation.*

See **Note 10**.

```
simulation=c.comets(layout,params)
simulation.run(delete_files=True)
```

32. *Create a dataframe for both strains and important metabolites regarding the pathway.*

We will extract the concentration of selected metabolites for each simulation cycle. This dataframe will serve as the basis for graphing the results in the next step. The metabolites chosen for the denitrification pathway are as follows:

- Glucose ("glc__D_e")
- Nitrate ("no3_e")
- Nitrite ("no2_e")
- Nitrogen ("n2_e")

```
media = simulation.media #We get the media composition results
dataframe=simulation.total_biomass #We get the biomass results
columns=['cycle','Nar_strain', 'Nir_strain']
dataframe.columns=columns
max_cycles=params.all_params['maxCycles']
for d in ['glc__D_e', 'no3_e', 'n2_e','no2_e']:
 met =media[media['metabolite'] == d]
 temp=np.zeros(max_cycles+1) #Create an array with all zeros
 dataframe[d]=temp #We add it to the dataframe
 j=1
 while j < (max_cycles+1): #For each cycle
 if (met.cycle==j).any(): #If the row exists
  dataframe.loc[j-1,d] = \ float(met[met['cycle']==j]['con-
c_mmol']) #Its dataframe value is changed
 j+=1

 if float(dataframe[str(d)].iloc[-1])==0:
 if dataframe[d].iloc[-2]==0:
 continue
 else:
 dataframe[d].iloc[-1]= \ dataframe[d].iloc[-2]
```

33. *Plot the results.*

The results from COMETS, which display metabolite concentrations and biomass levels across time cycles, can be illustrated using line graphs that plot these metrics over time. Utilize the provided code to generate graphs depicting the phenotype of the Nar (Fig. 1a) and Nir (Fig. 1b) strains separately, as well as for the consortia (Fig. 1c). In these graphs, dots represent the experimental data, while the lines denote the model's predictions. Referring to the original study, it is evident that their model failed to accurately predict the consortia phenotype, whereas the protocol outlined here successfully replicates the experimental results.

This type of representation is very suitable for temporal data to observe the growth dynamics of various organisms in a consortium, as well as the production and consumption of metabolites.

```
import plotly.express as px
import plotly.graph_objects as go
```

```
from plotly.subplots import make_subplots
fig = make_subplots(rows=1, cols=3,
 subplot_titles=['<b>Nar strain phenotype</b>', '<b>Nir strain
phenotype</b>', '<b> Consortia phenotype </b>'], x_title="time
(h)",
 y_title="<b>metabolite conc</b> (mM)")

fig.add_trace(
go.Scatter(x=dataframe_nar['cycle']*0.1, y=dataframe_nar
['no3_e'], name='Nitrate', marker=dict(color="blue")),row=1,
col=1
)
fig.add_trace(
go.Scatter(x=dataframe_nar['cycle']*0.1, y=dataframe_nar
['no2_e'], name='Nitrite', marker=dict(color="crimson")),
row=1, col=1
)
fig.add_trace(
go.Scatter(x=dataframe_nir['cycle']*0.1, y=dataframe_nir
['no3_e'], name='Nitrate', marker=dict(color="blue"), showle-
gend=False),row=1, col=2
)
fig.add_trace(
go.Scatter(x=dataframe_nir['cycle']*0.1, y=dataframe_nir
['no2_e'], name='Nitrite', marker=dict(color="crimson"),
showlegend=False), row=1, col=2
)

fig.add_trace(
go.Scatter(x=dataframe['cycle']*0.1, y=dataframe['no3_e'],
name='Nitrate',marker=dict(color="blue"), showlegend=False),
row=1, col=3
)
fig.add_trace(
go.Scatter(x=dataframe['cycle']*0.1, y=dataframe['no2_e'],
name='Nitrite',marker=dict(color="crimson"),showlegen-
d=False), row=1, col=3
)

fig.show()
```

Optimization and Analysis of the Consortium Using FLYCOP
The final step involves the optimization of the assembled community-scale model using FLYCOP. To accomplish this, we must define our input space and establish our desired objective, known as the fitness function, which we aim to maximize. To do that, the variables to be optimized should be selected, and their range of values has to be determined. The set of variables and their ranges is called input space,

and the objective is the identification of a set of values for those variables that maximize the fitness. Given the denitrification capabilities of this consortium, we will test its ability to produce N_2. Therefore, we will set the production of N_2 as the objective.

34. *Input space definition.*

The optimization process requires defining the variables to be optimized and their respective value ranges to prevent searching in an excessively large solution space. Two variables will be created for optimization: The initial biomass of the Nar (*biomass_nar*) and Nir (*biomass_nir*) strains, aiming to identify the optimized inoculum yielding optimal denitrification process. In this example, the number of searches was set to 100, but it can be increased for improved results.

```
from ConfigSpace import ConfigurationSpace,Float
from smac import Scenario
def n2_space():
 cs=ConfigurationSpace(seed=0)
 X0= Float("biomass_nar",bounds=(0.0001,0.2))
 X1= Float("biomass_nir",bounds=(0.0001,0.2))
 cs.add_hyperparameters([X0,X1])
 scenario=Scenario(cs, deterministic=True, n_trials=100)
 return(scenario)
```

35. *Integration of COMETS scenario into FLYCOP.*

To integrate the code from the previous step "Construction of Community-scale model" into FLYCOP, we need to create a function that encapsulates the simulation code. This function will then be used within the FLYCOP framework. In this FLYCOP optimization, the fitness function will be based on the production of the "products" of the denitrification pathway for each strain. The total searching trials will be stored in "detrinification_output.csv".

```
from FLYCOP.tools import make_df_and_graph
def n2_scenario(self,parameters):
 # Creation of the consortium with comets
 parameters=parameters
 configuration=','.join([str(value) for value in parameters.
values()])
 #Reading all the parameters and its variables
 #converting our models to comets format simulator.layout.
initial_pop=[[0.0,0.0,parameters['biomass_nar'],parameters
['biomass_nir']]]
 self.simulator.layout.add_typical_trace_metabolites()
self.simulator.layout.set_specific_metabolite("succ_e",25)
```

```
self.simulator.layout.set_specific_metabolite("no3_e",2)
self.simulator.layout.set_specific_metabolite("nh4_e",15)
self.simulator.layout.set_specific_metabolite("pqqA_t_c",2)

 self.simulator.simulate()
 simulation=self.simulator.get_results()
 print(simulation)
  results=make_df_and_graph([model_Nar_comets.id,model_Nir_co-
mets.id],["no3_e","no2_e","n2_e"],simulation,700)
print("The configuration is: "+configuration+"and the
result is: "+last_cycle)
# Writing the searches to nitrification_output.csv
with open("denitrification_output.csv","a") as outfile:
 outfile.write(configuration+','+last_cycle+'\n')
 n2=results.at[results.index[-2], 'n2_e']
 fitness=1/(n2+1)
 return fitness,0
```

36. *Create a list with the COMETS models generated.*

```
consortia=[model_Nar_comets,model_Nir_comets]
```

37. *Create a FLYCOP2 object.*

```
from FLYCOP import *
FLYCOP_optimization=FLYCOP()
```

38. *Select the COMETS simulator.*

```
FLYCOP_optimization.create_simulator(simulator_type="COMETS")
```

39. *Load the consortia into FLYCOP.*

```
FLYCOP_optimization.load_consortia(consortia)
```

40. *Set the COMETS parameters. The parameters from* **step 14** *are used.*

```
FLYCOP_optimization.set_simulator_params(params)
```

41. *Load the COMETS scenario created.*

The scenario defined in **step iv.b** is the one provided to FLYCOP for simulation.

```
FLYCOP_optimization.create_scenario()
```

```
FLYCOP_optimization.set_scenario(n2_scenario)
```

42. *Set the optimization space.*

Lastly, we need to assign the variables we want to optimize. *See* **Note 12**.

```
space=n2_space()
result=FLYCOP_optimization.optimize(space)
```

The result shows the best configuration identified. The maximum production is 1.848 mM of N_2.

See **Notes 13** and **14**.

Initial Biomass Nar strain (g/L)	Initial Biomass Nir strain (g/L)
0.185	0.08

The results of the inoculum optimization show that dinitrogen production is higher than that achieved in **step iii** with default inoculum (1 mM), demonstrating an increase.

3.2 Example 2: Violacein Production

In this section, we describe the process for optimizing SMC using FLYCOP. As an example, we will simulate a previously published synthetic consortia [12], focusing on overproducing Violacein. The optimization process consists of three main steps: Modifying the model to allow violacein production, conducting dynamic simulations of the assembled SMC with COMETS, and performing SMC optimization with FLYCOP.

The basic consortium is formed by a transformer strain and a producer strain. For the transformer strain, the authors utilized *E. coli* ET1. After reviewing the genetic changes in the strain, it was found that all were regulatory changes, aimed to prevent tryptophan biosynthesis repression by metabolites. We employ the available model *i*WFL_1372 [13], which represents the original strain of ET1 used in the study.

The *E. coli W* strain was used as the producer strain. It was engineered by adding the genes *vioA, vioB, vioE, vioD*, and *vioC* from *Chromobacterium violaceum*, while simultaneously deleting the genes *tnaA* and *trpR* to mitigate feedback inhibition from tryptophan.

Preliminary Evaluation of the Selected Models

For this example, neither the metabolites nor the reactions have BiGG identifiers [14]; thus, the assignment of new id's is required, as depicted in the tables below.

Table 3
Metabolites implicated in the synthesis of violacein and their associated id's according to BiGG notation

Metabolite name	BiGG id	New id
2-imino-3-(indol-3-yl)propanoate	None	2i3i3ppa
Indole-3-pyruvate imine dimer	None	3pyridm
Protodeoxyviolaceinic acid	None	ptdeovio
Protoviolaceinic acid	None	ptvio
Violacein	None	violacein

Table 4
Reactions implicated in the synthesis of violacein and their associated id's according to BiGG notation

Reaction name	BiGG id	New id
Flavin-dependent L-tryptophan oxidase	None	VIOA
2-imino-3-(indol-3-yl)propanoate dimerase	None	VIOB
Prodeoxyviolacein synthase	None	VIOE
Protodeoxyviolaceinate monooxygenase	None	VIOD
Violacein synthase	None	VIOC

See **Note 1** (Tables 3 and 4).

To modify the model used for the producer strain, *E. coli W* (*i*WFL_1372), according to the aforementioned adjustments, there is no need to verify if the metabolites already exist in the model, since they come from a heterologous pathway. Therefore, we can proceed directly to the construction of condition-specific models.

Construction of Condition-Specific Models

Construction of In Silico Producer Strain

1. *Load the metabolic model.*

   See **Note 2**.

```
import cobra as cb
model_producer = cb.io.load_model('iWFL_1372')
```

The model requires the integration of the heterologous violacein pathway along with its associated metabolites. Therefore, it is necessary to create and incorporate the reactions and metabolites accordingly. *See* **Note 6**.

2. *Add the reaction flavin-dependent L-tryptophan oxidase (VIOA) and the metabolite 2-imino-3-(indol-3-yl)propanoate (2i3i3ppa).*

```
reaction = cb.Reaction('VIOA')
reaction.name = 'Flavin-dependent L-tryptophan oxidase'
reaction.subsystem = ''
reaction.lower_bound = -1000
reaction.upper_bound = 1000
#metabolites associated with the reaction

trp__L_c=model_producer.metabolites.trp__L_c
o2_c=model_producer.metabolites.o2_c
h2o2_c=model_producer.metabolites.h2o2_c
# i3i3ppa_c metabolite
i3i3ppa_c=cb.Metabolite( \
 '2i3i3ppa_c', \
 formula='C11H9N2O2', \
 name='2-imino-3-(indol-3-yl)propanoate', \
 compartment='c', \
 charge=-1)

# Add the metabolites to the reaction and their stoichiometric
indexes
reaction.add_metabolites({ \
 trp__L_c: -1, \
 o2_c: -1, \
 h2o2_c: 1, \
 i3i3ppa_c:1})
reaction.gene_reaction_rule = ''
model_producer.add_reactions([reaction])
```

3. *Add the reaction 2-imino-3-(indol-3yl)propanoate dimerase (VIOB) and the metabolite indole-3-pyruvate imine dimer (i3pyridm).*

```
 reaction = cb.Reaction('VIOB')
reaction.name='2-imino-3-(indol-3yl)propanoate \ dimerase'
reaction.subsystem = ''
reaction.lower_bound = -1000
reaction.upper_bound = 1000

#metabolites associated creation
```

```
i3i3ppa_c= \
model_producer.metabolites.get_by_id('2i3i3ppa_c')
i3pyridm_c=cb.Metabolite( \
 'i3pyridm_c', \
 formula='C22H18N4O4', \
 name='indole-3-pyruvate imine dimer', \
 compartment='c', \
 charge=0)

reaction.add_metabolites({ \
 i3i3ppa_c: -2, \
 i3pyridm_c:1})
reaction.gene_reaction_rule = ''
model_producer.add_reactions([reaction])
```

4. *Add the reaction prodeoxyviolacein synthase (VIOE) and the metabolite protodeoxyviolaceinic acid (ptdeovio).*

```
reaction = cb.Reaction('VIOE')
reaction.name = 'Prodeoxyviolacein synthase'
reaction.subsystem = ''
reaction.lower_bound = 0 #
reaction.upper_bound = 1000 # This is the default

#metabolites associated with the reaction

i3pyridm_c= \
model_producer.metabolites.i3pyridm_c
ptdeovio_c=cb.Metabolite( \
 'ptdeovio_c', \
 formula='C21H15N3O2', \
 name='protodeoxyviolaceinic acid', \
 compartment='c', \
 charge=0)
reaction.add_metabolites({ \
 i3pyridm_c: -1, \
 ptdeovio_c:1})
reaction.gene_reaction_rule = ''
model_producer.add_reactions([reaction])
```

5. *Add the reaction protodeoxyviolaceinate monooxygenase (VIOD) and the metabolite protoviolaceinic acid (ptvio).*

```
reaction = cb.Reaction('VIOD')
reaction.name = 'Protodeoxyviolaceinate monooxygenase'
reaction.subsystem = ''
```

```
reaction.lower_bound = 0
reaction.upper_bound = 1000

#metabolites associated with the reaction

ptdeovio_c=model_producer.metabolites.ptdeovio_c
o2_c=model_producer.metabolites.o2_c
nadph_c=model_producer.metabolites.nadph_c
h_c=model_producer.metabolites.h_c
nadp_c=model_producer.metabolites.nadp_c
h2o_c=model_producer.metabolites.h2o_c
ptvio_c=cb.Metabolite( \
 'ptvio_c', \
 formula='C21H15N3O3', \
 name='protoviolaceinic acid', \
 compartment='c', \
 charge=0)

reaction.add_metabolites({ \
 ptdeovio_c: -1, \
 o2_c: -1, \
 nadph_c: -1, \
 h_c: -1, \
 nadp_c: 1, \
 h2o_c: 1, \
 ptvio_c: 1})

reaction.gene_reaction_rule = ''
model_producer.add_reactions([reaction])
```

6. *Add the reaction violacein synthase VIOC and the metabolite violacein (violacein).*

```
reaction = cb.Reaction('VIOC')
reaction.name = 'Violacein synthase'
reaction.subsystem = ''
reaction.lower_bound = 0
reaction.upper_bound = 1000

#metabolites associated with the reaction
ptvio_c=model_producer.metabolites.ptvio_c
o2_c=model_producer.metabolites.o2_c
nadph_c=model_producer.metabolites.nadph_c
h_c=model_producer.metabolites.h_c
nadp_c=model_producer.metabolites.nadp_c
h2o_c=model_producer.metabolites.h2o_c
```

```
violacein_c=cb.Metabolite( \
 'violacein_c', \
 formula='C20H13N3O3', \
 name='Violacein', \
 compartment='c', \
 charge=0)

reaction.add_metabolites({ \
 ptvio_c: -1, \
 o2_c: -1, \
 nadph_c: -1, \
 h_c: -1, \
 nadp_c: 1, \
 h2o_c:1, \
 violacein_c: 1})
reaction.gene_reaction_rule = ''
model_producer.add_reactions([reaction])
```

7. *Simulate a knockout of the gene tryptophanase/L-cysteine desulfhydrase (tnaA) with the locus ECW_m4007 to remove the tryptophan assimilation.*

To identify reactions associated with the genes, we must first consult the literature for their corresponding locus tags, as this is the identifier used in the models.

The locus tag of the gene *tnaA* is "ECW_m4007", whereas the protein encoded by *trpR* functions as an aporepressor and is not included in the model due to its regulatory nature.

```
model_producer.genes.ECW_m4007.knock_out()
```

8. *Open the exchange reaction of tryptophan (EX_trp__L_e).*

The producer phenotype requires the uptake of tryptophan produced by the transformer strain to synthesize the final compound, violacein.

See **Notes 4** and **5**.

```
model_producer.reactions.EX_trp__L_e.bounds=(-10,0)
```

9. *Add an exchange reaction (EX_violacein_e) for violacein, as well as a diffusion reaction (VIOtr).*

See **Note 15**.

```
#Adding transport reaction
reaction = cb.Reaction('VIOtr')
```

```
reaction.name = 'Violacein diffusion'
reaction.subsystem = ''
reaction.lower_bound = 0 #
reaction.upper_bound = 1000 # Only export

#metabolites associated with the reaction
violacein_c=model_producer.metabolites.violacein_c
violacein_e=cb.Metabolite( \
 'violacein_e', \
 formula='C20H13N3O3', \
 name='violacein', \
 compartment='e', \
 charge=0)

reaction.add_metabolites({ \
 violacein_c: -1,\
 violacein_e:1})
reaction.gene_reaction_rule = ''

model_producer.add_reactions([reaction])

#Adding exchange reaction
model_producer.add_boundary( \
model_producer.metabolites.get_by_id("violacein_e"), \
type="exchange")
```

10. *Modify violacein production (VIOA) flux to 0.022 mM h^{-1} $gCDW^{-1}$. Activate the galactose exchange reaction for galactose uptake. (EX_gal_e). Disable the uptake for the exchange reaction for glucose (EX_glc__D_e).*

See **Notes 5** and **11**.

In the optimization conducted in the study, it was found that galactose is the optimal carbon source for violacein production. Therefore, we will activate its exchange reaction to allow its uptake. Simultaneously, we disable the exchange reaction for glucose uptake, which is the default carbon source when loading the model.

```
model_producer.reactions.VIOA.lower_bound=0.022
model_producer.reactions.EX_gal_e.lower_bound=-2.5
model_producer.reactions.EX_glc__D_e.lower_bound=0
model_producer.summary()
```

Construction of In Silico Transformer Strain

11. *Load the metabolic model.*

```
model_transformer = cb.io.load_model('iWFL_1372')
```

12. *Replicate the overproduction of tryptophan by directing the model to carry flux through the reaction TRPAS2, which is responsible for tryptophan synthesis.*

See **Note 16**.

```
model_transformer.reactions.TRPAS2.bounds=(-1000,-0.6)
```

13. *Set the galactose uptake with the exchange of galactose (EX_gal_e). Close the exchange reaction for glucose (EX_glc__D_e), the carbon source by default, and tryptophan (EX_trp__L_e), so the model can only secrete it.*

```
model_transformer.reactions.EX_gal_e.lower_bound-2.25
model_transformer.reactions.EX_glc__D_e.lower_bound=0
model_transformer.reactions.EX_trp__L_e.lower_bound=0
```

Construction of Community-Scale Model

14. *Set up the COMETS layout for the simulation.*

In this instance, the culture media includes galactose at a concentration of 10 g/L as the carbon source, and the initial biomasses for the strains adhere to a 19:1 ratio for the transformer.
See **Note 8**.

```
import cometspy as c
#converting our models to comets format
model_transformer_comets=c.model(model_transformer)
model_producer_comets=c.model(model_producer)

#changing the ids of the models
model_transformer_comets.id='transformer_strain'
model_producer_comets.id='producer_strain'

layout=c.layout([model_transformer_comets, \ model_producer_-
comets])

layout.add_typical_trace_metabolites()
layout.set_specific_metabolite("gal_e", 55.5)
```

15. *Prepare the simulation parameters.*

See **Note 9**.

```
params=c.params()
params.all_params['maxCycles']=480
params.all_params['spaceWidth']=0.05
params.all_params['allowCellOverlap']= True
params.all_params['maxSpaceBiomass']= 1000
params.all_params['MediaLogRate']=1
params.all_params['exchangestyle']='Standard FBA'
params.all_params['writeTotalBiomassLog']=True
params.all_params['writeMediaLog']=True
```

16. *Run COMETS simulation.*

See **Note 10**.

```
simulation=c.comets(layout,params)
simulation.run(delete_files=True)
```

17. *Create a dataframe for both strains and important metabolites.*

```
import numpy as np
#Get the media composition results
media = simulation.media
#Get the biomass results
dataframe=simulation.total_biomass
# Rename column names with desired strings
columns=['cycle', \
'transformer_strain', \
'producer_strain']

dataframe.columns=columns
#Get the maximum cycles value
max_cycles=params.all_params['maxCycles']

for d in ['gal_e', 'trp__L_e', 'violacein_e']:
 met =media[media['metabolite'] == d]
 temp=np.zeros(max_cycles+1)
 dataframe[d]=temp #We add it to the dataframe
 j=1
 while j < (max_cycles+1): #For each cycle
 if (met.cycle==j).any(): #If the row exists
  dataframe.loc[j-1,d] = \ float(met[met['cycle']==j]['con-
c_mmol'])
 j+=1
 if float(dataframe[str(d)].iloc[-1])==0:
```

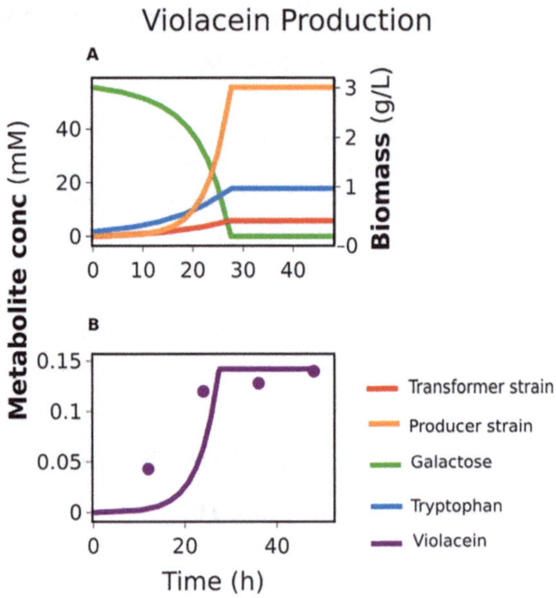

Fig. 2 (**a**) Graph depicting the biomass of the producer strain (orange) and the transformer strain (red) over time, along with the concentrations of galactose as a carbon source (green) and tryptophan (blue). (**b**) Graph illustrating the concentration of violacein produced over time (purple). The dots represent experimental observation of violacein accumulation

```
if dataframe[d].iloc[-2]==0:
continue
else:
dataframe[d].iloc[-1]=dataframe[d].iloc[-2]
```

18. *Plot the results.*

Utilize the Plotly graphical package to generate two line plots (as detailed in Fig. 2). The first plot will show the temporal evolution of biomass along with the concentrations of galactose and tryptophan. The second plot will illustrate the temporal variation in violacein concentration. Dots on the plots represent the experimental data for violacein production. Although the model predicts a longer lag phase, the final concentration of violacein achieved aligns closely with the experimental data.

```
import plotly.graph_objects as go
from plotly.subplots import make_subplots
fig = make_subplots(rows=2, cols=1, \
 subplot_titles=['<b>A</b>', '<b>B</b>'], \
 specs=[[{"secondary_y": True}], \
 [{"secondary_y": False}]])
```

```
for i in ['transformer_strain', 'producer_strain']:
# Add traces
 fig.add_trace( \
  go.Scatter(x=dataframe['cycle']*0.1,  \ y=dataframe[i],
name=i), \
 secondary_y=True, row=1, col=1)

for m in ['gal_e',"trp__L_e"]:
 fig.add_trace( \
  go.Scatter(x=dataframe['cycle']*0.1,  \ y=dataframe[m],
name=m), \
 secondary_y=False, row=1, col=1)

for j in ["violacein_e"]:
 fig.add_trace( \
  go.Scatter(x=dataframe['cycle']*0.1,  \ y=dataframe[j],
name=j), \
 row=2, col=1)

# Set x-axis title
fig.update_xaxes(title_text="time (h)")

# Set y-axes titles
fig.update_yaxes(title_text="<b>metabolite conc</b> \ (mM)",
secondary_y=False)
fig.update_yaxes(title_text="<b>biomass</b> (g/L)", \ secon-
dary_y=True)

fig.update_layout(legend=dict( \
 y=0.35, \
 orientation='v', \
 x=1.05 \
), template='simple_white',height=600, \ width=600,title_-
text="Violacein production")

fig.show()
```

Optimization and Analysis of the Consortium Using FLYCOP After the construction of the condition-specific models and the community model, the next step is the consortia optimization. To do that, the variables to be optimized should be selected, and their range of values has to be determined. In this case, the fitness will be the violacein final concentration.

19. *Input space definition.*

To reduce the searching space, the ranges are selected based on experimental values identified by *Gwon* et al. Three variables will be created to optimize the initial biomass of the transformer and producer strains to identify the optimized inoculum and the carbon source selected (glucose, galactose, fructose, sucrose, glycerol, xylose, maltose, and lactose). For this example, the number of searches is set to 100.

```
def violacein_space():
 cs=ConfigurationSpace()
 X0= Float("biomass_tranformer",\
0.0001,0.2,default_value=0.01)
 X1= Float("biomass_producer",\
0.0001,0.2,default_value=0.01)
 X2= CategoricalHyperparameter("carbon_source",\
["glc__D_e","gal_e","fru_e","sucr_e","gly_e",\
"xyl_e","mal__D_e","lcts_e"],default_value="glc__D_e")
 cs.add_hyperparameters([X0,X1,X2])
 scenario=Scenario({"deterministic":"true",
 "n_trials":100,
 "cs":cs})
 return(scenario)
```

20. *Integrate the COMETS scenario into FLYCOP.*

The code developed in the previous **step (iii)** is now integrated into FLYCOP. To accomplish this, a function needs to be established containing the relevant simulation code. The resulting configurations will be saved in a file named "violacein_output.csv".

```
from FLYCOP.tools import make_df_and_graph
```

```
def violacein_scenario(self,parameters):
 # Creation of the consortium with comets
 parameters=parameters
 configuration=','.join([str(value) for value in parameters.
values()])
 #Reading all the parameters and its variables
 #converting our models to comets format
 #layout=c.layout(self.consortia)
 self.simulator.layout.initial_pop=[[0.0,0.0,parameters["bio-
mass_transformer"],parameters["biomass_producer"]]] self.si-
mulator.layout.add_typical_trace_metabolites()
 self.simulator.layout.set_specific_metabolite(parameters
["carbon_source"],56)
 self.simulator.simulate()
 simulation=self.simulator.get_results()
 print(simulation)
 results=make_df_and_graph([model_transformer_comets.id,mod-
```

```
el_producer_comets.id],["violacein_e"],simulation,170)
print("The configuration is: "+configuration+"and the
result is: "+last_cycle)
# Writing the searches to violacein_output.csv
with open("violacein_output.csv","a") as outfile:
 outfile.write(configuration+','+last_cycle+'\n')
 violacein=results.at[results.index[-2], 'violacein_e']
 fitness=1/(violacein+1)
 return fitness,0
```

21. *Create a list with the modified models to create a microbial consortium.*

```
consortia=[model_transformer_comets,model_producer_comets]
```

22. *Create a FLYCOP object and select the COMETS simulator.*

```
from FLYCOP import *
FLYCOP_optimization=FLYCOP()
# FLYCOP set the dynamic FBA selected tool
FLYCOP_optimization.create_simulator(simulator_type="COMETS")
```

23. *Load the consortia into FLYCOP.*

```
FLYCOP_optimization.load_consortia(consortia)
```

24. *Set the previous COMETS parameters.*

```
FLYCOP_optimization.set_simulator_params(params)
```

25. *Set the COMETS scenario.*

```
FLYCOP_optimization.create_scenario()
FLYCOP_optimization.set_scenario(violacein_scenario)
```

26. *Set the optimization space.*

    See **Note 12**.

```
space=violacein_space()
result=FLYCOP_optimization.optimize(space)
```

27. *The best parameters found are stored in the "result" variable.*

    See **Notes 13** and **14**.

Initial Biomass Transformer	Initial Biomass Producer	Carbon source
0.104	0.011	Galactose

Another crucial consideration is exploring the correlation among variables to identify significant relationships. FLYCOP provides a large amount of search results that can be utilized to infer such relationships. For instance, correlations among the search values can be examined. A suitable tool for this analysis is the R language, along with the GGally package, which offers functionalities such as pairwise correlations.

```
library(tidyverse)
library(GGally)
density_color <- function(data, mapping,...){
 # Using default ggplot density function

 p <- ggplot(data = data, mapping = mapping) +
 stat_density2d(aes(fill=..density..), geom="tile", contour =
FALSE) +
 scale_fill_gradientn(colours=viridis::viridis(100, option="-
viridis"))
 p
}
FLYCOP_data <- read_csv("violacein_output.csv")
p=ggpairs(FLYCOP_data,lower=list(continuous=density_color),
aes(color=CarbonSource))
p
```

The result is a pair correlation plot (image not included due to its large dimension) with the correlation indices. The results indicate a positive correlation (0.2, p-value<0.005) between the inoculum size of the transformer strain and violacein production, as well as a negative correlation (−0.3, p-value<0.005) between the inoculum size of the producer strain and violacein production.

4 Notes

1. GEMs provide a comprehensive computational framework for describing the entire set of metabolic genes within an organism, along with the intricate gene–protein–reaction relationships. These models facilitate the simulation and prediction of metabolic fluxes through various optimization techniques, such as flux balance analysis. GEMs are fundamentally based

on stoichiometry and mass balance principles, derived from genome annotations and corroborated by empirical and experimental data.

2. When working with a SMC, it is essential that the GEMs from the different partners share a common namespace to facilitate effective communication among them. Additionally, it is desirable that the chosen namespace or nomenclature is comprehensible to ensure clarity and understanding. For example, the examples from this chapter have the BiGG namespace, but there are others quite common like KEGG (https://www.kegg.jp/), SEED (https://modelseed.org), MetaCyc (https://metacyc.org/), or MetaNetX (https://www.metanetx.org/).
3. Selecting an appropriate genome-scale metabolic model is a critical step, as its phylogenetic distance and quality can directly influence the outcomes of the protocol. Therefore, the chosen model must be of sufficient quality and closely related to the microorganism under study. If the model's quality is inadequate, it may contain various errors, such as missing metabolic or spontaneous reactions, incomplete pathways, or incorrect reaction directionality.
4. Exchange reactions are pseudo-reactions incorporated into GEMs to simulate the entry or exit of metabolites within the system. These reactions are employed to represent the theoretical chemical environment of the model.
5. When referring to the boundaries of reactions, we denote the theoretical flux limitations imposed on a given reaction. A "closed" or disabled reaction has its boundaries set to 0, thereby preventing any flux through the reaction. The boundaries also ensure the directionality of reactions in GEMs. Typically, one-directional reactions will have a lower bound of 0 and an upper bound greater than 0, allowing flux only toward the product side. Open or activate reactions, conversely, will have a lower bound less than 0 and a higher upper bound.

   The "default" boundary settings for a one-directional reaction are as follows:

```
model.reactions.ReactionID.lower_bound = 0
model.reactions.ReactionID.upper_bound = 1000
```

This configuration permits flux only toward the products, up to a theoretical maximum flux of 1000 mmol gDW^{-1} $h^{-1°}$, a value significantly higher than actual flux rates observed in microorganisms.

Boundaries are also critical for exchange reactions, as they constrain the metabolites a model can import and export to the extracellular compartment. For exchange reactions, the lower bound, indicated by a negative value, represents the capacity to incorporate a metabolite.

Furthermore, a model can be constrained to carry flux through a specific reaction by setting identical values for both the lower and upper boundaries. This approach is particularly useful for integrating experimental data into the model.

6. When incorporating a reaction into the model, an error may occur if the reaction id already exists within the model, as each id must serve as a unique identifier for a reaction or metabolite. In such cases, it is advisable to review the reaction with the same ID to determine if it is identical to the reaction being added. Additionally, it is important to note that if a specific metabolite is not already present in the model, it must be defined prior to adding the reaction.
7. The COBRApy method `summary` is very useful in assessing the functionality of an orthologous construction. By scrutinizing the exchange reactions within the model that exhibit flux, we can effectively identify any potential errors in the construction process prior to proceeding with the protocol.
8. The layout in COMETS serves as the foundational configuration for the simulation, encompassing elements such as initial biomass(es) and culture media composition. Various methodologies exist for constructing this layout, with the preferred approach outlined in this protocol by the authors. However, alternative methods can be found in the COMETSPY documentation.
9. COMETS simulation parameters play a crucial role in determining the behavior of the simulation. Below, we provide further insight into the parameters modified in the protocol:
   - *maxCycles*: This parameter sets the maximum number of cycles to simulate. Typically, one cycle represents approximately 1 h, and adjustments are made in the protocol based on information from reference papers.
   - *spaceWidth*: The total width designated for microbial growth (in centimeters). The default value of this parameter is 0.1, which is often insufficient and should be adjusted accordingly.
   - *allowCellOverlap*: A Boolean variable indicating whether cell overlaps are permitted in the simulation.
   - *maxSpaceBiomass*: The maximum total biomass allowed. If there are no limitations, this parameter should be set to a high value.

- *MediaLogRate*: A Boolean variable enabling the creation of a media table by COMETS, which is essential for plotting results and other analyses.
- *exchangestyle*: Determines the method used for calculating maximum exchange rates for all extracellular metabolites (Standard FBA, Monod Style, or Pseudo-Monod Style).
- *writeTotalBiomassLog*: A Boolean variable allowing COMETS to generate a biomass table for all organisms over time.
- *writeMediaLog*: A Boolean variable enabling COMETS to generate a media table for all metabolites over time.

10. When using COMETS for the first time, encountering certain issues is not uncommon. One of the most prevalent challenges is COMETS failing to locate the necessary paths for operation. In such instances, it is advisable to updatethe.bashrc file with the following environmental variables:

```
export COMETS_HOME= "Your path to COMETS directory"
export GUROBI_HOME= "Your path to Linux64 folder inside your
Gurobi directory"
export GUROBI_COMETS_HOME =$(echo $(ls -d "path from GUROBI_-
HOME") | awk {'print $NF'})
LD_LIBRARY_PATH=$LD_LIBRARY_PATH:$GUROBI_COMETS_HOME/lib/
export PATH=$PATH:$COMETS_HOME:$GUROBI_HOME:$GUROBI_COMET-
S_HOME
```

When working with COMETS, it is important to recognize that it functions slightly differently from running a flux balance analysis with COBRA. Initially, we need to configure the layout, which includes specifying the models to use, their respective initial biomasses, and the composition and quantities of the culture media. Subsequently, we must configure the parameters before initiating the simulation with COMETS.

One common issue encountered when using COMETS is remembering to define the culture media for the experiment and ensuring that the exchange and transport reactions for the relevant metabolites are open.

Another recurring challenge is encountering an expired Gurobi license. In such cases, refer to the Gurobi webpage to obtain a new license.

11. As indicated in the study [12], flux traverses violacein synthesis reactions. Referring to Figure 4.D from [12], we can obtain the requisite data to approximate the flux through the VIOA reaction. Subsequently, we can calculate the flux and assign it to the lower boundary of VIOA. In this instance, violacein production is estimated to be approximately 42 mg/L at 24 h with

an OD600 of 6.5. The approximated flux amounts to 0.022 mM h^{-1} $gCDW^{-1}$.

12. The execution time of the FLYCOP software's optimization step can vary significantly, spanning from a few minutes to several hours. This variation depends on factors such as the complexity of the problem and the computational resources allocated to the task.
13. FLYCOP identifies the optimal configuration within the explored search space. However, this does not guarantee it as the best solution across the entire solution space. Increasing the maximum number of searches enhances the likelihood of discovering the optimal solution.
14. The search space enables the identification of good solutions, but FLYCOP does not always follow the same search paths, leading to potentially different results in some cases. Nonetheless, the maximization of fitness is guaranteed for the search paths.
15. Ensuring the inclusion of metabolite release through transport and exchange reactions is crucial. This necessity arises from COMETS' coding limitations, which exclusively quantify metabolites within the extracellular compartment, disregarding those contained within the cell.
16. Replicate the overproduction of tryptophan by guiding the model to channel flux through reaction TRPAS2, the reaction responsible for tryptophan synthesis.

Acknowledgments

This research received funding from the European Union's Horizon 2020 research and innovation program under grant agreement number no. 870294 (MixUp) and HE no 101112378 (Promiseang), and 101000733 (Promicon); Funds from the Spanish Ministry of Science, Innovation and Universities (AEI/10.13039/501100011033) through research grant Rob3D (PID2022-139247OB-I00) is also acknowledged.

Conflict of Interest Statement This study was funded by [PROMOSEANG] [GA n° 101112378].

References

1. Brenner K, You L, Arnold FH (2008) Engineering microbial consortia: a new frontier in synthetic biology. Trends Biotechnol 26(9): 483–489
2. Rafieenia R, Atkinson E, Ledesma-Amaro R (2022) Division of labor for substrate utilization in natural and synthetic microbial communities. Curr Opin Biotechnol 75:102706

3. Tarzi C, Zampieri G, Sullivan N, Angione C (2024) Emerging methods for genome-scale metabolic modeling of microbial communities. Trends Endocrinol Metab 35:533–548
4. García-Jiménez B, Torres-Bacete J, Nogales J (2021) Metabolic modelling approaches for describing and engineering microbial communities. Comput Struct Biotechnol J 19:226–246
5. San León D, Nogales J (2022) Toward merging bottom–up and top–down model-based designing of synthetic microbial communities. Curr Opin Microbiol 69:102169
6. García-Jiménez B, García JL, Nogales J (2018) FLYCOP: metabolic modeling-based analysis and engineering microbial communities. Bioinformatics 34(17):i954–i963
7. Dukovski I, Bajić D, Chacón JM et al (2021) A metabolic modeling platform for the computation of microbial ecosystems in time and space (COMETS). Nat Protoc 16:5030–5082
8. Ebrahim A, Lerman JA, Palsson BO, Hyduke DR (2013) COBRApy: constraints-based reconstruction and analysis for python. BMC Syst Biol 7:1–6
9. Lindauer M, Eggensperger K, Feurer M et al (2022) SMAC3: a versatile Bayesian optimization package for hyperparameter optimization. J Mach Learn Res 23(54):1–9
10. Gowda K, Ping D, Mani M, Kuehn S (2022) Genomic structure predicts metabolite dynamics in microbial communities. Cell 185(3): 530–546
11. Nogales J, Mueller J, Gudmundsson S et al (2020) High-quality genome-scale metabolic modelling of Pseudomonas putida highlights its broad metabolic capabilities. Environ Microbiol 22(1):255–269
12. Gwon DA, Seo E, Lee JW (2023) Construction of synthetic microbial consortium for violacein production. Biotechnol Bioprocess Eng 28(6):1005–1014
13. Monk JM, Charusanti P, Aziz RK, Lerman JA, Premyodhin N, Orth JD, Feist AM, Palsson BØ (2013) Genome-scale metabolic reconstructions of multiple Escherichia coli strains highlight strain-specific adaptations to nutritional environments. Proc Natl Acad Sci USA 110(50):20338–20343
14. King ZA, Lu J, Dräger A et al (2016) BiGG Models: a platform for integrating, standardizing and sharing genome-scale models. Nucleic Acids Res 44(D1):D515–D522

Chapter 15

Dynamic Simulation of Growth and Cross-Feeding in Microbiomes with μbialSim

Ali Nawaz, Jessye L. Schaefer, and Florian Centler

Abstract

Microbial cells in natural environments are typically embedded in microbial communities consisting of few to many different species. Close proximity and high diversity of neighboring cells facilitate manifold interactions on several layers, from substrate competition, exchange of genetic material, to metabolic cross-feeding. The complexity of these ecological interaction networks makes microbiomes notoriously difficult to study. While microbiome dynamics can routinely be elucidated by meta-omics technologies, pinpointing mechanisms driving these observed dynamics remains a challenge. Mechanistic mathematical modeling with its ability to focus on individual interactions and exploring their isolated impact on overall dynamics has emerged as a suitable tool in this context. Here, we use μbialSim, an open-source simulator that extends the Flux Balance Analysis approach to microbial communities, considering substrate competition and metabolic cross-feeding but neglecting any other microbial interactions. Assuming a well-mixed bioreactor environment, simulated trajectories enable the analysis of growth behavior of individual microbiome members, dynamics of intracellular enzymatic fluxes across all species, as well as the analysis of cross-feeding behavior and how it changes over time. The MATLAB implementation of μbialSim is available from https://github.com/fcentler/microbialSim.

Key words Microbiome modeling, Microbial cross-feeding, Microbiome dynamics, Microbial communities, Flux balance analysis, Genome-scale metabolic networks, Systems biology

1 Introduction

Naturally occurring microbial cells are typically not exclusively surrounded by cells of the same species, but they are embedded in microbial communities consisting of few to many other species. The advent of culture-independent methods such as Next-Generation Sequencing (NGS) has enabled microbiological research to move beyond the traditional approach of studying culturable species in monoculture to directly assessing the true diversity of naturally occurring microbial communities and tracking how these change over time or differ between conditions. This has led to gaining fundamental insights into the functioning and the

Karthik Raman and Gayathri Sambamoorthy (eds.), *Modelling the Microbiome*, Methods in Molecular Biology, vol. 3006, https://doi.org/10.1007/978-1-0716-5080-6_15,

importance of microbial communities, or microbiomes, in the context of environmental science, biotechnology, and human health. While the majority of microbiome studies remains in principle observational in nature, for example, by assessing microbiome states and correlating changes thereof to external parameters, mathematical modeling has recently been adopted to study complex microbiomes to elucidate mechanisms driving observed dynamics. Interactions of microbial species within a community can take many forms, from competition for substrate, antibiotics warfare, exchange of genetic material, to the beneficial exchange of metabolites for growth. In the generalized Lotka–Volterra modeling (gLV), such interactions are reduced to a single parameter for each species pair, indicating a growth-promoting or inhibiting effect of one species on the other which remains constant throughout the simulation. While the mathematical simplicity of gLV models enables qualitative model analysis as a big advantage, this approach might not be applicable under all circumstances [1]. As an alternative, constraint-based approaches, such as Flux Balance Analysis which predicts microbial growth behavior based on genome-scale metabolic network models of individual species and the availability of nutrients, have been expanded to microbial communities. Such models are able to capture substrate competition and microbial cross-feeding as important microbial interactions [2]. Several simulation tools have been proposed (reviewed in [3]) that either focus on steady-state behavior or consider dynamics, and either consider space explicitly or treat the system as well-mixed, neglecting any spatial gradients. In the following, we will use µbialSim [4], Version 1.2.1, an open-source simulator which is able to simulate microbiome dynamics assuming a well-mixed reactor environment and demonstrate how the activity of individual species and microbial cross-feeding can be explored.

2 Materials

The simulation software µbialSim is written in MATLAB (MathWorks Inc.) which is available for all major operating systems (Linux, MacOS, and Windows). The source code of µbialSim can be downloaded from GitHub at https://github.com/fcentler/microbialSim. It depends on several other software packages besides MATLAB which need to be installed before it can be properly used. The following instructions describe all required installation steps. For more detailed installation instructions on individual software components, the reader is referred to the documentation of the respective software packages.

1. *MATLAB.* A MATLAB license is required for installing the software. Many scientists and students might have access to such a license through institutional license agreements. For

simulating complex microbiomes and making use of parallel execution on multi-core CPUs, the *Parallel Computing Toolbox* is required as well. For this chapter, MATLAB for Linux, Version R2021a, Update 8, was mainly used while also testing the execution on MacOS and Windows. Other versions of MATLAB are expected to work as well without any modifications of the code.

2. *COBRAToolbox v3.0.* For handling SBML models and performing Flux Balance Analysis computations, µbialSim relies on the popular constraint-based modeling toolbox COBRAToolbox [5], which is available at https://opencobra.github.io/cobratoolbox/stable/index.html. Follow the installation instructions (additional tools including `curl` and `git` are required for the installation) at https://opencobra.github.io/cobratoolbox/stable/installation.html and perform these steps:
   1. Using the shell, clone the repository with "`git clone ---depth=1` https://github.com/opencobra/cobratoolbox.git".
   2. In MATLAB, change to folder "`cobratoolbox`", and run "`initCobraToolbox`".
   3. In order to (optionally) test the proper installation of COBRAToolbox, run "`testAll`".

   If errors occur referring to "`TranslateSBML`", move to the `cobratoolbox` folder in the shell and run two git commands:

```
git submodule init
git submodule update
```

   and proceed with **step 2** above.

3. *µbialSim.* Installation of µbialSim is done by cloning the repository via

   "`git clone` https://github.com/fcentler/microbialSim.git". The repository includes the AGORA 1.01 collection of 773 gut microbes [6]. In order to use them in simulations, unzip the file "`microbialSim/models/AGORA-1.01-Western-Diet.zip`" at the location of the ZIP file. Make sure that the new subfolder "`1.01-Western-Diet`", containing the individual models, resides in the "`models`" directory, alongside the ZIP file.
4. For the visualization of cross-feeding activity as networks, additional software is required such as the free tools Gephi (https://gephi.org) and Cytoscape (https://cytoscape.org). We will use Cytoscape in this chapter; its installation instructions are available at the respective website.

3 Methods

3.1 Running Preconfigured Scenarios

Four simulation scenarios are preconfigured in µbialSim and can be run out of the box to test the proper installation of µbialSim and all required software packages. Scenario 1 represents a batch culture of *Methanococcus maripaludis* performing hydrogenotrophic methanogenesis. Syntrophic growth of a co-culture of *Syntrophobacter fumaroxidans* and *Methanospirillum hungatei* on propionate is implemented in Scenario 2. Scenario 3 simulates batch growth of a complex microbiome consisting of all 773 microbial species contained in the AGORA 1.01 model collection. And finally in Scenario 4, batch growth on a western diet of the eight-species SIHUMIx community [7] representing a minimal human gut microbiome is considered. *See* [4] for a more detailed description of the scenarios, including selected initial conditions. The scenarios can be simulated by providing the scenario ID (1, 2, 3, or 4) as an argument to the function "microbialSimMain()". For example, for running Scenario 2, change to the µbialSim folder in MATLAB and run:

```
simulationOutput = microbialSimMain(2).
```

µbialSim will report on the progress during the numerical simulation and store simplified simulation results in the specified MATLAB variable simulationOutput and the full simulation results as a series of files in the filesystem. The naming of these files follows a common scheme: "simulatedTrajectory_<-date> <time>.<ending>". Both date and time refer to the start of the simulation. Three MATLAB files are generated with the following endings:

- .mat: This file contains the full trajectory of the simulation, including both biomass and compound concentrations in the reactor over time, and intracellular flux distributions across all species for all time steps.
- mat_dynamics.mat: This is a reduced file only containing biomass and compound concentrations in the reactor over time but not internal flux distributions. This file is considerably smaller than the full trajectory file.
- mat_restartInit.mat: The final state of the simulation is stored in this file and can be used to seamlessly continue the simulation in a subsequent simulation run.

Additionally, three figures are automatically produced (*see* Figs. 1, 2, and 3), including:

1. Temporal dynamics of biomasses of all species in one graph, showing the difference to the initial biomass concentration for

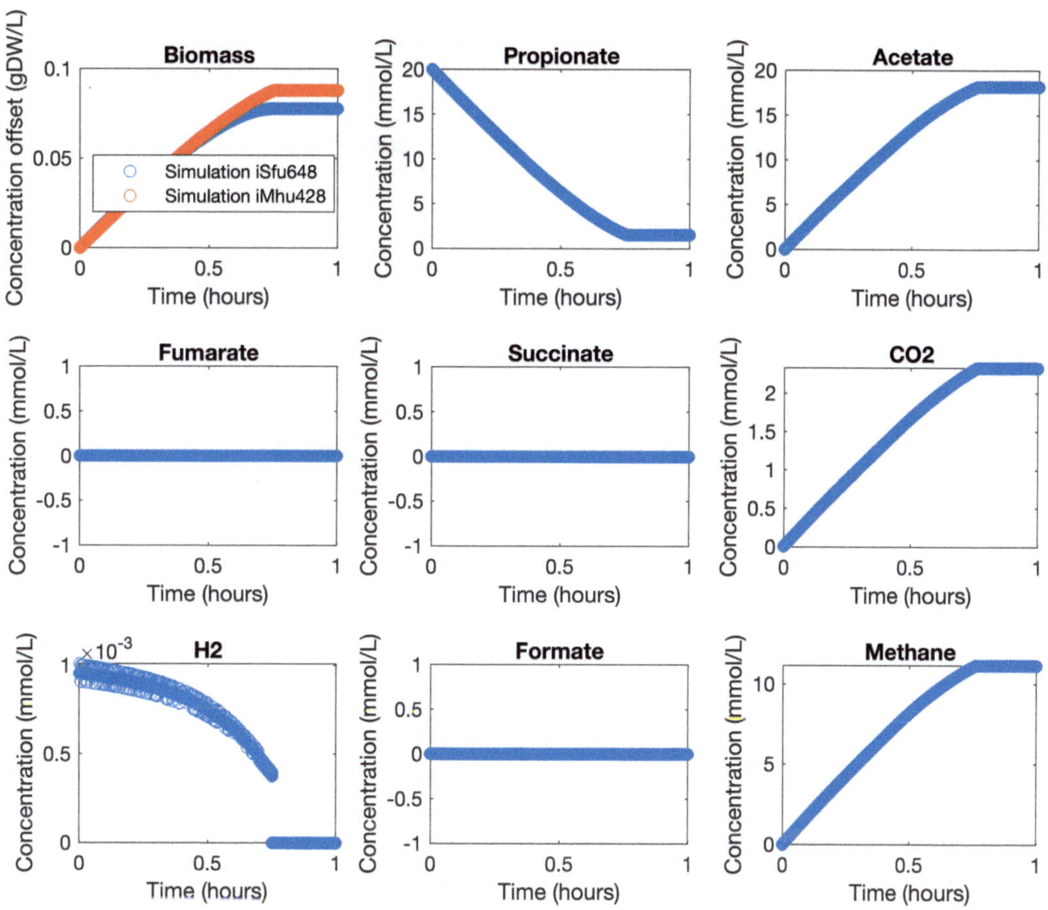

Fig. 1 Automatically generated figure detailing the evolution of biomass concentrations for all species (top left) as deviation from initial biomass concentrations ("offset") and evolution of reactor compound concentrations during the simulation. This figure was generated by preconfigured Scenario 2, syntrophic growth of a binary community consisting of propionate utilizing *S. fumaroxidans* (in silico model iSfu648) and methanogen *M. hungatei* (in silico model iMhu428)

each species ("offset"), and individual graphs for each reactor compound considered in the simulation (produced using function `plotTrajectoryCmp()`).

2. Temporal dynamics of exchange fluxes indicate the exchange of compounds across the cell membrane for each species. Two graphs are generated per species. In the first, exchange fluxes are shown which are coupled to reactor compounds. It is these fluxes that change the compound concentrations in the reactor. In the second graph, the remaining exchange fluxes are shown. These fluxes do not influence the system state as these refer to compounds which are expected to be available in sufficient amounts and hence not being growth-limiting. In both plots, exchange reactions are excluded which do not carry any flux

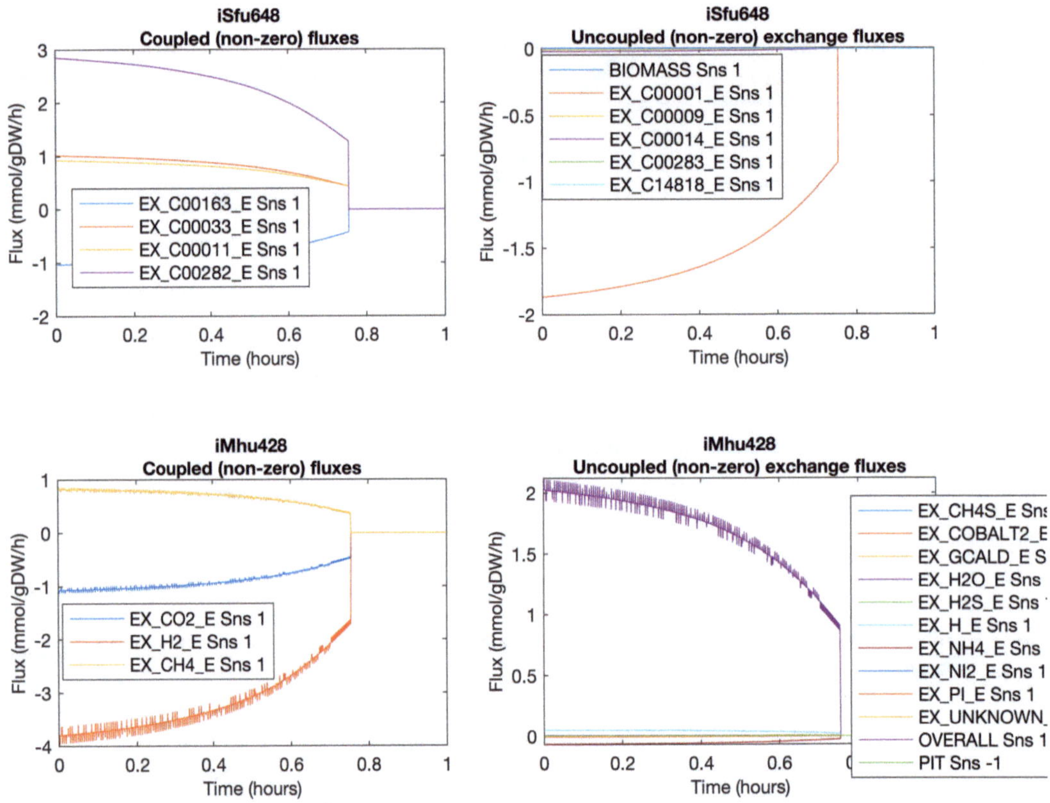

Fig. 2 Automatically generated figure detailing the evolution of exchange fluxes indicating the transfer of compounds across the cell membrane. Both coupled (left) and uncoupled exchange reactions are shown (right). Only coupled exchange fluxes lead to changes in reactor compound concentrations. Each row contains both graphs for a particular species. "Sns 1" in legend, short for "sense = 1", indicates that negative flux values indicate compound uptake, and positive flux values indicate compound secretion. This figure was generated by preconfigured Scenario 2

throughout the full simulation period (produced using function `plotExchangeFluxes()`).

3. Temporal dynamics of biomass concentration in absolute values in one graph, temporal evolution of all reactor compounds in a second graph, and finally the temporal evolution of all specific growth rates μ for all species in one graph (based on the reduced trajectory, using function `plotTrajectoryDynamics()`).

The automated generation of figures can be deactivated by uncommenting the respective code lines 321, 322, and 338 in `microbialSimMain.m` by preceding them with a percent sign ("%").

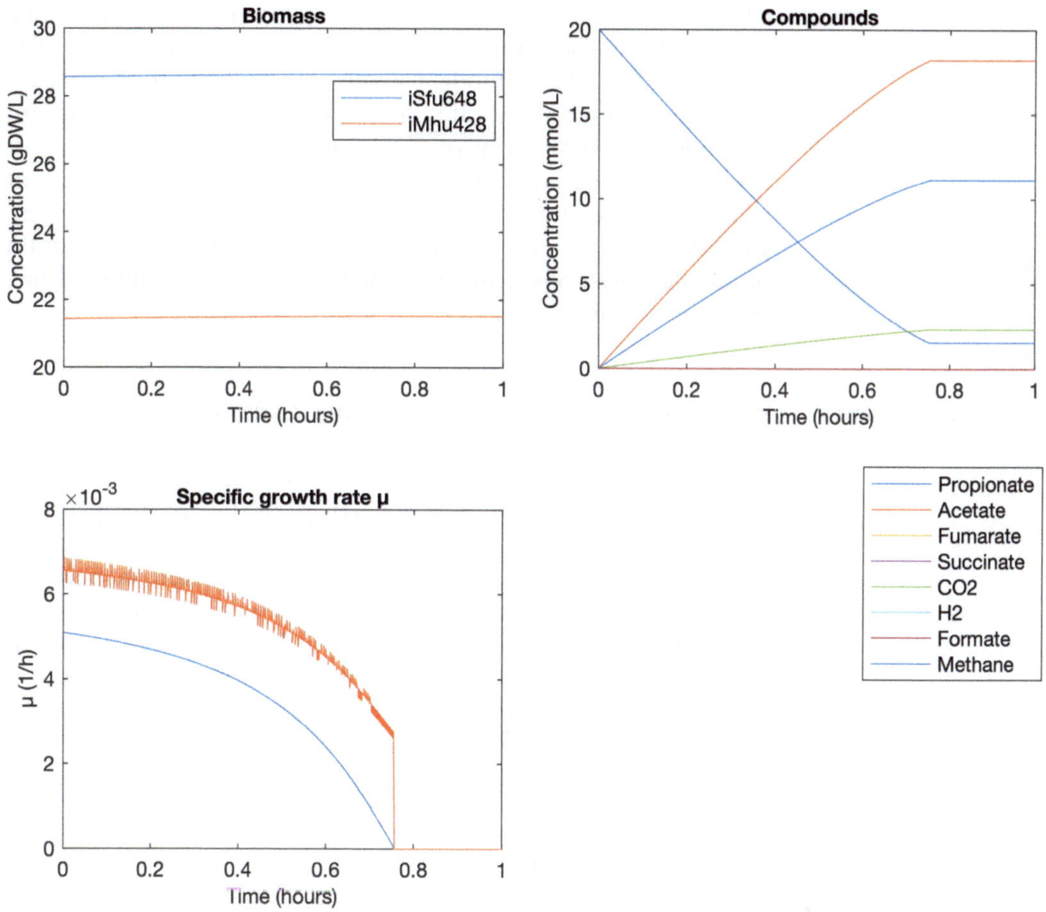

Fig. 3 Automatically generated figure detailing the evolution of biomass concentration for all species in absolute values (top left), evolution of all reactor compound concentrations (top right), and evolution of specific growth rate μ for all microbial species (bottom left). This figure was generated by preconfigured Scenario 2

3.2 Analyzing Simulated Data

When running a simulation using the function `microbialSim-Main()`, an object is returned containing the simulated trajectory but only including the temporal evolution of all species biomasses and chemical compounds contained in the reactor. These data, together with the temporal evolution of internal metabolic flux distributions for all microbial species, are saved to simulation files as described above. In the returned object, the field `time` refers to the time points at which data was recorded, `biomass` and `com-pounds` contain the evolution of biomass and compound concentrations, respectively, and `mu` the evolution of the specific growth rate for each species. Names for species and compounds are provided in fields `modelNames` and `compoundNames`, respectively, defining the order in which data is presented in the abovementioned fields. The same data can be obtained from the simulation file ending in `mat_dynamics.mat` which additionally contains the used reactor flow rate. Files ending in `mat` contain three fields `elapsedHours`, `reactor`, and `trajectory`. The first field records the elapsed wall clock time for the simulation run. The second field records all model parameters pertaining to the reactor, including volume, flow rate, compounds, initial concentrations, and inflow concentrations of all compounds and species. The final field `trajectory` contains fields already discussed above, but additionally all simulation parameters used (field `solverPars`), and information on each model (field `FBA`) including the selection of coupled reactions, and the temporal evolution of internal flux distributions (field `fluxes`).

3.2.1 Inspecting Internal Fluxes

While exchange fluxes indicating the transport of metabolites across the cell membrane can be plotted by the provided function `plotExchangeFluxes()`, internal fluxes must be visualized manually. As an example, we plot the activity of the enzyme pyruvate kinase which transforms phosphoenolpyruvate to pyruvate in the last step of glycolysis across all species considered in Scenario 4. After loading the full simulation trajectory file ending in "`.mat`", the following code in which variable `fluxToPlot` is set to the reaction name which is to be plotted (here: PYK,) produces Fig. 4. Admissible names for this code can be retrieved by inspecting the column names of `trajectory.FBA.fluxes`.

```
fluxToPlot = 'PYK';
plotData = zeros(length(trajectory.time) - 1, length(tra-
jectory.modelNames));
for i = 1:size(plotData,2)
 plotData(:,i) = trajectory.FBA(i).fluxes.(fluxToPlot);
end
plot(trajectory.time(1:end-1), plotData)
title(fluxToPlot)
xlabel('Time (hours)')
ylabel('v (mmol/gDW/h)')
legend(trajectory.modelNames, 'Interpreter', 'none')
```

3.2.2 Analyzing Microbial Cross-Feeding

The exchange of chemical compounds between microbial cells embedded in a phylogenetically diverse microbial community is likely an important factor determining microbiome dynamics. Chemical compounds secreted by one microbial species during growth as a by-product might serve as a growth substrate for another species. Although this interaction involves a temporal and spatial gap—first requiring the compound to be secreted, then transported to the receiver, and finally to be taken up—we employ the following simplification here. As space is not considered in the model, we assume that a cross-feeding interaction occurs whenever

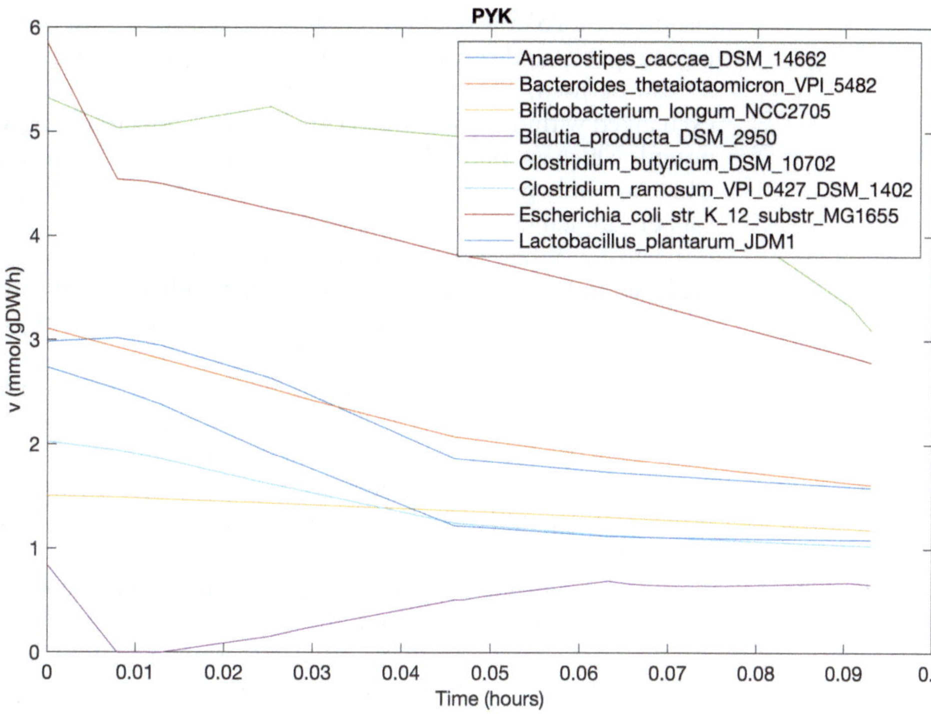

Fig. 4 Plotting the temporal evolution of flux activity of pyruvate kinase as the last step in glycolysis across all species of Scenario 4

one species secretes a compound which is consumed by another at the same time, assuming that cross-feeding interactions occur instantaneously. To identify such interactions from a simulated trajectory, load the respective simulation result file (ending in `.mat`) by double-clicking it in MATLAB's file pane. A table containing all exchange fluxes at the time point `time` can then be obtained by "`[exchangeTable, reportTime] = getCmpndExchangeTable(trajectory, time);`". Make sure to inspect `reportTime`, as the closest time step from the trajectory will be selected for reporting and might deviate from the requested time point `time`. "-1" will be returned if the requested time point is outside the simulated time span. The variable `exchangeTable` contains all microbial species in columns and compounds in rows and indicates the production and consumption fluxes in unit "mM/h", with negative values indicating uptake. The obtained exchange table can be filtered for interesting compounds using the function `filterCmpndExchangeTable(exchangeTable, mode, threshold)`. The parameter `threshold` can be used to define a minimal meaningful value for exchange fluxes. Fluxes below this value will be set to zero. The parameter `mode` determines the type of filtering to be applied. Removing compounds which are neither secreted nor taken up can be eliminated from the table by choosing "mode = 0". To only retain compounds which are at least produced by one species and consumed by at least one other species, "mode = 1" can be used. This will deliver compounds which are currently exchanged within the microbial community. The selection of compounds which are only taken up but not produced by any species, or which are only produced but never taken up by any other species, can be implemented by chosing "mode = 2" and "mode = 3", respectively. While the former indicates compounds which need (at least) to be provided externally for all members of the community to be able to grow, the latter indicates waste products, which are not further metabolized within the community. Filtered exchange tables can be written to an adjacency matrix in an external file using function `writeCmpndExchangeTableToFile(exchangeTable, filename)`. This enables a quick visualization of the compound exchange network, for example, in Cytoscape [8] using the app aMatReader [9]. Select the file just written by MATLAB in Cytoscape (`Apps/aMatReader/Import Matrix File`), and choose "Directed" as style. As an example, compound exchange halfway throughout the simulation of the SIHUMIx microbiome simulation of Scenario 4 is shown in Fig. 5, using 0.05 as the time parameter in `getCmpndExchangeTable()` and using `mode = 1` and a threshold value of 0.05 when calling `filterCmpndExchangeTable()`.

3.3 Setting Up a Simulation

The preconfigured scenario simulations can be used as a template to set up new scenarios including metabolic models of other species in SBML format. If species are to be simulated for which models are available in the AGORA 1.01 model collection, one can simply change the vector `speciesToConsider` in Scenario 4, either specifying the number of the respective model (*see* top of `prepareFBAmodel_Agora.m` for complete list) or specifying model file names directly. In general, the reactor and the microbial models are set up in separate files: `reactorDefinition_*.m` and `prepareFBAmodel_*.m`. The parameters to be set up in both files are collected in Table 1. Both files define functions that are called from the main program `microbialSimMain.m`. If using different file-/function names, these must hence be updated in the main program as well. The model setup file can also be used to set additional flux constraints for the model. Due to the numerical implementation in µbialSim, genome-scale metabolic models must only contain irreversible reactions. The function `convertToOnlyIrrevReac()` can be used to ensure this to be the case for SBML models also containing reversible reactions. After the bioreactor and microbial models are configured, the simulation end time and desired time stepping can be set using the variables `solverPars.tend` and `solverPars.timeStepSize`. Note that several additional parameters controlling the numerical solver of µbialSim can be set at the top of the file `microbialSimMain.m`. Refer to the comments

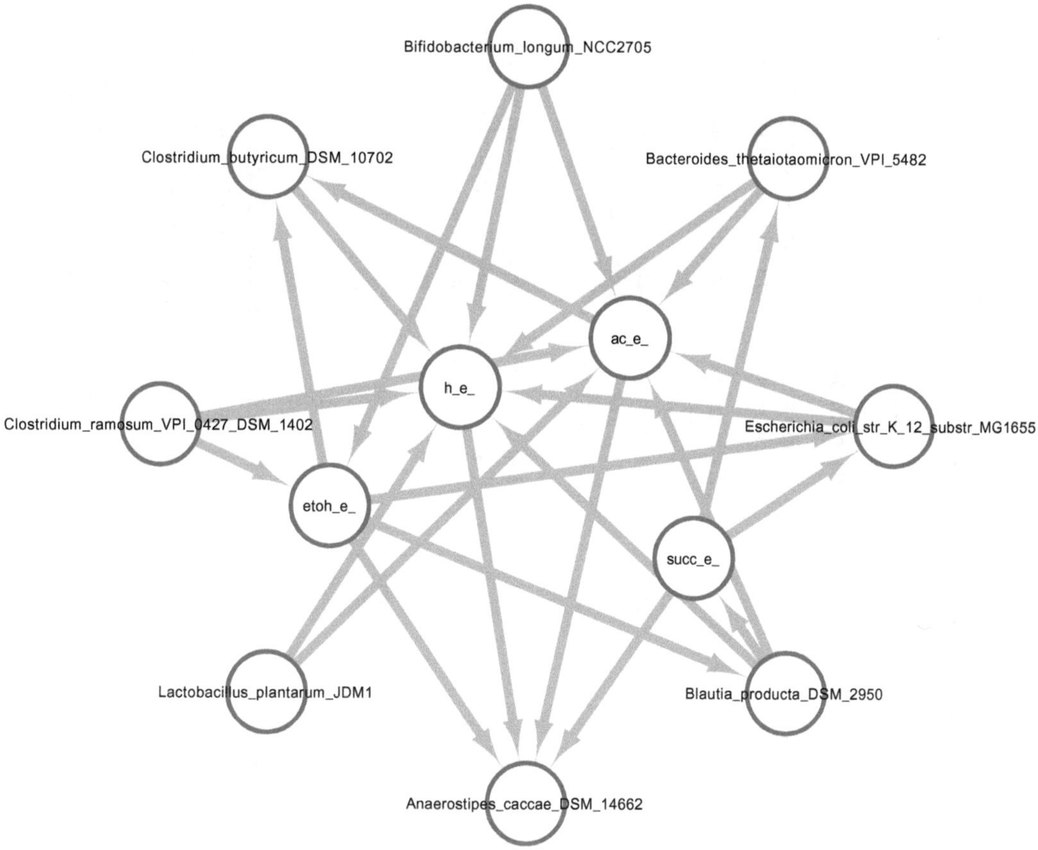

Fig. 5 Visualizing compound exchange in the eight-species SIHUMIx microbiome (Scenario 4) with Cytoscape, only showing uptake and secretion fluxes above 0.05 mM/h at $t = 0.0491$ h halfway through the simulation. Compound "Water" was removed for clarity

contained therein for details. Once the simulation is started, progress will be continuously reported on. Note that the adaptive decrease of the time step size will transiently also decrease simulation speed, making the prediction of the remaining simulation time somewhat unreliable. Depending on the chosen scenario and the unfolding exchange dynamics, simulations can take a long time. *See* **Notes 1–3** for options to speed-up the simulation. After a simulation is done and evaluated, it sometimes becomes evident that a longer simulation time would have been beneficial in order to capture interesting dynamics in full or ensuring that a steady state is reached. *See* **Note 4** on how this can be done in µbialSim without the need to rerun the full simulation. Setting up simulations by modifying parameters directly in the file `microbialSimMain.m` becomes cumbersome if scenario simulations are run in which parameter values are systematically varied. For such cases, parameters can be set in an external file, as explained in detail in **Note 5**.

Table 1
Setting up a simulation in µbialSim requires the definition of bioreactor parameters (left columns) and microbial species models making up the simulated microbiome (right columns)

Defining the bioreactor (file `reactorDefinition_*.m`)		Defining microbial species models (file `prepareFBAmodel_*.m`)	
Parameter [unit] variable	**Description**	**Parameter/variable**	**Description**
Reactor volume [L] `reactor.volume`	The liquid volume of the bioreactor (required for chemostat simulations only)	`ngamReac`	Index of the reaction defining energy demand due to non-growth associated maintenance
Flow rate [L/h] `Reactor.flowRate`	Flow rate for chemostat simulations, set to 0 for simulating a batch system	`biomassReac`	Index of the reaction defining biomass composition
Compound names `reactor.compounds`	List with the names of all chemical compounds present in the reactor. The chosen order defines the order in which values must be given in parameters below	`IDs` (first occurrence)	List with indices of all reactions that provide or remove chemical compounds from the vicinity of the cell
Initial compound concentrations [mM] `reactor.compoundsInit`	List with initial compound concentrations in the bioreactor; the order follows the compound names list	`sense`	Indication for each reaction in `IDs`, whether positive flux indicates compound secretion (value 1) or uptake (value −1).
Initial biomass in reactor [gDW/L] `reactor.biomassInit`	List with initial biomass concentrations. The order is given by the order of species in variable `models` as set up in `microbialSimMain.m`	`IDs` (second occurrence)	Only retain indices referring to compounds which are explicitly modeled in the bioreactor
Compound concentrations in inflow [mM] `reactor.compoundsInflow`	Concentration of compounds in inflow during chemostat simulations. For order, *see* above	`vmax, ks`	Maximal uptake flux and K_S value for converting reactor concentrations to uptake constraints using Monod kinetics for each reactor compound, order as in `IDs`
Biomass concentrations in inflow [gDW/L] `reactor.biomassInflow`	Concentrations of species biomasses in inflow during chemostat simulations. For order, *see* above	`reactorCompoundIDs`	Mapping compounds to the reactor: order as in `IDs`, value refers to position of respective compound in `reactor.compounds`

4 Notes

1. *Choosing numerical solver.* Several commercial and free numerical solvers have been surveyed for their performance for FBA calculations [10]. Besides glpk, µbialSim can make use of IBM's CPLEX and the Gurobi solver. In order to use them, they must be made available to the COBRAToolbox. Follow the respective instructions and set variable `solverPars.COBRAsolver` to 2 for using CPLEX and to 3 for Gurobi.
2. *Parallel execution.* In case MATLAB's Parallel Computing Toolbox is available, parallel execution on multicore CPUs can be activated by setting `solverPars.parallel = 1` and assigning the maximum number of parallel processes to `solverPars.maxWorkers`.
3. *Preloading species models.* The loading of SBML model files takes some time and becomes limiting when simulating microbiomes of high diversity. Instead of re-loading individual SBML files at the beginning of each simulation run, the MATLAB object containing all models can be stored during the first simulation run and then used to speed up consecutive runs. To use this feature, use `solverPars.saveLoadedModelsToFile = 1` in the first simulation run which will store the species models to file `loadedModels_*.mat`. In consecutive simulation runs, reset `solverPars.saveLoadedModelsToFile = 0`, and instead of calling the function `prepareFBAmodel_*()`, simply use `load('./loadedModels_*.mat');`.
4. *Continuing previous simulations.* To easily facilitate the continuation of the simulation, the file `simulatedTrajectory_*.mat_restartInit.mat` stores the final system state at the end of each simulation. This file can be used to set the initial conditions for a consecutive simulation run. Simply set variable `solverPars.readInitialStateFrom` to the restart filename and make sure to increase the simulation end time `solverPars.tend` to a new end time.
5. *Setting simulation parameters* via *file.* Instead of configuring simulation parameters and selecting microbiome composition, etc., directly in `microbialSimMain.m`, the choices can be stored in a YAML file. *See* `config.yaml` for an example. A simulation taking the parameters from this file can be started by `microbialSimMain(config.yaml)`. Note that, in this case, the name of the YAML file will be applied to the output files instead of a time stamp. If the respective output file already exists, an error will be reported, preventing the accidental overwriting of existing simulation files.

Conflict of Interest Statement

The authors have no conflicts of interest to declare that are relevant to the content of this chapter.

References

1. Dedrick S, Warrier V, Lemon KP, Momeni B (2023) When does a Lotka-Volterra model represent microbial interactions? Insights from in vitro nasal bacterial communities. mSystems 8:e00757–e00722. https://doi.org/10.1128/msystems.00757-22
2. Culp EJ, Goodman AL (2023) Cross-feeding in the gut microbiome: ecology and mechanisms. Cell Host Microbe 31:485–499. https://doi.org/10.1016/j.chom.2023.03.016
3. Scott WT, Benito-Vaquerizo S, Zimmermann J et al (2023) A structured evaluation of genome-scale constraint-based modeling tools for microbial consortia. PLoS Comput Biol 19: e1011363. https://doi.org/10.1371/journal.pcbi.1011363
4. Popp D, Centler F (2020) μbialSim: constraint-based dynamic simulation of complex microbiomes. Front Bioeng Biotechnol 8: 574. https://doi.org/10.3389/fbioe.2020.00574
5. Heirendt L, Arreckx S, Pfau T et al (2019) Creation and analysis of biochemical constraint-based models using the COBRA Toolbox v.3.0. Nat Protoc 14:639–702. https://doi.org/10.1038/s41596-018-0098-2
6. Magnúsdóttir S, Heinken A, Kutt L et al (2017) Generation of genome-scale metabolic reconstructions for 773 members of the human gut microbiota. Nat Biotechnol 35:81–89. https://doi.org/10.1038/nbt.3703
7. Becker N, Kunath J, Loh G, Blaut M (2011) Human intestinal microbiota: characterization of a simplified and stable gnotobiotic rat model. Gut Microbes 2:25–33. https://doi.org/10.4161/gmic.2.1.14651
8. Shannon P, Markiel A, Ozier O et al (2003) Cytoscape: a software environment for integrated models of biomolecular interaction networks. Genome Res 13:2498–2504. https://doi.org/10.1101/gr.1239303
9. Settle B, Otasek D, Morris JH, Demchak B (2018) aMatReader: importing adjacency matrices via cytoscape automation. F1000Res 7:ISCB Comm J-823. https://doi.org/10.12688/f1000research.15146.2
10. Machado D (2024) A benchmark of optimization solvers for genome-scale metabolic modeling of organisms and communities. mSystems:e0083323. https://doi.org/10.1128/msystems.00833-23

Chapter 16

LambdaPy and LambdaR: Thermodynamics-Based Biogeochemical Reaction Modeling Packages for Integrating High-Resolution Mass Spectrometry Data

Manokaran Veeramani, Sanjog Kharel, Hugh C. McCullough, Xingyuan Chen, Jianqiu Zheng, James C. Stegen, Timothy D. Scheibe, and Hyun-Seob Song

Abstract

Microorganisms are key drivers of biogeochemical cycles in natural environments. Microbially mediated biogeochemical reactions are influenced by both biotic and abiotic factors, including microbe–microbe interactions, enzyme kinetics, and chemical traits. To address the impact of chemical substrates on biogeochemical reactions, this chapter provides guidance on substrate-explicit thermodynamic modeling (SXTM) (also known as lambda modeling). SXTM enables automatic formulation of stoichiometric and kinetic models of biogeochemistry from the chemical formulas of organic matter (OM). This approach is particularly useful for formulating biogeochemical reaction models from ultra-high-resolution mass spectrometry data that identify thousands of compounds with distinct molecular formulas. Regardless of the complexity of OM data, SXTM requires only two parameters, maximum growth rate (μ_{max}) and harvest volume (V_h), thereby avoiding the issue of overparameterization caused when one attempts to include a large set of chemical compounds. While the original formulation has been demonstrated with a focus on aerobic respiration, recent work has extended its scope to various other forms of electron acceptors. Here, we provide a tutorial on formulating biogeochemical reaction models using SXTM, with river corridor OM data collected by the Worldwide Hydrobiogeochemistry Observation Network for Dynamic River Systems (WHONDRS) consortium as an example. The software packages for implementing SXTM are available in two languages, Python and R, and are referred to as LambdaPy and LambdaR, respectively. These tools will significantly facilitate modeling of complex OM pools in microbially driven biogeochemical cycling and will help improve our understanding of the interplay among microbes, enzymes, and OM when integrated with complementary approaches, including microbial- and enzyme explicit modeling.

Key words Substrate-explicit modeling, Biogeochemical modeling, Organic matter, Thermodynamics, FTICR-MS

Karthik Raman and Gayathri Sambamoorthy (eds.), *Modelling the Microbiome*, Methods in Molecular Biology, vol. 3006, https://doi.org/10.1007/978-1-0716-5080-6_16,

1 Introduction

Degradation of organic matter (OM) significantly influences biogeochemical cycling in terrestrial and aquatic systems, affecting nutrient availability in the environment, ecosystem functioning, and ecological resilience. Microorganisms play a pivotal role in these processes by producing extracellular enzymes to break down and transform complex OM, influencing feedback cycles in biogeochemical processes through ecological interactions. Accurate prediction of biogeochemical cycling, therefore, requires a comprehensive consideration for key factors affecting this process, including (1) characteristics of microbial growth and interactions (microbial traits), (2) synthesis and degradation of extracellular enzymes (enzymatic traits), and (3) chemical and thermodynamic properties of OM (chemical traits).

Current ecosystem models have oversimplified descriptions of microbially driven biogeochemical reactions, significantly limiting their predictive power. Therefore, the consensus regarding the need for a high-resolution representation of microbial dynamics and enzymatic reactions in ecosystem models led to the development of microbial- [1–4] and enzyme-explicit models [5–7]. Despite these advancements, the limited complexity of substrate pools serves as a significant bottleneck to enhance the accuracy of these models [8, 9].

To address this, Song et al. (2020) developed substrate-explicit thermodynamic modeling (SXTM), a novel thermodynamics-based biogeochemical reaction modeling approach. This framework provides an unprecedented capability to directly incorporate high-resolution OM data for the prediction of biogeochemical reaction rates [10]. Based on thermodynamic principles, SXTM derives stoichiometric equations for metabolic reactions from chemical formulas of organic compounds. This is achieved by combining catabolic and anabolic reactions using a parameter λ such that energy generation and consumption through these two reaction pathways are balanced out. The parameter λ denotes the number of times catabolic reactions must occur to provide the energy required to synthesize a unit of carbon mole of biomass. Thus, λ is frequently used for quantifying thermodynamic efficiency of OM degradation reactions, which is why this formulation is also referred to as the *lambda model.*

SXTM further formulates the kinetic expressions for OM degradation rates using only two parameters: The maximal growth rate and harvest volume. This significantly simplifies the complexity associated with the identification of a large number of parameters. This feature enables incorporating complex OM datasets from various analytical instruments (including GC-MS, LC-MS, HPLC-MS, and FTICR-MS) into biogeochemical modeling.

SXTM was initially demonstrated with oxygen as the terminal electron acceptor (EA) [10] and recently extended to a few selected non-oxygenic EAs [11]. SXTM has been actively applied for various purposes, including: Prediction of watershed dynamics under environmental perturbations [12], identification of the key factors determining OM bioavailability in river corridors [13], evaluation of the impact of bioavailability of pyrogenic OM [14], understanding how nutrient limitations influence aerobic respiration in hyporheic zones [15], and developing reactive transport models [16]. To facilitate model building, simulations, and post-processing, we previously developed software packages using Software Development Kits in the U.S. Department of Energy's Systems Biology Knowledgebase (KBase). The developed KBase apps include: "Run Lambda Analysis", "Build Stoichiometric Reaction Models From Chemical Formulas", "Simulate Batch Biogeochemical Reaction Model", and "Simulate CSTR Biogeochemical Reaction Model" [17].

In this work, we significantly extend the scope of SXTM simulations by accounting for 21 EA equations under six different categories classified based on the family to which the EA belongs (shown in Table 1). We present the resulting simulation packages written in two commonly used languages, Python and R, and term them LambdaPy and LambdaR, respectively. The inclusion of different EAs allows SXTM to effectively assess diverse energy flows through multiple metabolic pathways activated during OM decomposition. With a focus on LambdaPy, we illustrate the implementation of these simulation packages using ultra-high-resolution OM data from FTICR-MS, collected by the Worldwide Hydrobiogeochemistry Observation Network for Dynamic River Systems (WHONDRS) consortium [18, 19].

2 Materials

LambdaPy and LambdaR are generalized packages to formulate SXTM with different electron donors and acceptors and can run on all platforms supporting Python and R codes. Other than the choice of language, these two packages are identical in all aspects, including the structure of input files and the range of output variables they produce. Both packages are freely available for download from https://github.com/hyunseobsong/lambda. In this chapter, we will demonstrate the application of LambdaPy and LambdaR using Jupyter Notebook, an open-source web-based interactive platform that supports both Python and R through installation of appropriate kernels. Jupyter Notebook can be installed using Anaconda, a distribution of Python and R, which simplifies package management and deployment. If we prefer a minimal installation, we can use Miniconda (available at https://

Table 1
List of different electron-acceptor reactions supported by LambdaPy and LambdaR

Classification	Family identifier	Subcategories	Electron-acceptor reaction
Oxygen-based	0	–	$O_2 + 4H^+ + 4e^- \rightarrow 2H_2O$
Nitrogen-based	1.	1.1	$NO_3^- + 10H^+ + 8e^- \rightarrow NH_4^+ + 3H_2O$
		1.2	$NO_3^- + 2H^+ + 2e^- \rightarrow NO_2^- + H_2O$
		1.3	$NO_3^- + 6H^+ + 5e^- \rightarrow (\frac{1}{2})N_2 + 3H_2O$
		1.4	$NO_2^- + 8H^+ + 6e^- \rightarrow NH_4^+ + 2H_2O$
		1.5	$NO_2^- + 4H^+ + 3e^- \rightarrow (\frac{1}{2})N_2 + 2H_2O$
		1.6	$N_2 + 8H^+ + 6e^- \rightarrow 2NH_4^+$
Sulphur-based	2.	2.1	$SO_4^{2-} + (\frac{9}{2})H^+ + 8e^- \rightarrow (\frac{1}{2})H_2S + (\frac{1}{2})HS^- + 4H_2O$
		2.2	$SO_4^{2-} + 2H^+ + 2e^- \rightarrow SO_3^{2-} + H_2O$
		2.3	$SO_4^{2-} + 8H^+ + 6e^- \rightarrow S + 4H_2O$
		2.4	$SO_4^{2-} + 5H^+ + 4e^- \rightarrow (\frac{1}{2})S_2O_3^{2-} + (\frac{5}{2})H_2O$
		2.5	$SO_4^{2-} + 9H^+ + 8e^- \rightarrow HS^- + 4H_2O$
		2.6	$SO_3^{2-} + (\frac{15}{2})H^+ + 6e^- \rightarrow (\frac{1}{2})H_2S + (\frac{1}{2})HS^- + 3H_2O$
Iron-based	3.	3.1	$Fe(OH)_3 + 3H^+ + e^- \rightarrow Fe^{2+} + 3H_2O$
		3.2	$FeOOH + 3H^+ + e^- \rightarrow Fe^{2+} + 2H_2O$
		3.3	$Fe_3O_4 + 8H^+ + 2e^- \rightarrow 3Fe^{2+} + 4H_2O$
		3.4	$Fe^{3+} + e^- \rightarrow Fe^{2+}$
Hydrogen-based	4.	4.1	$HCO_3^- + 9H^+ + 8e^- \rightarrow CH_4 + 3H_2O$
		4.2	$H^+ + e^- \rightarrow (\frac{1}{2})H_2$
Acetate-based	5.	–	$CH_3COO^- + 9H^+ + 8e^- \rightarrow 2CH_4 + 2H_2O$
Manganese-based	6.	–	$MnO_2 + 4H^+ + 2e^- \rightarrow Mn^{2+} + 2H_2O$

docs.conda.io/projects/miniconda/en/latest/). In Subheading 2.1, we describe the steps involved in the installation of Anaconda and Jupyter Notebook, followed by instructions for using LambdaPy in Jupyter Notebook and LambdaR in Jupyter Notebook and RStudio. Subheadings 2.2 and 2.3 provide discussion on the structural requirements for input and output csv files necessary for conducting thermodynamic analysis, as well as the thermodynamic variables obtained from the analysis.

2.1 Software Installation

2.1.1 Anaconda and Jupyter Notebook

1. Download Anaconda from https://www.anaconda.com/download according to the operating system (Windows, Mac, or Linux). Clicking on the provided link will take us to the download page. Select "Download" from that page.
2. An installation file will begin to download for our operating system (Mac, Windows, or Linux). Make sure to have at least 4 GB of disk space available to install it in the system.

3. Once the installation file is downloaded, run it from the appropriate folder. Follow the instructions from the installation prompt to complete the installation.
4. Launch the Jupyter Notebook using Anaconda Navigator, a desktop graphical interface included in Anaconda distribution, by clicking the "Launch" button.
5. To create a new notebook, click on the "New" button in the top right corner of the Jupyter Notebook interface and select "Python [Root]" under the "Notebooks" section. This will create a new notebook with the version of Python kernel installed in our system.

2.1.2 LambdaPy

The following steps will guide us on how to run it using Jupyter Notebook.

1. First, launch the Jupyter Notebook, as described in Subheading 2.1.1. This will open a command prompt window. Subsequently, a new webpage will be automatically launched in our system's browser, as shown in Fig. 1. If the webpage does not open automatically, we can manually copy and paste the local host/link (for example, it will resemble something like "http://localhost:8888/?token=....") into our browser's address bar. Alternatively, we can hold down the "Ctrl" key while left-clicking on either of the two links.
2. After this, we will be taken to a page that will show our system folders. Go to the folder where "LambdaPy.ipynb" was downloaded. Then, click on it to open the notebook and start running the LambdaPy code.

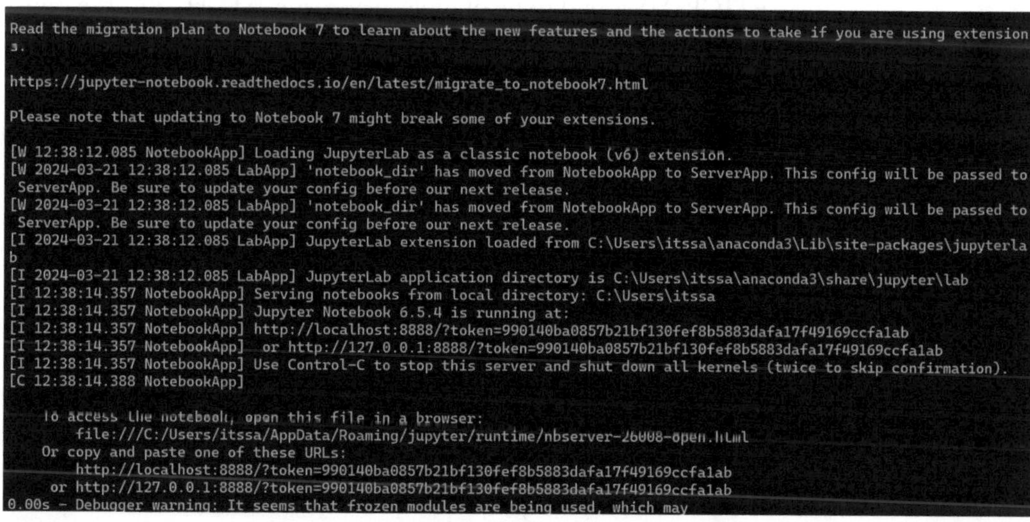

Fig. 1 Screenshot of command prompt window when Jupyter Notebook is launched

2.1.3 LambdaR in Jupyter Notebook

1. Installation of Jupyter Notebook is the same as above, with the only additional requirement being the installation of R in our Conda environment via Anaconda Navigator.
2. This can be done by going to the environment tab in the Anaconda Navigator.
3. Then, we can create a new environment and select the box for R under Packages.
4. Next, click the play button on the new environment and choose "Open with Jupyter Notebook".
5. Now, we can open LambdaR. Ensure the kernel selected is R and not Python 3.

2.1.4 LambdaR in RStudio

1. To use LambdaR in RStudio's RMarkdown format, first download and install R from https://www.r-project.org/.
2. Next, download and install RStudio from https://posit.co/download/rstudio-desktop/.
3. After installation, launch RStudio, open the LambdaR RMarkdown, and modify the input file path and specify the parameters.

2.2 Structure of the Input File

To perform thermodynamic analysis of organic compounds using LambdaPy and LambdaR, a csv file is required as the input containing molecular compositions of organic compounds obtained from FTICR-MS analysis.

Ensure the csv file includes columns for the chemical composition, such as carbon (C), hydrogen (H), nitrogen (N), oxygen (O), phosphorus (P), and sulphur (S). It is important to note that naming conventions (C, H, N, O, P, S) followed in the csv file reflect the LambdaPy code. For instance, if the OM identified from FTICR-MS is said to be of molecular formula $C_6H_{12}O_6$, the values in the columns of C, H, and O are 6, 12 and 6, and 0 in the columns of N, P, and S.

Formularity package (https://github.com/PNNL-Comp-Mass-Spec/Formularity) can also be used to preprocess the raw MS data and obtain the desired csv file required for performing thermodynamic analysis.

2.3 Structure of the Output File

Both LambdaPy and LambdaR generate the output file in csv format, which includes a set of relevant thermodynamic properties and stoichiometries of all individual compounds listed in the input file. The full list of output variables from LambdaPy and LambdaR and their short description is provided in Table 2. Detailed explanation and mathematical expressions of all the thermodynamic properties listed in Table 2 can be found in [10, 13]. Note that all the thermodynamic properties provided in Table 2 are computed based on the Gibbs energy of formation in Table 3 for all species included in the package.

Table 2
List of output variables obtained from executing the LambdaPy/LambdaR code and their descriptions

Output variable	Description
CUE	Carbon use efficiency denoting how much of carbon for OC is assimilated as biomass
NUE	Nitrogen use efficiency denoting how much of nitrogen for OC is assimilated as biomass
TER	Threshold element ratio denoting balanced requirement of C/N ratio for growth and maintenance of microbiome
delGd0	Gibbs free energy of electron donor reaction computed at pH = 0
delGd	Gibbs free energy of electron donor reaction computed at user-specified pH
delGcat0	Gibbs free energy of catabolic reaction computed at pH = 0
delGcat	Gibbs free energy of catabolic reaction computed at user-specified pH
delGan0	Gibbs free energy of anabolic reaction computed at pH = 0
delGan	Gibbs free energy of anabolic reaction computed at user-specified pH
delGdis0	Gibbs free energy of dissipation reaction computed at pH = 0
delGdis	Gibbs free energy of dissipation reaction computed at user-specified pH
lambda0	Parameter λ of OC computed at pH = 0
lambda	Parameter λ of OC computed at user-specified pH
stoichD_*	Stoichiometric coefficient of different compounds involved in the electron donor reaction. _* denotes individual compounds. For example, stoichD_OC denotes the stoichiometric coefficient of OC in electron donor reaction
stoichA_*	Stoichiometric coefficient of different compounds involved in the electron acceptor reaction
stoichcat_*	Stoichiometric coefficient of different compounds involved in the catabolic reaction
stoichAn_*	Stoichiometric coefficient of different compounds involved in the anabolic reaction
stoichMet_*	Stoichiometric coefficient of different compounds involved in the metabolic reaction formed by coupling the catabolic and anabolic reactions

It should be noted that all thermodynamic properties are computed under the assumption that the biomass composition is $CH_{1.8}N_{0.2}O_{0.5}$ [24]

3 Methods

3.1 Workflow of SXTM and Scope of LambdaPy and LambdaR

The SXTM builds stoichiometric and kinetic reaction models from molecular formulas of organic carbons (OCs) based on thermodynamic theories (Fig. 2). The derivation of stoichiometric equations require implementing the following steps: (1) Write e-donor and e-acceptor half-reactions, merging them into the catabolic equation (R_{Cat}) to meet electron balance; (2) write an anabolic equation by taking OC as the C source and NH_4^+ as the N source (if electron balance is not achieved in the previous equation); perform

Table 3
Key chemical species associated with the oxidative degradation of OC and their standard Gibbs free energy of formation (ΔG_i^0) at 298.15 K and 1 bar

S.no	Compound	ΔG_i^0 (kJ/mol)	References
1.	OC	Dependent on chemical formula	
2.	H_2O	−237.2	[25]
3.	HCO_3^-	−586.9	[25]
4.	NH_4^+	−79.5	[25]
5.	HPO_4^{2-}	−1089.1	[25]
6.	HS^-	12	[25]
7.	H^+	0[a]	
8.	e^-	0[a]	
9.	O_2	16.5	[25]
10.	NO_3^-	−111.3	[26]
11.	NO_2^-	−32.2	[26]
12.	N_2	0	
13.	Fe^{3+}	−4.6	[26]
14.	Fe^{2+}	−78.9	[26]
15.	H_2	0	
16.	SO_4^{2-}	−744.6	[26]
17.	H_2S	−33.4	[26]
18.	SO_3^{2-}	−486.6	[26]
19.	S	0	
20.	$S_2O_3^{2-}$	−522.5	[26]
21.	$Fe(OH)_3$	−690	[27]
22.	FeOOH	−489.8	[28]
23.	Fe_3O_4	−1012.6	[29]
24.	CH_4	−34.06	[25]
25.	CH_3COO^-	−392	[26]
26.	MnO_2	−465.14	[30]
27.	Mn^{2+}	−228	[30]
28.	Biom[b]	−67	[30]

The standard-state Gibbs energies of formation of several compounds were calculated using the CHNOSZ R package [25] at 298.15 K (25 °C) and 1 bar. For other compounds, ΔG_i^0 was taken from the corresponding literature as listed in Table 2. Here, OC denotes organic carbon, and Biom is biomass

[a]All thermodynamic properties of H^+ and e^- including Gibbs energies are 0 at all pressures and temperatures, and all other aqueous species properties are relative to this

[b] ΔG_i^0 for Biom was taken from [24, 30], assuming a constant biomass composition ($CH_{1.8}N_{0.2}O_{0.5}$)

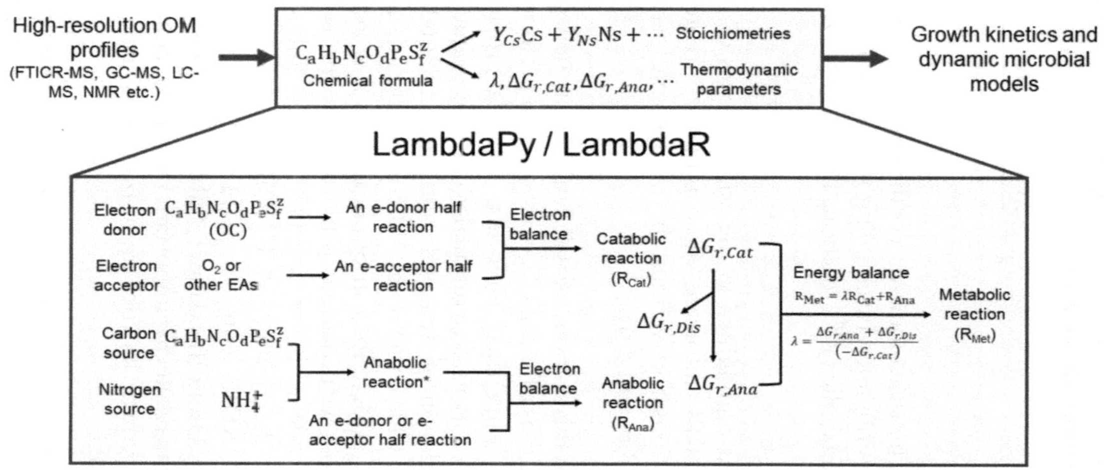

Fig. 2 Schematic illustrating the workflow of SXTM. The two boxes highlighted in the middle represent the scope of LambdaPy and LambdaR packages. (Adapted from Song et al. [10] with modifications)

additional integration with the e-donor or e-acceptor half-reactions to obtain R_{Ana}; (3) integrate R_{Cat} and R_{Ana} to obtain a metabolic reaction (R_{Met}) such that the total amount of energy generated from catabolism ($-\Delta G_{r,\ Cat}$) matches with the energy requirement for anabolism ($\Delta G_{r,\ Ana}$) and maintenance ($\Delta G_{r,\ Dis}$). The final step defines λ, a key parameter that quantifies the overall thermodynamic favorability of metabolic reactions: Lower (higher) λ value means that the corresponding metabolic reaction is thermodynamically more (less) favorable.

$$\lambda = \frac{\Delta G_{r,\text{Ana}} + \Delta G_{r,\text{Dis}}}{\left(-\Delta G_{r,\text{Cat}}\right)}$$

The stoichiometric coefficient vector $\boldsymbol{y}$ of the metabolic reaction is obtained by coupling the stoichiometries of catabolic ($\boldsymbol{y}^{\mathbf{Cat}}$) and anabolic ($\boldsymbol{y}^{\mathbf{An}}$) reactions by parameter λ, i.e.,

$$y = \lambda y^{\text{Cat}} + y^{\text{An}}$$

As highlighted in the middle boxes of Fig. 2, LambdaPy and LambdaR automate these processes to generate stoichiometric equations and various thermodynamic parameters (Subheadings 3.4, 3.5, and 3.6).

3.2 Executing the LambdaPy Code

From this section, we will demonstrate how to build stoichiometric equations of metabolic reactions and perform thermodynamic analyses using their molecular formula provided from FTICR-MS (with a focus on LambdaPy). At this point, it is expected that the installation of Anaconda and Jupyter Notebook is complete, and the csv file containing OM composition data with naming conventions as described in Subheading 2.2 is available. In the current section, we provide step-by-step description for executing LambdaPy code with user-specified inputs. Subheading 3.3 provides a brief description of FTICR-MS OM data collected by the WHONDRS consortium [18, 19] used for illustration in this work. Using this dataset, Subheadings 3.4, 3.5, 3.6, and 3.7 present the results for important thermodynamic properties and microbial growth kinetics for various EAs included in the LambdaPy package.

1. Import the LambdaPy code into Jupyter Notebook by following the instructions from Subheading 2.1.2.
2. After loading the template for LambdaPy, click the "Cell" option in the toolbar and from there click "Run All" as illustrated in Fig. 3. We will then be asked for an input for a csv file.
3. The first input needed is the file path for the csv file containing the composition of compounds for analysis. To get the file's path correctly, press "Shift + right click" on the input csv file

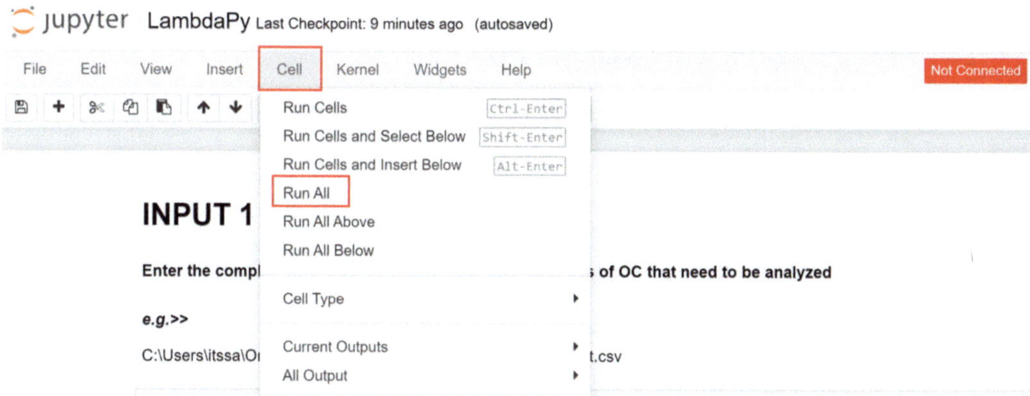

Fig. 3 Screenshot of executing LambdaPy in Jupyter Notebook

```
In [*]: import pandas as pd
        while True:
            file_pathway = input("Enter the pathway of the CSV file to be analyzed as shown in above example: ")
            try:
                data = pd.read_csv(file_pathway)
                print("File successfully loaded.")
                break
            except FileNotFoundError:
                print("\nError: File not found. Please check the pathway and try again.\n")
            except pd.errors.EmptyDataError:
                print("\nError: The file is empty.\n")
            except pd.errors.ParserError:
                print("\nError: Unable to parse the CSV file. Please ensure it is formatted correctly.\n")
            except Exception as e:
                print("\n An unexpected error occurred:", e, '\n')

        Enter the pathway of the CSV file to be analyzed as shown in above example:
```

Fig. 4 Screenshot of the input field for loading input csv file with the file pathway into LambdaPy

```
In [*]: while True:
            reaction_pathway = input("Enter a number from the provided list for an electron acceptor reaction: ")
            try:
                reaction_pathway = float(reaction_pathway)
                if reaction_pathway in [0, 1.1, 1.2, 1.3, 1.4, 1.5, 1.6, 2.1, 2.2, 2.3, 2.4, 2.5, 2.6, 3.1, 3.2, 3.3, 3.4, 4.1, 4.2, 5, 6
                    break
                else:
                    print("Please enter a valid value for an electron acceptor reaction pathway from the provided list")
            except ValueError:
                print("Invalid input. Please enter a number.")

        Enter a number from the provided list for an electron acceptor reaction:
```

Fig. 5 Screenshot of the input field in LambdaPy for specifying the reaction pathway (electron acceptor reaction)

that we want to analyze and select the option as "Copy as path." Then, paste it in the first input box of Fig. 4 and remove the quotation marks at the beginning and end of the pathway and press enter. If this step is completed correctly, we will receive a message that the file is loaded successfully. This will automatically lead us to the next input pop up for electron acceptor reaction, as depicted in Fig. 5.

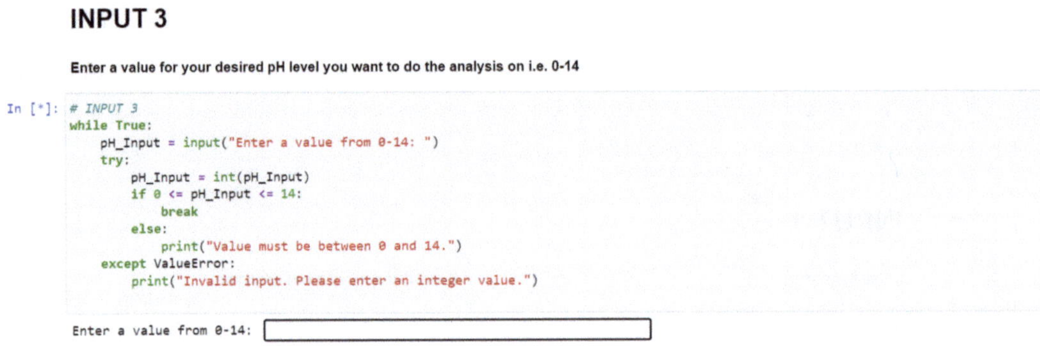

Fig. 6 Screenshot of the input field for pH value input in LambdaPy

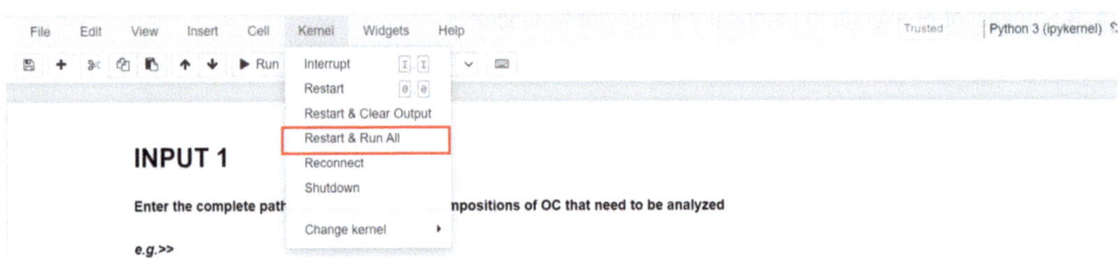

Fig. 7 Screenshot to restart and run the entire LambdaPy code

4. The second input required is a number ranging from 0 to 6 (some number may have subcategories) denoting the family identifier (and subcategory) of EA reaction to be included for performing thermodynamic analysis (refer to Table 1; *see* Fig. 5 for the input box). For instance, selecting 0 for the reaction pathway signifies that oxygen will be used as the EA. It is important to note that all compounds in the csv file will be analyzed using the specified reaction pathway. Enter a number referring to Table 1 to choose a particular EA reaction and press enter.
5. The third input is the pH value denoting the conditions under which thermodynamic properties are evaluated, as shown in Fig. 6. Enter a value for pH from 0 to 14 and press enter.
6. After providing all the three inputs, the code will run automatically, and we will get a table as output in the Jupyter Notebook after the code cell at the end of the notebook. In addition, the output from the SXTM analysis is also saved as a csv file (as described in Subheading 2.3) in the current working folder.

   Note: Sometimes the code might freeze when the inputs are not properly provided. In those cases, press "Kernel" in the toolbar and choose "Restart & Run All", as illustrated in Fig. 7.

Note that LambdaPy code must be run several times if thermodynamic analysis is required for multiple EAs. The output csv file from each EA can be aggregated later to perform in-depth analysis, as shown in Subheadings 3.4, 3.5, 3.6, and 3.7.

3.3 Description of FTICR-MS Dataset

WHONDRS, a global research consortium led by researchers from PNNL, was established to facilitate the research community with a scientific basis for understanding the linkages between the dynamics of hydrology, biogeochemistry, and microbiology in river corridors across different scales. The dataset used in this study was from WHONDRS summer 2019 sampling campaign (S19S) collected across 97 global river corridor systems between July and September 2019 (https://doi.org/10.15485/1729719). High-resolution dissolved OM data from surface water was obtained via 12 Tesla Fourier transform ion cyclotron resonance mass spectrometry (FTICR-MS) through the Environmental Molecular Sciences Laboratory (EMSL; https://www.pnnl.gov/environmental-molecular-sciences-laboratory). The raw FTICR-MS data was processed using Formularity package (https://github.com/PNNL-Comp-Mass-Spec/Formularity) to assign elemental formulas for the detected organic compounds using a compound identification algorithm [20]. This dataset also includes additional measurements such as non-purgeable organic carbon (NPOC), stable water isotopes (O and H), specific conductivity (SpC), total nitrogen (TN) and anions, bacterial abundance data, dissolved inorganic carbon (DIC) data, metadata, including international generic-sample number (IGSN) mapping file, file-level metadata, and data dictionary. For this illustration, we have only used the FTICR-MS data.

3.4 Gibbs Free Energy of Catabolic Reaction

For different EAs included in LambdaPy package, we compute $\Delta G_{r,cat}$ for various OM compounds characterized by FTICR-MS. From Fig. 8a, we find that the computed $\Delta G_{r,cat}$ is in the order of O_2 > nitrogen-based electron acceptors > MnO_2 > sulphur-based electron acceptors > iron-based electron acceptors > CH_3COO^-, which is in line with the predictions by the redox ladder [21] for most cases. It should be noted that for EAs with different subcategories (nitrogen-, sulphur-, and iron-based electron acceptors), we have aggregated all the $\Delta G_{r,cat}$ values and plotted their distribution as in Fig. 8a. However, we also find that some of the Fe- and S-based EAs are less energetically favorable than CH_3COO^-. These energetically less favorable EAs are identified based on $\Delta G_{r,cat}$ plots for N, S, and Fe family (Fig. 8b–d). We also performed the sensitivity analysis of $\Delta G_{r,cat}$ estimations from SXTM across different pH values ranging from 3 to 7. For this analysis, we have considered one EA from each family of EA included in the package. It is observed that $\Delta G_{r,cat}$ shifts considerably for $Fe(OH)_3$ denoting higher dependency with pH, as shown in Fig. 9 (*see* **Note 1**).

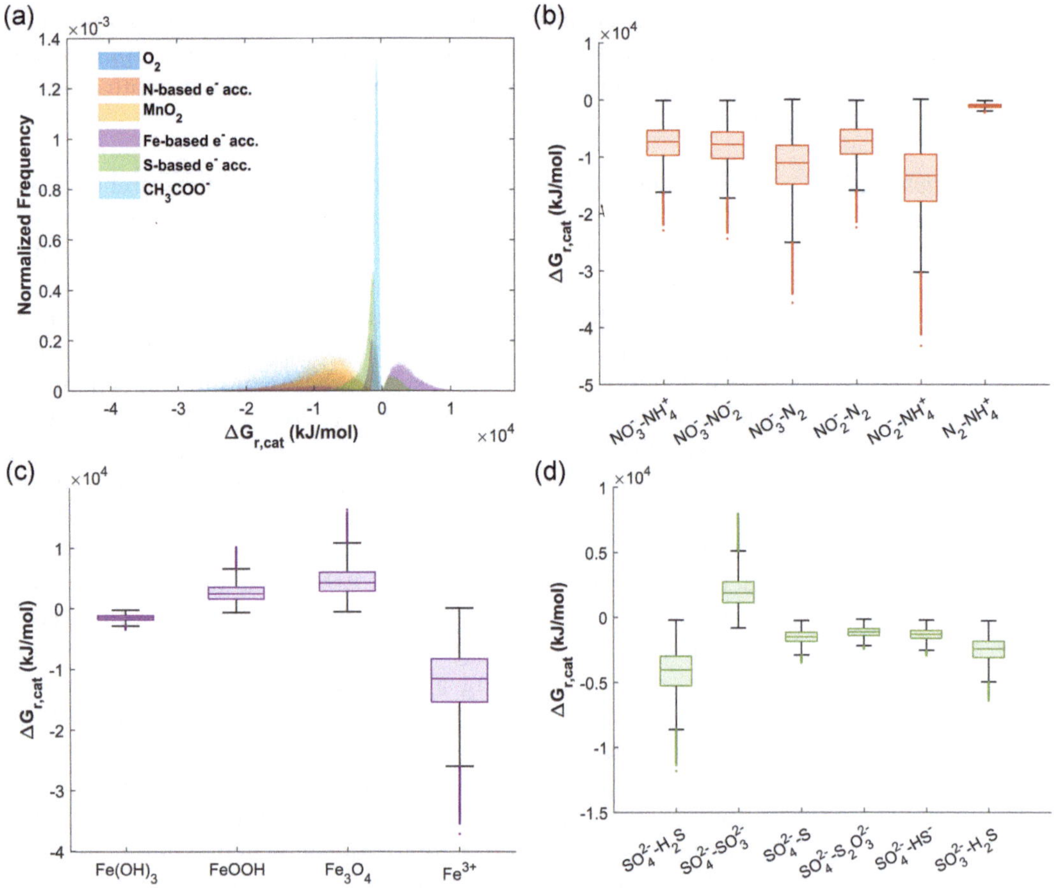

Fig. 8 Gibbs free energy of catabolic reactions ($\Delta G_{r,\text{cat}}$) for electron acceptors included in the LambdaPy package. (**a**) Distribution of $\Delta G_{r,\text{cat}}$ at pH = 7 for different electron acceptors. For N-, Fe-, and S-based electron acceptors, the distributions aggregate results across all electron acceptors within each family as defined in LambdaPy. (**b**) Boxplots of $\Delta G_{r,\text{cat}}$ at pH = 7 for N-based electron acceptor families, (**c**) Fe-based families, and (**d**) S-based families. In panels (**b–d**), *x*-axis labels are shown as *c1–c2*, where *c1* denotes the electron acceptor and *c2* denotes the major reaction product

3.5 Parameter λ

Figure 10 shows the scatter plot of λ as a function of number of C atoms in OC for the same set of EAs considered as in Fig. 9. It can be observed from Fig. 10 that $O_2, MnO_2, NO_3^{2-} - NH_4^+, SO_4^{2-} - H_2S$ are thermodynamically more favorable in the transformation of OC when compared to $Fe(OH)_3$, CH_3COO^-.

3.6 Growth Rate

Based on the thermodynamic theory [22], the microbial growth rate related to the consumption of a particular OC_i is formulated as a function of stoichiometric coefficients of metabolic reaction as follows [10]:

$$\mu_i = \mu_{\max} \exp\left(-\frac{|y_{OC,i}|}{V_h[OC_i]}\right) \exp\left(-\frac{|y_{EA}|}{V_h[EA]}\right)$$

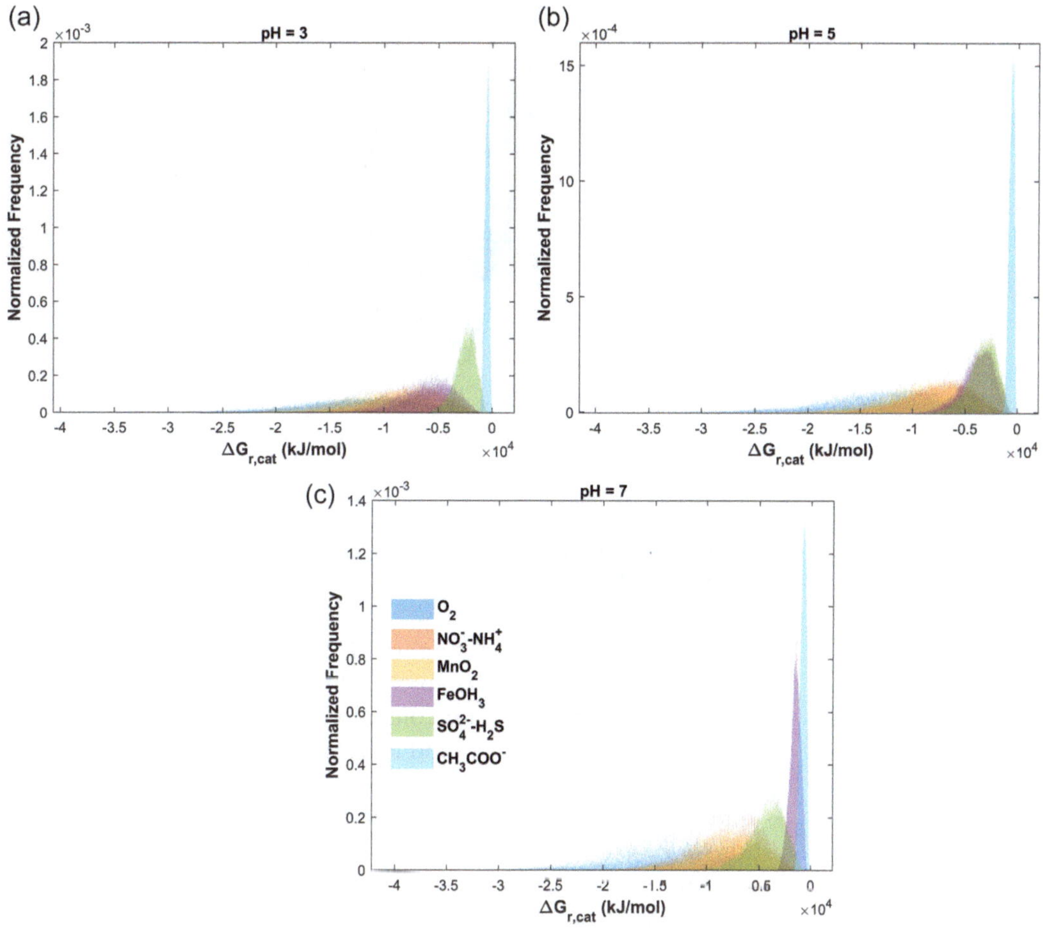

Fig. 9 Sensitivity of $\Delta G_{r,\mathrm{cat}}$ computed from SXTM for different electron acceptors as a function of pH. (**a**) pH $=3$, (**b**) pH $=5$, and (**c**) pH $=7$

Here, $y_{\mathrm{OC},\,i}$ is the stoichiometric coefficient of ith OC, y_{EA} is the stoichiometric coefficient of EA used in the analysis, $[\mathrm{OC}_i]$ and $[\mathrm{EA}]$ are concentrations of OC_i and EA, $\mu_{\max}$ is the maximum growth rate, and V_{h} is the harvest volume denoting the volume accessible by microbes for their growth in the environment. The expression for microbial growth rates can be further simplified to emulate different substrate limiting conditions such as

$$\text{OC is limiting}: \mu_i = \mu_{\max}\ \exp\left(-\frac{|y_{\mathrm{OC},i}|}{V_{\mathrm{h}}[OC_i]}\right)$$

$$\text{EA is limiting}: \mu_i = \mu_{\max}\ \exp\left(-\frac{|y_{\mathrm{EA}}|}{V_{\mathrm{h}}[\mathrm{EA}]}\right)$$

Using the above equations, microbial growth rates across different EAs are computed under different substrate-limiting

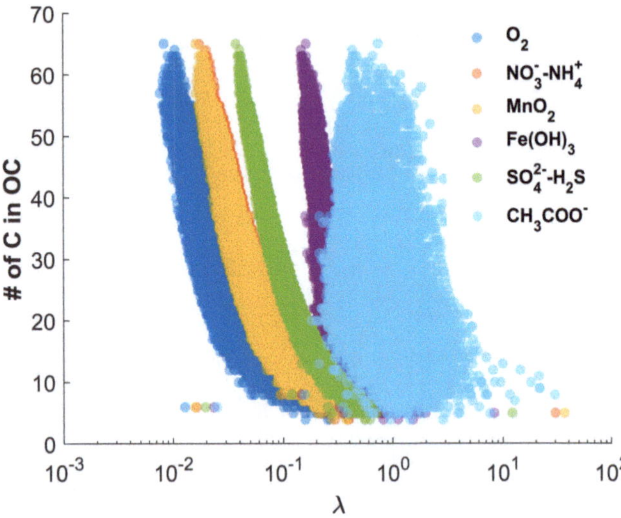

Fig. 10 λ as a function of number of C atoms in OC across different electron acceptors at pH = 7

conditions: (a) OC is limiting, (b) EA is limiting, and (c) OC and EA are limiting. For each substrate-limiting condition, two different levels of limitations are considered: (a) OC is limiting, severe limitation—$V_h[OC] = 0.5$ and moderate limitation—$V_h[OC] = 2$; (b) EA is limiting, severe limitation—$V_h[EA] = 1$ and moderate limitation—$V_h[EA] = 5$; (c) OC and EA are limiting, severe limitation—($V_h[OC] = 0.5$, $V_h[EA] = 1$) and moderate limitation—($V_h[OC] = 2$, $V_h[EA] = 5$). Fixing the value of the maximum growth rate μ_{max} to be 1 (*see* **Note 2**), growth rates for different EAs are computed at various substrate-limiting conditions. In case of OC limiting conditions, we find that predicted growth rates μ are in the order of $O_2 > MnO_2 > NO_3^{2-} - NH_4^+ > SO_4^{2-} - H_2S > Fe(OH)_3 > CH_3COO^-$ for both levels of limitations (Fig. 11a, d), especially for OC with high C atoms (C > 40), with μ reaching the maximum value of 1 at moderate limiting conditions for $O_2, NO_3^{2-} - NH_4^+, MnO_2, SO_4^{2-} - H_2S$. However, in the case of both EA-limiting (Fig. 11b, e) and OC–EA-limiting conditions (Fig. 11c, f), we find that predicted μ is in the order of $NO_3^{2-} - NH_4^+ > O_2 > SO_4^{2-} - H_2S > MnO_2 > CH_3COO^- > Fe(OH)_3$ at both levels of limitation. This indicates that under moderate limiting conditions, less energetic EAs such as CH_3COO^- have higher growth rates compared to $Fe(OH)_3$. We also predicted growth rates for different EAs by considering μ_{max} to be constrained by catabolic activity, i.e., $\mu_{max} = \frac{3}{N_e}$ [23] (*see* **Note 2**), where N_e is the number of electrons released during the catabolic reaction. For this μ_{max} setting, the predicted growth rates follow the same trend for different substrate-limited conditions at different levels of limitation (Fig. 12) when μ_{max} was fixed to be 1, and

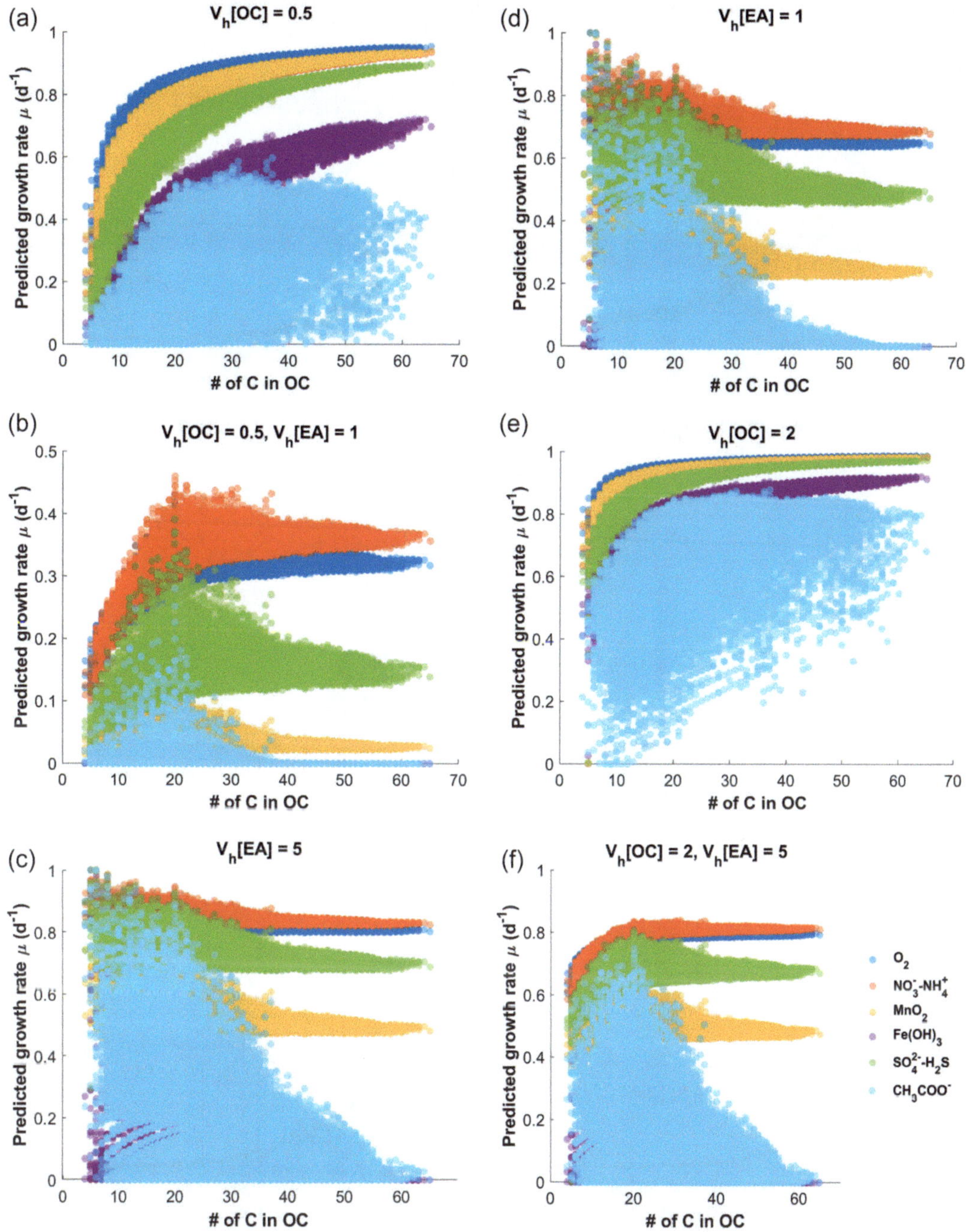

Fig. 11 Predicted growth rates for various electron acceptors (pH = 7) at different substrate-limiting conditions when $_{max}$ was fixed to be 1. Columns represent substrate-limiting regimes: The first column shows predictions under severe limitation, and the second column shows predictions under moderate limitation. Panels (**a–c**) correspond to severe limitation of organic carbon (OC), electron acceptor (EA), and combined OC + EA, respectively, while panels (**d–f**) show the corresponding growth rate predictions under moderate limitation

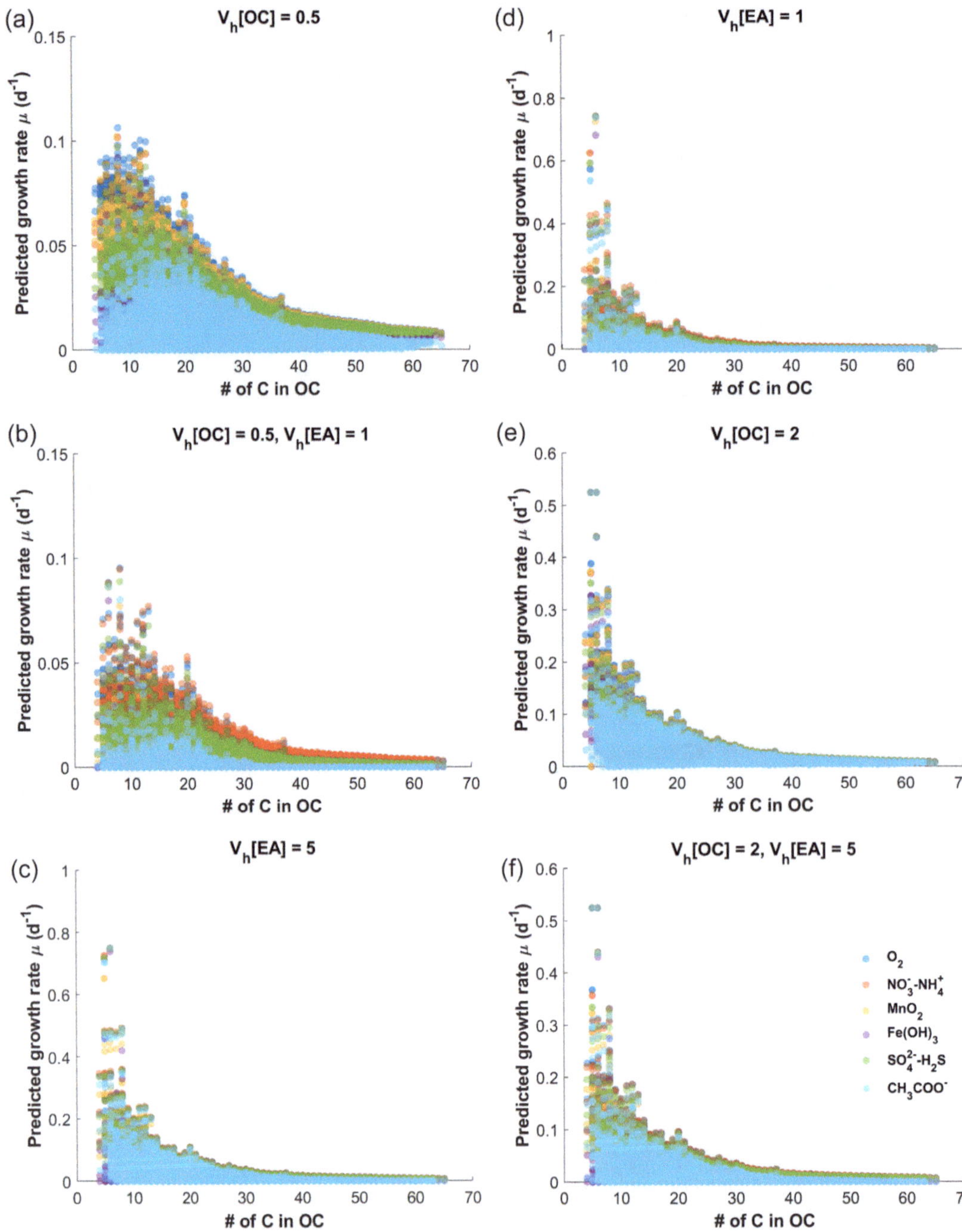

Fig. 12 Predicted growth rates for various electron acceptors (pH $=$ 7) at different substrate-limiting conditions when $_{\max}$ was constrained by catabolic activity ($_{\max} = 3/N_e$, where N_e is the number of electrons transferred during catabolic reaction). Columns represent substrate-limiting regimes: The first column shows predictions under severe limitation, and the second column shows predictions under moderate limitation. Panels (**a–c**) correspond to severe limitation of organic carbon (OC), electron acceptor (EA), and combined OC + EA, respectively, while panels (**d–f**) show the corresponding growth rate predictions under moderate limitation

maximum μ value is obtained when the number of C atoms in OC is approximately equal to 10 (*see* **Note 3**).

3.7 Nutrition Use Efficiencies

Carbon and nitrogen use efficiencies (CUE and NUE) denote the efficiency of a microbe assimilating the carbon and nitrogen in OC into their biomass. Threshold elemental ratio (TER) is another key metabolic trait often used in literature to quantify the actual C/N ratio required for biomass production. In this section, CUE, NUE, and TER for different OC are computed for the same set of EAs used in Subheading 3.5. From Fig. 13a, we find that the order of CUE is $O_2 > MnO_2 > NO_3^{2-} - NH_4^+ > SO_4^{2-} - H_2S > Fe(OH)_3 > CH_3COO^-$, with O_2 being the efficient EA in carbon assimilation. For NUE, we find that EAs $O_2, MnO_2, SO_4^{2-} - H_2S$ approach NUE equal to 1 when C/N ratio is approximately 15 (Fig. 13b). Also, we find that $NO_3^{2-} - NH_4^+$ EA is least efficient with respect to NUE as additional NH_4^+ are produced from the EA reaction. From Fig. 13c, we find that

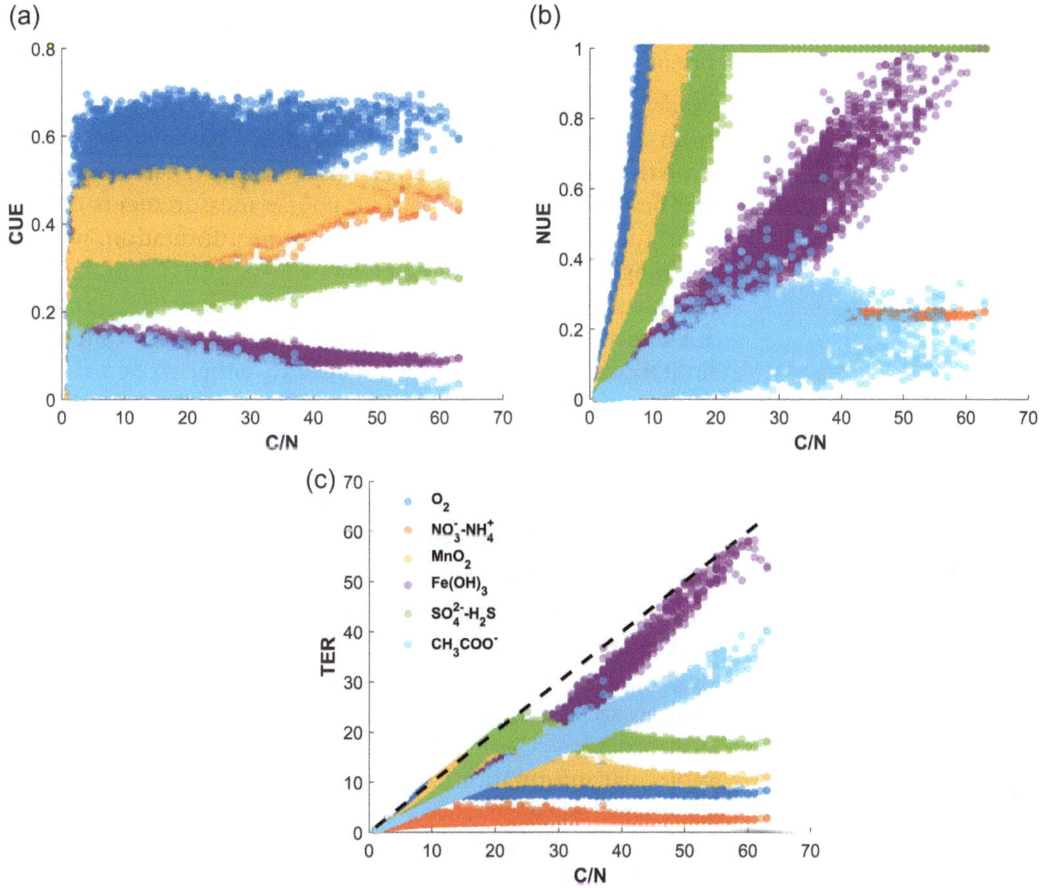

Fig. 13 Distribution of nutrient use efficiencies and threshold element ratio for different electron acceptors as a function of C/N ratio. (**a**) Carbon use efficiency (CUE), (**b**) nitrogen use efficiency (NUE), and (**c**) threshold element ratio (TER). Black dotted lines in (**c**) denote the region when C/N ratio is equal to TER

$O_2, MnO_2, NO_3^{2-} - NH_4^+, SO_4^{2-} - H_2S$ EAs are favorable for balanced growth when C/N ratio is less than 20 and for higher C/N ratio, and $Fe(OH)_3$ is more favorable for balanced growth (*see* **Note 3**).

4 Notes

1. *Uncertainty in the estimation of thermodynamic parameters.* Estimated Gibbs free energy of catabolic, anabolic, and metabolic reactions from SXTM plays an important role in the computation of reaction stoichiometries, λ, growth rates, and predicted respiration rates. Although we are not generally concerned with the variations of Gibbs free energy as it is the function of logarithm of the species activity, our analysis indicated the sensitivity of $\Delta G_{r,\text{cat}}$ when $Fe(OH)_3$ was used as EA (shown in Fig. 9), denoting the higher dependency with pH. It should be noted that there is also uncertainty associated with the estimation of Gibbs free energy of oxidation ΔG_{Cox}, as an inverse relationship between the nominal oxidation state of organic carbon (NOSC) was used [10].
2. *Model parameters.* In this study, the growth rates are computed from a simple model involving only two parameters, specific growth rate ($\mu_{\max}$) and harvest volume (V_{h}). These parameters can be estimated from the time-course measurements of concentrations of OC and EA. Without this information, we have set ranges for the parameter values V_{h}, [OC] and [EA] to include different substrate-limiting conditions, OC-limiting, EA-limiting, and both OC and EA-limiting with two levels of limitations. We have also assumed $\mu_{\max}$ either to be fixed to a constant value of 1 across all OM compositions or fixed based on the catabolic activity. While the above conditions can be utilized for calculating model-predicted growth rates and respiration rates facilitating a qualitative comparison with the measured respiration rates, this does not represent the actual respiration/growth rates from the environment.
3. *Bioavailability of OM.* The respiration rate computed from SXTM assumes that thermodynamics is the sole driver of OM degradation, and all the organic compounds from the FTICR analysis are assumed to be bioavailable. Although this assumption provides considerable results [10], using the full set of molecular information of organic compounds identified from the characterization studies for SXTM resulted in poor correlation between the experimental and model predicted respiration rates [13]. To circumvent this problem, a data-driven study was performed to find a subset of OM which is assumed to be bioavailable [13]. Also, multiple factors, such as C/N ratio,

number of C atoms, and λ, are found to play a vital role in the transformation of OM. Incorporation of this information can improve the consistency of model predictions with experimental data.

Acknowledgments

This research was supported by the U.S. Department of Energy (DOE), Office of Biological and Environmental Research (BER), and Environmental System Science (ESS) program through the River Corridor Scientific Focus Area project at Pacific Northwest National Laboratory. This work was also supported by the partnership with the IDEAS-Watersheds.

Conflict of Interest Statement The authors have no conflicts of interest to declare that are relevant to the content of this chapter.

References

1. Allison SD (2012) A trait-based approach for modelling microbial litter decomposition. Ecol Lett 15(9):1058–1070
2. Wieder W, Grandy A, Kallenbach C, Bonan G (2014) Integrating microbial physiology and physio-chemical principles in soils with the MIcrobial-MIneral Carbon Stabilization (MIMICS) model. Biogeosciences 11(14): 3899–3917
3. Wang G, Jagadamma S, Mayes MA, Schadt CW, Megan Steinweg J, Gu L, Post WM (2015) Microbial dormancy improves development and experimental validation of ecosystem model. ISME J 9(1):226–237
4. Sulman BN, Phillips RP, Oishi AC, Shevliakova E, Pacala SW (2014) Microbe-driven turnover offsets mineral-mediated storage of soil carbon under elevated CO2. Nat Clim Chang 4(12):1099–1102
5. Moorhead DL, Rinkes ZL, Sinsabaugh RL, Weintraub MN (2013) Dynamic relationships between microbial biomass, respiration, inorganic nutrients and enzyme activities: informing enzyme-based decomposition models. Front Microbiol 4:56126
6. Song H-S, Liu C (2015) Dynamic metabolic modeling of denitrifying bacterial growth: the cybernetic approach. Ind Eng Chem Res 54(42):10221–10227
7. Song H-S, Thomas DG, Stegen JC, Song X, Fredrickson JK, Scheibe TD (2017) Regulation-structured dynamic metabolic model provides a potential mechanism for delayed enzyme response in denitrification process. Front Microbiol 8:275696
8. Garayburu-Caruso VA, Stegen JC, Song H-S, Renteria L, Wells J, Garcia W, Resch CT, Goldman AE, Chu RK, Toyoda J (2020) Carbon limitation leads to thermodynamic regulation of aerobic metabolism. Environ Technol Lett 7(7):517–524
9. Stegen J, Johnson T, Fredrickson J, Wilkins M, Konopka A, Nelson W, Arntzen E, Chrisler W, Chu R, Fansler S (2018) Influences of organic carbon speciation on hyporheic corridor biogeochemistry and microbial ecology. Nat Commun 9:585
10. Song HS, Stegen JC, Graham EB, Lee JY, Garayburu-Caruso VA, Nelson WC, Chen X, Moulton JD, Scheibe TD (2020) Representing organic matter thermodynamics in biogeochemical reactions via substrate-explicit modeling. Front Microbiol 11:531756. https://doi.org/10.3389/fmicb.2020.531756
11. Zheng J, Scheibe TD, Boye K, Song H-S (2024) Thermodynamic control on the decomposition of organic matter across different electron acceptors. Soil Biol Biochem 193: 109364
12. Chen X, Lee RM, Dwivedi D, Son K, Fang Y, Zhang X, Graham E, Stegen J, Fisher JB, Moulton D (2021) Integrating field observations and process-based modeling to predict

watershed water quality under environmental perturbations. J Hydrol 602:125762

13. Ahamed F, You Y, Burgin A, Stegen JC, Scheibe TD, Song H-S (2023) Exploring the determinants of organic matter bioavailability through substrate-explicit thermodynamic modeling. Front Water 5. https://doi.org/10.3389/frwa.2023.1169701
14. Graham EB, Song H-S, Grieger S, Garayburu-Caruso VA, Stegen JC, Bladon KD, Myers-Pigg AN (2023) Potential bioavailability of representative pyrogenic organic matter compounds in comparison to natural dissolved organic matter pools. Biogeosciences 20(16): 3449–3457
15. Tureţcaia AB, Garayburu-Caruso VA, Kaufman MH, Danczak RE, Stegen JC, Chu RK, Toyoda JG, Cardenas MB, Graham EB (2023) Rethinking aerobic respiration in the hyporheic zone under variation in carbon and nitrogen stoichiometry. Environ Sci Technol 57(41):15499–15510
16. Song H-S, Ahamed F, Lee J-Y, Henry CS, Edirisinghe JN, Nelson WC, Chen X, Moulton JD, Scheibe TD (2025) Coupling flux balance analysis with reactive transport modeling through machine learning for rapid and stable simulation of microbial metabolic switching. Sci Rep:6042. https://doi.org/10.1038/s41598-025-89997-9
17. Song H-S, Lee J-Y (2020) Substrate-explicit modeling tutorial. United States. https://doi.org/10.25982/65526.69/1755438
18. Stegen JC, Goldman AE (2018) WHONDRS: a community resource for studying dynamic river corridors. mSystems 3(5). https://doi.org/10.1128/mSystems.00151-18
19. Toyoda JG, Goldman AE, Arnon S, Bar-Zeev E, Chu RK, Danczak RE, Daly RA, Forbes B, Garayburu-Caruso VA, Graham EB, Lin X, Moran JJ, Ren H, Renteria L, Resch CT, Tfaily M, Tolic N, Torgeson JM, Wells J, Wrighton KC, Stegen JC; WHONDRS Consortium T (2020) WHONDRS Summer 2019 sampling campaign: global river corridor surface water FTICR-MS, NPOC, TN, anions, stable isotopes, bacterial abundance, and dissolved inorganic carbon (v6). United States. https://doi.org/10.15485/1603775
20. Tolić N, Liu Y, Liyu A, Shen Y, Tfaily MM, Kujawinski EB, Longnecker K, Kuo L-J, Robinson EW, Paša-Tolić L (2017) Formularity: software for automated formula assignment of natural and other organic matter from ultrahigh-resolution mass spectra. Anal Chem 89(23):12659–12665
21. Bethke CM, Sanford RA, Kirk MF, Jin Q, Flynn TM (2011) The thermodynamic ladder in geomicrobiology. Am J Sci 311(3):183
22. Desmond-Le Quemener E, Bouchez T (2014) A thermodynamic theory of microbial growth. ISME J 8(8):1747–1751. https://doi.org/10.1038/ismej.2014.7
23. Gonzalez-Cabaleiro R, Ofiteru ID, Lema JM, Rodriguez J (2015) Microbial catabolic activities are naturally selected by metabolic energy harvest rate. ISME J 9(12):2630–2641. https://doi.org/10.1038/ismej.2015.69
24. Stephanopoulos G, Aristidou AA, Nielsen J (1998) Metabolic engineering: principles and methodologies. Academic Press, San Diego
25. Dick JM (2019) CHNOSZ: thermodynamic calculations and diagrams for geochemistry. Front Earth Sci 7:180
26. Rittmann BE, McCarty PL (2001) Environmental biotechnology: principles and applications. McGraw-Hill Education, New York
27. Majzlan J, Navrotsky A, Schwertmann U (2004) Thermodynamics of iron oxides: part III. Enthalpies of formation and stability of ferrihydrite (~ Fe (OH) 3), schwertmannite (~ FeO (OH) 3/4 (SO4) 1/8), and ε-Fe2O3. Geochim Cosmochim Acta 68(5): 1049–1059
28. Majzlan J, Grevel K-D, Navrotsky A (2003) Thermodynamics of Fe oxides: part II. Enthalpies of formation and relative stability of goethite (α-FeOOH), lepidocrocite (γ-FeOOH), and maghemite (γ-Fe2O3). Am Min 88(5–6):855–859
29. Zegeye A, Ruby C, Jorand F (2007) Kinetic and thermodynamic analysis during dissimilatory γ-FeOOH reduction: formation of green rust 1 and magnetite. Geomicrobiol J 24(1): 51–64
30. Kleerebezem R, Van Loosdrecht MCM (2010) A generalized method for thermodynamic state analysis of environmental systems. Crit Rev Env Sci Tec 40(1):1–54. https://doi.org/10.1080/10643380802000974

Index

A

Adapter trimming 35
AGORA 192, 193, 223, 224, 235–240, 247–249, 251, 253, 319, 320, 325
AGORA2 .... 177, 235, 237, 239, 241, 242, 246, 249
Amino acid auxotrophs 263, 266, 267
Amplicon sequencing 35, 39–42, 123, 135, 145, 146, 149
Antimicrobial resistance (AMR) 48–51
Assembly metrics
  L50 .... 103, 104
  L90 .... 103, 104
  N50 .... 42, 103, 104, 108, 109, 125
  N90 .... 103, 104
Association networks 143–161
Average Nucleotide Identity (ANI) 125, 178, 179

B

Bacteriophages 84
  See also Phages; Viromes
BBTools 92, 94, 95
BiGG databases 177, 196, 203, 223
Biogeochemical modeling 331–351
Bioinformatics analysis 35, 40, 58–61
Binning algorithms 52, 58, 59, 64–65, 84, 106, 107, 118, 119, 128
Bowtie2 .... 58, 99–102

C

Catabolic reactions 332–351
CheckM 35, 53, 64, 109–111, 123
CheckV 118, 122, 123, 132
Co-Culture/Community Analyses for Metabolite Production (CAMP) 221–230
COMETS 197, 198, 277–314
Community metabolic modeling 191–214, 221–230, 233–257, 261–275, 277–314, 317–329
Community modelling 181, 192, 194, 197, 199–201, 205–214, 222, 225–227, 229, 230, 235–237, 248–257, 267–270, 275, 307
  dynamic approaches 170, 197
  genome-scale approaches 192, 194, 197, 200, 201, 205, 211, 262, 278, 318
Community shuffling 156, 157, 159, 160
Community structures 6, 39, 57, 154, 168–172, 185
Competition 192, 318
CompNet 143–161
CONCOCT 64, 106
Conda 40, 86, 178, 200, 279, 336
Constraint-based modeling 191–214, 221–230, 233–257
  COBRA 192, 193, 195, 197, 198, 200, 201, 203, 209, 214, 222–224, 226, 234, 235, 237, 238
  flux balance analysis (FBA) 195, 209–210, 234, 318, 319
  flux variability analysis (FVA) 192, 200, 201, 210–211, 223–225, 249, 255
Constraint-based reconstruction and analysis (COBRA) 118, 192, 193, 195, 200, 223, 234, 236–238, 249, 279, 285, 290, 313
Constructed wetlands (CWs) 47–69
Contamination control 44, 51
Contigs 35, 42, 59, 60, 64, 65, 67, 68, 98, 103, 104, 146
Co-occurrence networks 67, 145–147
Cross-feeding 176, 192, 317–329
Cytoscape 130, 134, 319, 326, 327

D

Data preprocessing 336
Decontamination 6–8, 20, 91
DEMETER 235, 237, 240–248
Dereplication 123–128
DNA extraction 2, 3, 26–28, 34–35, 38–39, 49, 52, 55–57, 103
Driver microbes 143–161, 163
Dynamic FBA (dFBA) 192, 197, 199, 222
Dynamic simulation 277–314, 317–329

E

Electron acceptors 281, 287, 332–351
Electron donors 332–351

Karthik Raman and Gayathri Sambamoorthy (eds.), *Modelling the Microbiome*, Methods in Molecular Biology, vol. 3006, https://doi.org/10.1007/978-1-0716-5080-6,

Enrichment cultures.....75–80
EPICS.....167–173
European Nucleotide Archive (ENA).....44, 89
Exchange reactions.....182–184, 214, 222, 226, 234, 256, 257, 267–269, 284, 290, 302–304, 311, 312, 314, 321, 322

F

FastQC.....35, 40, 52, 58, 91
Filtration.....44, 47, 51, 52, 55–57, 63, 69, 77, 149
Flux balance analysis (FBA).....191–214, 221–230, 233–257, 261–275, 277–314, 317–329
  biomass objective function.....206, 235, 236
  flux distributions.....262, 324
  steady-state assumption.....235
Flux couplings.....261–275
Flux sampling (FS).....192, 200, 201, 211, 212
Flux variability analysis (FVA).....192, 200, 201, 210, 223–225, 249, 255
FLYCOP.....197, 277–314
Fourier transform ion cyclotron resonance mass spectrometry (FTICR-MS).....332–351

G

Galaxy.....84
Gap filling (metabolic models).....177, 195, 196, 246, 247
GenBank.....89, 195, 196
Genome binning.....35, 105, 119
Genome-scale community metabolic model.....201
Genome-scale metabolic models (GEMs, GSMMs).....175–189, 191–214, 221–230, 233–257, 261–275, 277–314, 317–329
Genome-scale metabolic networks.....176, 212
Gibbs free energy.....332–351
Growth kinetics.....277–314, 317–329
GTDB-Tk.....53, 114, 130–132

H

High-order interactions.....167–173
Host predictions.....130, 133, 134
Host–virus interactions.....130

I

Integrated Phage-Host Prediction (iPHoP).....130, 132–134

K

KBase.....177, 222–224, 241, 333
Keystone species.....153

L

LambdaPy.....331–351
LambdaR.....331–351
Lignocellulose.....76, 79, 80, 194
Linear programming.....210, 223, 237, 238, 264, 269, 270, 272
Low biomass samples.....38, 44

M

MAGs, *see* Metagenome-assembled genomes (MAGs)
Marine microbiomes.....33–45, 163
MaxBin2.....42, 106
MEGAHIT.....59, 60, 96, 98
MetaBAT2.....35, 42, 106–111
Metabolic flux.....192, 194, 196, 206, 210, 214, 234, 310, 323
Metabolic interactions.....187, 192, 197, 201, 206, 214, 221, 234, 262
Metabolite exchange.....181, 262
Metagenome-assembled genomes (MAGs).....35, 42–43, 50, 53, 58–60, 64–66, 106–111, 113–115, 117, 118, 123–126, 130–135, 146, 240
Metagenome assembly.....6, 96–100, 102–106, 109, 118
Metagenomics.....1–22, 25–45, 47–69, 83–136, 143–149, 192, 201, 206, 233–257, 280
MetaQUAST.....103, 104, 106
MetaSPAdes.....35, 42, 96
MetQuest.....175–188
mgPipe.....233–257
MICOM.....191–214, 222
Microbial consortium.....75–80, 192, 194, 201, 278, 309
Microbial co-occurrence.....146
Microbial cross-feeding.....318, 324–326
Microbial ecology.....3, 39, 51, 75, 83–136
Microbial interactions.....144, 145, 147, 157, 160, 176, 194, 200, 318
Microbial networks.....143–163, 167–173
Microbiome dynamics.....250, 318, 324
Microbiome engineering.....277–314, 317–329
Microbiome modeling.....191–214, 221–230, 233–257, 261–275, 277–314, 317–329, 331–351
Microbiome Modeling Toolbox (MMT; MMT 2.0).....199, 233–257
µbialSim.....198, 317–329
Multi-sample mapping.....135
Mutualism.....176, 227

N

NCBI Sequence Read Archive (SRA).....44, 53, 86, 88, 90
NetShift.....143–163

Network centrality
betweenness centrality 153, 159
degree centrality 153, 154, 160
Next-generation sequencing (NGS) 25–32, 45, 91, 146, 317

O

Objective function 203, 206, 223, 235, 236, 256
Optimization 2, 25, 31, 50, 195, 198, 200, 206, 212, 223, 229–230, 235, 237, 278, 294–297, 303, 307–310, 314
Organic matter (OM) 34, 44, 163, 332–351
Ordinary differential equations (ODEs) 199, 277–314, 317–329

P

Paired-end reads 7, 8, 91, 96
Parameter estimation 350
Pathogen detection 65–68
Personalized medicine 234
Phages 92, 93, 117, 130, 162, 163, 194
See also Bacteriophages; Viromes
Plastics 75–80
Population dynamics 170
Pymgpipe 236

Q

Quality control (QC) 4, 8–12, 26, 28, 35, 38, 40, 42, 44, 50, 56, 58, 84, 91–96, 99–106, 108, 114, 118, 133, 179, 244–246

R

Reaction knockouts 229–230
Read QC 102
Read trimming 92
Reference databases 4, 41, 45, 58, 59, 65, 195, 196

S

Sampling protocols 6, 34, 38, 39, 146
Scaffold (assembly) 98–100, 102–109, 113, 115–119, 121–123, 126–128, 132, 133, 146
Secreted metabolites 242, 253
Sequencing depth 3, 6, 11, 44, 58, 62, 89, 94, 103, 105, 238
Shadow prices 256–257
Shotgun metagenomics 1–22, 25–45, 47–69
Simulation frameworks 191–214, 221–230, 233–257, 261–275, 277, 317–329
16S rRNA gene sequencing 34, 39
Snakemake 8, 22, 86
Species abundances 135, 168, 169, 171, 227, 253
Statistical analyses 2, 3, 5, 6, 35, 69, 236, 237, 240, 254–257
SteadyCom 197, 222, 223, 226, 261–275
Stoichiometric matrix 222, 264, 271
Sustainable bioprocesses 144
Synthetic microbial consortia 277–314, 317–329
Syntrophy 320, 321
Systems biology 192, 195, 196, 209, 234, 333
Systems Biology Markup Language (SBML) 177, 180, 193, 243, 244, 263–265, 319, 325, 329

T

Taxonomic and functional annotations 52, 58, 59
Taxonomic profiles 3, 5–8, 10, 12, 19
Thermodynamics 244, 245, 331–351
Time-course modeling 277–314, 317–329
Trimmomatic 35, 40, 91

V

Viral ecology 83–136
Viral metagenome-assembled genomes (vMAGs) 117–135
Viromes 83–136
See also Bacteriophages; Viral ecology; Viromics
Viromics 84, 123, 135
vRhyme 118, 119, 122, 123, 132

W

Wastewater treatment 47–69

MIX
Papier aus verantwortungsvollen Quellen
Paper from responsible sources
FSC® C105338

If you have any concerns about our products,
you can contact us on
ProductSafety@springernature.com

In case Publisher is established outside the EU,
the EU authorized representative is:
Springer Nature Customer Service Center GmbH
Europaplatz 3, 69115 Heidelberg, Germany

Printed by Libri Plureos GmbH
in Hamburg, Germany